射频识别：
应用中的 MIFARE 和非接触式智能卡
（影印版）

Gerhard H. Schalk，Renke Bienert 著

南京　东南大学出版社

图书在版编目(CIP)数据

射频识别:应用中的 MIFARE 和非接触式智能卡:英文/(德)沙尔克(Schalk,G.H.),(德)比奈特(Bienert,R.)著. —影印本. —南京:东南大学出版社,2015.9

书名原文:RFID:MIFARE and Contactless Smartcards in Application

ISBN 978-7-5641-5947-4

Ⅰ.①射… Ⅱ.①沙… ②比… Ⅲ.①射频-无线电信号-信号识别-英文 Ⅳ.①TN911.23

中国版本图书馆 CIP 数据核字(2015)第 170009 号

© 2013 by Elektor International Media BV

Reprint of the English Edition, jointly published by Elektor International Media BV and Southeast University Press, 2015. Authorized reprint of the original English edition, 2015 Elektor International Media BV, the owner of all rights to publish and sell the same.

All rights reserved including the rights of reproduction in whole or in part in any form.

英文原版由 Elektor International Media BV 出版 2013。

英文影印版由东南大学出版社出版 2015。此影印版的出版和销售得到出版权和销售权的所有者—— Elektor International Media BV 的许可。

版权所有,未得书面许可,本书的任何部分和全部不得以任何形式重制。

射频识别:应用中的 MIFARE 和非接触式智能卡(影印版)

出版发行:东南大学出版社
地　　址:南京四牌楼 2 号　邮编:210096
出 版 人:江建中
网　　址:http://www.seupress.com
电子邮件:press@seupress.com
印　　刷:常州市武进第三印刷有限公司
开　　本:787 毫米×980 毫米　16 开本
印　　张:30.25
字　　数:592 千字
版　　次:2015 年 9 月第 1 版
印　　次:2015 年 9 月第 1 次印刷
书　　号:ISBN 978-7-5641-5947-4
定　　价:89.00 元

本社图书若有印装质量问题,请直接与营销部联系。电话(传真):025-83791830

Contents

Preface **13**

1 RFID Fundamentals **17**
   1.1 Introduction to RFID 17
        1.1.1 RF 17
        1.1.2 ID 18
        1.1.3 RFID system classification 19
              1.1.3.1 Frequencies and transmission principles 19
              1.1.3.2 Applications 20
   1.2 RFID system components 22
        1.2.1 Card (PICC) 22
        1.2.2 Reader (PCD) 23
   1.3 ISO/IEC 14443 application example 24
        1.3.1 Public transport ticketing 24
        1.3.2 Employee identification cards 25
        1.3.3 Electronic passports and identity cards 26
        1.3.4 Other applications 27
   1.4 Physical fundamentals 27
        1.4.1 Energy transmission 27
        1.4.2 Data transmission from PCD to PICC 29
              1.4.2.1 Amplitude modulation 29
              1.4.2.2 Standard data rate (106 Kb/s) 29
              1.4.2.3 Higher data rates (up to 848 Kb/s) 31
        1.4.3 Data transmission from PICC to PCD 32
              1.4.3.1 Load modulation 33
              1.4.3.2 Subcarrier modulation:
                        Manchester encoding using ASK 33
              1.4.3.3 Subcarrier modulation:
                        NRZ encoding with BPSK 34

2 Overview of the Relevant Standards **35**
   2.1 ISO/IEC 14443 35
        2.1.1 Part 1: physical properties 36
        2.1.2 Part 2: RF properties and signals 37
        2.1.3 Part 3: Card selection and activation 38
              2.1.3.1 Type A: UIDs 38
              2.1.3.2 Type A: card activation 40
              2.1.3.3 Type A: SAK encoding 42
              2.1.3.4 Type A: collision-detection and
                        conflict resolution 44
              2.1.3.5 Type B card activation 46
              2.1.3.6 Type B: Card activation parameters 48

- 2.1.4 Part 4: communication protocol 49
  - 2.1.4.1 Protocol activation 49
  - 2.1.4.2 T=CL protocol block structure 56
- 2.1.5 Information Block (I Block) 56
  - 2.1.5.1 Receive-ready Blocks (R Blocks) 58
  - 2.1.5.2 Supervisory Blocks (S Blocks) 59
- 2.1.6 Electromagnetic disturbance (EMD) 61
  - 2.1.6.1 Rest period 62
  - 2.1.6.2 Rest level 63
  - 2.1.6.3 Distinction between invalid card response and EMD 63
  - 2.1.6.4 The MFRC522 reader IC and EMD 63
- 2.2 ISO/IEC 10373-6 test methods 64
  - 2.2.1 Test equipment 65
    - 2.2.1.1 Calibration coil 65
    - 2.2.1.2 Test PCD assembly 65
    - 2.2.1.3 ReferencePICC 66
  - 2.2.2 Tuning and calibration 68
    - 2.2.2.1 Tuning 68
    - 2.2.2.2 Calibration 68
  - 2.2.3 Measurements at the reader 69
    - 2.2.3.1 Range measurement 69
    - 2.2.3.2 Measurement effort 69
  - 2.2.4 Tests for Layer 3 and 4 70
- 2.3 Near Field Communication (NFC) 70
  - 2.3.1 Introduction 70
  - 2.3.2 NFC air interface 71
    - 2.3.2.1 NFC device as card 71
    - 2.3.2.2 NFC device as a reader 71
    - 2.3.2.3 NFC device in 'active' mode 72

3 RFID Antenna Design 73
- 3.1 Theoretical Fundamentals 73
  - 3.1.1 Antenna as Resonant Circuit 73
  - 3.1.2 Transformer Model 77
  - 3.1.3 The Biot-Savart Law 79
  - 3.1.4 Optimal Antenna Size 80
- 3.2 Reader Antennas 83
  - 3.2.1 Antenna Quality 83
    - 3.2.1.1 Data Transmission Bandwidth 83
    - 3.2.1.2 Stability Against Detuning 86
  - 3.2.2 Electrically Conductive Surfaces in the Vicinity of the Antenna 89
    - 3.2.2.1 Ferrites 90
  - 3.2.3 Balanced and Unbalanced Antennas 92
    - 3.2.3.1 Balanced Antenna 93
    - 3.2.3.2 Unbalanced Antenna 96

3.3 Card Antennas 98
      3.3.1 Standard Cards (ID-1) 98
      3.3.2 Tokens with Smaller Inlays (Smaller than ID-1) 100
  3.4 Impedance Measurements with the miniVNA 101
      3.4.1 miniVNA User Software 102
      3.4.2 Disadvantages and Limitations of the miniVNA 102
          3.4.2.1 Lack of Sign for Imaginary Numbers 102
          3.4.2.2 Lack of Calibration and Compensation 102
      3.4.3 How Do I Find the Correct Compensation? 103
      3.4.4 Coil Inductance Measurement 105

4 Security and Cryptography 109
  4.1 Protection Objectives 109
      4.1.1 Keeping Data Secure 109
      4.1.2 Data Integrity 111
      4.1.3 Privacy Protection 112
  4.2 Attacks on Smartcards 112
      4.2.1 Logical Attacks 112
          4.2.1.1 Unauthorized Reading of Data 112
          4.2.1.2 Unauthorized Manipulation of Data 113
          4.2.1.3 'Replay' Attack 114
          4.2.1.4 'Relay' Attack 114
          4.2.1.5 'Man-in-the-Middle' Attack 116
          4.2.1.6 Denial-of-Service Attack 116
      4.2.2 Physical Attacks 116
          4.2.2.1 Side Channel Attacks and Power Analysis 116
          4.2.2.2 Reverse Engineering 118
          4.2.2.3 Light and Laser Attacks 119
          4.2.2.4 Temperature and Frequency 120
      4.2.3 Combined Attacks 120
  4.3 Cryptography 120
      4.3.1 Asymmetric Cryptography 120
      4.3.2 Symmetric Cryptography 122
      4.3.3 Block and Stream Cipher 123
      4.3.4 Encryption Standards: DES and AES 124
      4.3.5 DES Cascading 125
      4.3.6 Operation mode 127
          4.3.6.1 Electronic Code Book (ECB) 127
          4.3.6.2 Cipher Block Chaining (CBC) 127
  4.4 Application of Cryptography 129
      4.4.1 Mutual Authentication 129
      4.4.2 Data Encryption 131
      4.4.3 Message Authentication Code (MAC) 131
      4.4.4 Key Management 132
          4.4.4.1 Dynamic Keys 132
          4.4.4.2 Key Diversification 132
      4.4.5 Secure Application Module (SAM) 135
      4.4.6 Security Assessment 135

CONTENTS

5 Introduction to Cards and Tags **137**
  5.1 Overview 137
      5.1.1 Memory Cards and Microcontroller Cards 137
      5.1.2 Advantages and Disadvantages of Contactless Cards 138
          5.1.2.1 Robustness 138
          5.1.2.2 Longevity 138
          5.1.2.3 Usability 138
          5.1.2.4 Infrastructure 138
          5.1.2.5 Contact Between Card and Reader 139
      5.1.3 Dual-Interface Cards 139
  5.2 MIFARE 139
      5.2.1 MIFARE Overview 139
          5.2.1.1 Success Story 140
          5.2.1.2 MIFARE Clone 140
          5.2.1.3 MIFARE Hack 140
          5.2.1.4 MIFARE Product Overview 140
      5.2.2 MIFARE Ultralight 141
          5.2.2.1 Instruction Set 142
          5.2.2.2 Memory Organization 144
          5.2.2.3 Security Functions 144
      5.2.3 MIFARE Ultralight C 145
          5.2.3.1 Instruction Set 146
          5.2.3.2 Memory Organization 147
          5.2.3.3 Security Functions 147
      5.2.4 MIFARE Classic 148
          5.2.4.1 Instruction Set 150
          5.2.4.2 Memory Organization 151
          5.2.4.3 Security Functions 152
      5.2.5 MIFARE Plus 152
          5.2.5.1 Memory Organization 152
          5.2.5.2 MIFARE Plus S and MIFARE Plus X 152
          5.2.5.3 Security Levels 153
      5.2.6 MIFARE DESFire (EV1) 156
          5.2.6.1 Memory Organization 156
          5.2.6.2 File Types 157
          5.2.6.3 Data File 157

6 Reader Antenna Design **159**
  6.1 MF RC522 Reader Module 159
      6.1.1 Digital Interfaces 160
          6.1.1.1 UART 160
          6.1.1.2 SPI 160
          6.1.1.3 I²C 160
      6.1.2 Oscillator 161
      6.1.3 Analog Interfaces 161
          6.1.3.1 Transmitter Outputs 161
          6.1.3.2 Receive Input 165

CONTENTS

   6.1.4 Test Signals 169
      6.1.4.1 MFOUT 170
      6.1.4.2 AUX1 and AUX2 171
   6.1.5 Miscellaneous 173
      6.1.5.1 Power Supply and GND 173
      6.1.5.2 Tolerances 173
 6.2 Antenna Design 173
   6.2.1 Coil Design 174
      6.2.1.1 Measuring the Coil Parameters 175
      6.2.1.2 Determine the Q Factor and the Series Resistance ... 176
   6.2.2 Matching: Calculating the Initial Values 176
      6.2.2.1 Parallel Equivalent Circuit 176
      6.2.2.2 Partitioning and Simplifying the Circuit Diagram 177
      6.2.2.3 Low-Pass Filter 177
      6.2.2.4 Matching Network 178
   6.2.3 Matching: Simulation and Measurement 179
   6.2.4 Measurements on the Transmitted Pulse 181
   6.2.5 Measurement and Adjustment of the Receive Path 183
   6.2.6 Eliminating Interference 183
   6.2.7 Range Checking 185

7 The Elektor RFID Reader 187
 7.1 Introduction 187
 7.2 Reader Hardware 190
   7.2.1 Power Supply 192
   7.2.2 The P89LPC936 Microcontroller 193
   7.2.3 The MF RC522 Reader IC 194
   7.2.4 The FT232R USB/RS-232 Converter 197
      7.2.4.1 Configuring the FT232R 199
      7.2.4.2 USB Driver Modification 201
 7.3 Construction and Operation 202
   7.3.1 Installing the USB Driver 202
   7.3.2 Reader Firmware Update 203
   7.3.3 Firmware Version Control 205
 7.4 Reader Modes 205
   7.4.1 Terminal Mode 205
   7.4.2 PC Reader Mode 207
      7.4.2.1 Activating the PC Reader Mode 207
 7.5 The Firmware 208
   7.5.1 The Software Architecture 208
   7.5.2 The Main Program 208
   7.5.3 The PC_ReaderMode() Function 210
      7.5.3.1 The RS-232 Communication Protocol 210
 7.6 The PC Development Tools 212
   7.6.1 Elektor RFID Reader Programming in .NET 212
   7.6.2 Smart Card Magic.NET 214
      7.6.2.1 It's Usable without Programming 214

CONTENTS

	7.6.2.2	A Scripting Tool or a C# Compiler?	215
	7.6.2.3	Our First Program: "Hello World".	216
	7.6.2.4	Compiling and Running.	219
	7.6.2.5	User Input from the Console Window	220
	7.6.2.6	Are There Really No Breakpoints?	222
7.6.3	Visual C# 2012 Express Edition		223
	7.6.3.1	Creating a Simple Console Application	223
	7.6.3.2	Integrating the Elektor RFID Reader Library	225

8 Cards and Tags in Application 227

- 8.1 ISO/IEC 14443 Type A Card Activation 228
  - 8.1.1 Card Types from the Perspective of Card Activation 228
  - 8.1.2 The Activation Sequence. 230
    - 8.1.2.1 The Request and Wake-Up Commands 230
    - 8.1.2.2 The Anti-collision and Select Commands 233
    - 8.1.2.3 The HALT Command 234
  - 8.1.3 Elektor RFID Reader Library: Card Activation 236
  - 8.1.4 Program Examples 241
    - 8.1.4.1 Card Activation 241
    - 8.1.4.2 Reader Selection 243
    - 8.1.4.3 Polling for Cards 244
    - 8.1.4.4 Simplified Card Activation 247
    - 8.1.4.5 Testing the Reading Range 249
    - 8.1.4.6 Listing All Cards in the Reader's Field 250
- 8.2 MIFARE Card-Type Detection 254
  - 8.2.1 Program Example. 255
- 8.3 The MIFARE Ultralight Card 258
  - 8.3.1 Memory Organization 258
  - 8.3.3 Function of the One-Time-Programmable (OTP) Bytes 261
    - 8.3.3.1 Lock Bits Functionality. 262
  - 8.3.4 Elektor RFID Reader Library: MIFARE Ultralight. 263
  - 8.3.5 Program Examples 264
    - 8.3.5.1 Writing and Erasing Data 264
    - 8.3.5.2 Reading the Entire Memory Contents 265
    - 8.3.5.3 Reading and Writing Strings 267
    - 8.3.5.4 A Simple Ticket Application 268
    - 8.3.5.5 Cloning the Memory Content 271
    - 8.3.5.6 Secure Data Storage 273
- 8.4 The MIFARE Classic Card 281
  - 8.4.1 MIFARE Classic 1K Card Memory Organization. 281
  - 8.4.2 MIFARE Classic 4K Card Memory Organization. 281
  - 8.4.3 MIFARE Mini Card Memory Organization. 282
  - 8.4.4 Instruction Set. 282
  - 8.4.5 The MIFARE Value Format. 286
  - 8.4.6 Decrement, Increment, Restore and Transfer 289
  - 8.4.7 Changing the Keys and Access Condition 290
  - 8.4.8 Elektor RFID Reader Library: MIFARE Classic 294

CONTENTS

- 8.4.8.1 The MifareClassicUtil Class 294
- 8.4.8.2 The IMifareClassic Interface 297
- 8.4.9 Program and Case Studies 300
  - 8.4.9.1 Writing and Erasing Data 300
  - 8.4.9.2 Reading the Entire Memory Contents. 301
  - 8.4.9.3 Optimizing the Read and Write Speeds 303
  - 8.4.9.4 Optimized Reading of the Entire Memory Contents . . . 307
  - 8.4.9.5 The Problem of Data Corruption 310
  - 8.4.9.6 The MIFARE Value Format Methods 314
  - 8.4.9.7 Electronic Purse with Backup Management. 316
- 8.5 The MIFARE Ultralight C Card. 322
  - 8.5.1 Memory Organization 322
  - 8.5.2 Instruction Set. 324
  - 8.5.3 Triple-DES Authentication 324
  - 8.5.4 Elektor RFID Reader Library: MIFARE Ultralight C 327
  - 8.5.5 Programming Examples 327
    - 8.5.5.1 The MIFARE Ultralight C Authentication Sequence . . . 327
    - 8.5.5.2 MIFARE Ultralight C Card Personalization 333
- 8.6 The T=CL Transmission Protocol. 337
  - 8.6.1 T=CL Protocol Activation and Deactivation 338
    - 8.6.1.1 Multi-Card Activation 341
  - 8.6.2 Data Exchange. 341
    - 8.6.2.1 Smart Card Magic.NET – Exchange Mode. 342
    - 8.6.2.2 Block Chaining 343
    - 8.6.2.3 Waiting Time Extension 344
    - 8.6.2.4 Error Detection and Correction. 344
  - 8.6.3 Elektor RFID Reader Library: T=CL 345
  - 8.6.4 Example Programs. 350
    - 8.6.4.1 T=CL Protocol Activation and Deactivation 350
    - 8.6.4.2 Multi-Card Activation 352
- 8.7 The MIFARE DESFire EV1 Card 355
  - 8.7.1 MIFARE DESFire EV1 Commands 356
  - 8.7.2 DESFire Native Command Structure 357
    - 8.7.2.1 Card Command Structure. 357
    - 8.7.2.2 Card Response Structure 357
    - 8.7.2.3 DESFire Block Chaining 358
  - 8.7.3 The DESFire File System 359
    - 8.7.3.1 File Types 359
    - 8.7.3.2 Data File Structure 360
    - 8.7.3.3 Directory Names 361
    - 8.7.3.4 File Names 362
  - 8.7.4 Data Structure. 362
  - 8.7.5 Elektor RFID Reader Library: MIFARE DESFire EV1. 362
  - 8.7.6 Example Programs 363
    - 8.7.6.1 Creating a DESFire Application 363
    - 8.7.6.2 Standard Data File: Reading and Writing Data 366
- 8.8 Application Protocol Data Units (APDUs) 368
  - 8.8.1 Command APDU Data Structure. 368

CONTENTS

      8.8.1.1 Class Byte (CLA) 369
      8.8.1.2 Instruction Byte (INS) 369
      8.8.1.3 Parameter Bytes P1 and P2 369
      8.8.1.4 Coding of Length Fields Lc and Le 369
   8.8.2 Response APDU Data Structure 371
   8.8.3 Examples of ISO/IEC 7816-Compatible APDUs 372
      8.8.3.1 The SELECT Command 372
      8.8.3.2 The READ BINARY Command 373
      8.8.3.3 The Update Binary Command 375
   8.8.4 Elektor RFID Reader Library: APDU 376
   8.8.5 Accessing an ISO/IEC 7816 File System 378
      8.8.5.1 Example Program 380

9 Elektor RFID Projects 385

 9.1 Programming the MF RC522 Reader IC 385
   9.1.1 Elektor RFID Reader Library: MF RC522 386
   9.1.2 Program Examples 386
      9.1.2.1 Changing the RF Parameter Configuration 386
      9.1.2.2 MF RC522 SFR Programming — Card Activation . . . 387
 9.2 RFID Access Control Systems. 396
   9.2.1 Online Systems 396
   9.2.2 Offline Systems 396
   9.2.3 Elektor RFID Reader as Access Control System 396
      9.2.3.1 Functional Description 397
      9.2.3.2 Access Control Manager 398
      9.2.3.3 Microcontroller Firmware 399
      9.2.3.4 Reading and Deleting from the
          P89LPC936 EEPROM 404
 9.3 An Electronic ID Card 406
   9.3.1 Personalization. 406
   9.3.2 Reading the ID Card Data 407
 9.4 Launching a Windows Application 408

10 Smart Card Reader API Standards. 411

 10.1 Introduction 411
 10.2 Card Terminal API (CT-API) 412
 10.3 Open Card Framework (OCF) 412
 10.4 Personal Computer/Smartcard (PC/SC) 413
   10.4.1 The PC/SC Architecture. 413
      10.4.1.1 Integrated Circuit Card (ICC) 414
      10.4.1.2 Interface Device (IFD) 414
      10.4.1.3 Interface Device Handler (IFD Handler) 414
      10.4.1.4 ICC Resource Manager (RM) 415
      10.4.1.5 Service Provider 416
      10.4.1.6 ICC-Aware Applications 417

11 PC/SC Readers 419
11.1 Contactless Cards 419
11.1.1 Contactless Microcontroller Smartcards 419
11.1.2 Contactless Memory Cards 419
11.1.2.1 PC/SC-Compliant APDUs. 420
11.1.3 Answer To Reset (ATR) 422
11.1.3.1 Contact-type Card Activation Sequence 422
11.1.3.2 ATR Structure of a Contact-Type Smartcard 424
11.1.3.3 Contactless Smartcard Pseudo-ATR Structure. 425
11.2 The Microsoft WinSCard API. 427
11.2.1 WinSCard API Programming 428
11.2.1.1 Programming the WinSCard API in C 430
11.3 Java and PC/SC. 438
11.3.1 JPC/SC Java API 438
11.3.2 Java Smartcard I/O API. 440
11.4 The CSharpPCSC Wrapper for .NET 441
11.4.1 How Does One Create an API Wrapper? 441
11.4.2 The `WinSCard` Class 442
11.4.3 The `PCSCReader` Class. 445
11.4.4 Program Examples 447
11.4.4.1 "Hello Contactless Card". 447
11.4.4.2 Determine All Installed PC/SC Drivers 449
11.4.4.3 Getting the Reader and Card Properties 451
11.4.4.4 Testing the Reading Range 452
11.4.4.5 Determining the Type of Contactless Memory Card. 455
11.4.4.6 MIFARE Classic 1K/4K and MIFARE Ultralight. 456
11.4.4.7 MIFARE DESFire EV1 459

12 List of Abbreviations 465

13 Bibliography 469

14 Index 471

Preface

Dr. Thomas Wille

RFID technology, which uses radio frequency to identify people or objects, opens up pathways to newer and more expansive applications. From easier access control using contactless (RFID) cards to public transport fare tickets to bank cards to use in passports as well as in the new German identity cards, the applications are becoming more complex and require ever more computing power.

One of the reasons for contactless technology's success is the cards' ease of use and convenience. Simply placing an RFID device, without any special orientation, within a few centimeters of the reader suffices, and a transaction is completed in a fraction of a second.

Another advantage of this technology is its robustness. RFID devices are now highly integrated and measure just a few square millimeters in area. The coupling antennas are connected directly to the ICs' antenna terminals, resulting in minimal anisotropic connections. The antenna and IC may then be housed very stably within a card enclosure. By minimizing the required electrical connections in this manner, there are no further mechanical-electrical connections that, from a quality perspective, could present weaknesses. It is only possible by means of this robust construction, to achieve life spans of over ten years, and to use the technology in extreme environments.

All of these advantages are enabled with the design of more complex contactless interfaces on the cards and readers. In contrast with serial contact card interfaces with their 8 contacts, data coding for RFID carriers is more complex and the high positioning tolerance in regard to the reader device is made possible by complex and robust analog interface circuitry design. Optimization of the reader–card system is also crucial for optimal implementation and robust dimensioning of RFID technology.

This book gives the reader a simple and understandable introduction to the fundamentals of RFID technology and explains the relevant international standards. Then, in a separate chapter, the security aspects of the technology are discussed. Further, all RFID system components are systematically presented and explained in detail, from RFID antenna design to the different card and tag types to the design of an RFID reader, including the antenna design.

The design of a reader, particularly the Elektor RFID Reader, is very carefully described to the user, from software/firmware to user interface. However, for the user, practical application and associated problems are the key. This is where extensive description of some Elektor RFID Reader use cases and useful tips, tricks and solutions for problem situations comes in. The chapters that follow fully describe the included reader software for an access control system functionally. The book also explains the user interface standards.

This book is thus a systematic introduction to the RFID arena that leads the user through the design of a practical card-reader system. In addition, the book is an indispensible guide for any user who wants to conceptualize, design and optimize RFID systems, and offers the crucial assistance required for the often complex problems inherent in RFID system design.

With this book, the authors have closed the loop for users, hopefully aiding the continuous march of RFID technology to further, unimagined extents. For this, we owe the authors a large debt of gratitude.

Hamburg, January 2011

Dr. Thomas Wille
Senior Director Architecture & Technology

Business Unit Identification, NXP Semiconductors

About the authors

Gerhard Schalk has been an application engineer at NXP Semiconductors (formerly Philips Semiconductors) since 2001. In this role, he supports global clients in developing smartcard operating systems. He is also a freelance lecturer at the FH-Hagenberg and FH-Campus 02 (Graz) technical colleges.

Renke Bienert has also been an application engineer at NXP Semiconductors since 2001. He supports global clients in developing readers and systems that use contactless smartcards. He was involved, among other things, with the introduction of electronic tickets for the 2006 FIFA World Cup, as well as electronic travel passes.

Acknowledgments

When we began this project, we had no idea how much blood, sweat and tears would be required to physically nail down all of these concepts in writing. It's perhaps good that we didn't know that in advance.

For technical assistance, we want to thank all of our colleagues and former colleagues in the Identification Team at NXP, from Hamburg (Germany), Gratkorn (Austria), Caen (France), Eindhoven (The Netherlands), Leuven (Belgium), Singapore, Shanghai (China) and all other locations!

We also thank all collaborators in standardization, especially the members of SC17/WG8 and of the DIN working group, as well as all of our customers throughout the world. Without NXP's support, this book would not have been possible.

Special thanks go to Dr. Thomas Wille, Dipl.-Ing. Reinhard Szoncsó, Ing. Martin Möstl, Dr. Ute Merkel and Mag. Annemarie Brunner for intensive proofreading, as well as Raimund Krings for the supervision and support of Elektor Publishing.

February 2011, Gerhard H. Schalk and Renke Bienert

My special thanks go to my dear wife, Annemarie Brunner, and my two sons, Maximilian and Simon. Without the patience and support of my family, my contribution to this book would not have been possible.

February 2011, Gerhard H. Schalk

My sincere thanks go to all who have tolerated and supported me during the writing of this book, especially my wife, Ute Merkel.

February 2011, Renke Bienert

1
RFID Fundamentals

Renke Bienert

1.1 Introduction to RFID

The term 'RFID' is widely used, but not everyone knows how many different applications and techniques it covers. It's found in the media, where it's often used in an alarmist context, with analogies made to the "Transparent Man," and "Big Brother". Yet, when it comes to automated systems that bring comprehensive solutions to logistical chaos and complicated paperwork, RFID is often presented as a wide-ranging solution to everyday hassle. What, exactly, is RFID?

RFID is an abbreviation, consisting of two distinct concepts: RF and ID.

Figure 1.1.
RFID label on a roll (Source: NXP Semiconductors, 2007).

1.1.1 RF

RF stands for radio frequency.

'Radio frequency' typically refers to all electromagnetic waves radiated into free space and recaptured for the purpose of the wireless transmission of electrical energy or in-

1 RFID FUNDAMENTALS

formation. Today, frequencies of between 100 kHz and 10 GHz are widely used. Over a range this wide, transmission characteristics and propagation conditions are as diverse as the technology used for each situation. A variety of RFID applications can be found throughout the entire range.

1.1.2 ID

'ID' stands for identification. Who or what identifies whom or what?

Many RFID applications are differentiated by exactly this question, and, in answering it, there will often be significant differences in the techniques used, as well as in the details — for example, data rates, data volumes, safety, performance, range, etc.

Basically, RFID is the technology used to identify objects, operations or people by means of wireless transmission of data (and energy).

Figure 1.2.
RFID tag examples.

We can't take a closer look without examining the different classification criteria in order to organize and better understand different RFID systems.

This book deals specifically with a single, widely-used technology. It is thus quite interesting to clarify the differences and overlaps with other systems.

1.1.3 RFID system classification

From a technology point of view, RFID systems may be sorted by operating frequency, operating range, data rates, energy usage and security. For some applications, a clear distinction may be drawn, while in others overlapping cannot be avoided.

1.1.3.1 Frequencies and transmission principles

As mentioned in the previous section, the term 'radio frequency' is so general that it applies to many different technologies. The fact that RFID is constrained to specific frequencies is less about the physical parameter limitations and more about the regulation of frequency use. Manufacturers obviously want their applications to run in frequency bands that permit as much worldwide use as possible. The simpler the approval, the faster and easier it is to get an RFID system to market.

The following table provides an overview of the most-used frequency bands, as determined by frequency usage plans:

Table 1.1. RFID frequencies

Frequency	Transmission principle	Technology	Geographic area
125 kHz (LF)	magnetic coupling	passive	worldwide
13.56 MHz (HF)	magnetic coupling	passive	worldwide
433 MHz (UHF)	electromagnetic wave	active	worldwide
868 MHz (UHF)	electromagnetic wave	passive	Europe
915 MHz (UHF)	electromagnetic wave	passive	USA
2.45 GHz (SHF)	electromagnetic wave	passive / active	worldwide

Here, we'll consider just the 13.56 MHz technology, which is used worldwide. This technology uses passive tags, or tags without their own power source, which are 'dead' in the absence of a reader.

A 'tag' is a generic term that could just as easily apply to a sticker. The corresponding verb, 'to tag,' literally means to mark something, label it, or attach a tag to it. So, an RFID tag could also be a contactless smartcard, as described in this book.

For energy and data transmission, a 13.56 MHz magnetic coupling is used, details of which are below.

1 RFID FUNDAMENTALS

1.1.3.2 Applications

Depending on the technological specifications, RFID can be used (or abused) in many applications. Let's distinguish between the two main application areas: the identification of objects and the identification of people.

Identification of objects

There are many ways to identify objects electronically. The prolific barcode is one currently very widely used technology, while RFID is gaining popularity.

Essentially, to identify an object, such as a package or packaged commodity, it is assigned a unique identity, which can then be read automatically. Barcodes are read optically, while RFID makes use of radio waves.

RFID object-identification applications typically require as wide as possible a reach, so that, for example, an entire cart or palette of goods can be scanned at automatically. As many tags as possible must be read 'simultaneously', although very little data need be transmitted: To identify a product, a few bytes are usually sufficient.

Also, in these applications, normally no encryption or other security measures are required, i.e. typically, the unencrypted 'identification number' is read out (as with a barcode). The tags need not perform any cryptographic processing, but are usually just read and seldom written to.

The tags must be very cheap, so they can be deployed in large quantities.

Objects seldom move by themselves, but are moved when necessary, so the reader must be able to identify objects regardless of their position and location, as long as they're in range.

From the above specifications, the approximate requirements are outlined below.
For the tags:

- lowest possible energy consumption;
- very low cost;
- very few bytes storage capacity;
- no complicated processing necessary.

For the reader:

- long range (high power with high sensitivity);
- direction-independent reading;
- read many tags at a time;
- specific, location-dependent antennas often required.

Identification of people

Different requirements apply to the identification of people.

Contactless smartcard technology (a term we use here to make the distinction from RFID tags in general) are used:

- in public transportation (as E-tickets or smart paper tickets);
- for access control at concerts or stadiums (e.g. at the Football World Cup 2006 in Germany);
- for access to secured premises or in secure areas of buildings;
- for employee time recording ('electronic time card');
- for payment for goods (e.g. for payment for cafeteria food or 'coffee cards' for vending machines);
- for public payments (e.g. in credit cards);
- for monitoring persons (as used in the new German identity card and passport).

It's easy to see that different requirements apply to contactless smartcard applications for identifying people than those for the identification of objects.

For the tag:

- short range (helps to avoid unauthorized reading);
- high data rates (to transfer large amounts of data, for example saved images);
- increased energy consumption (to drive the microcontroller on the card, in order to enable complex processing operations);
- greater security (for example, encryption, so that unauthorized parties can't eavesdrop on information or manipulate it).

For the reader:

- short range, but high data rates;
- no direction-independent reading (because the user's tag is held to the reader);
- 'robust design' for large quantities (helps reduce environment-specific antenna selection).

What follows will be only the consideration of the technology suitable for identification of persons: contactless smartcards.

1 RFID FUNDAMENTALS

1.2 RFID system components

Figure 1.3 shows the general concept of an RFID system, consisting typically of many RFID tags (in our case smartcards), the reader devices, as well as the background system.

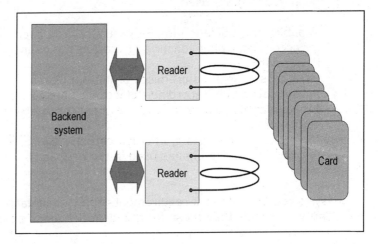

Figure 1.3. RFID system components.

Often, the readers are connected directly to a background system, but sometimes they operate stand-alone. In our examples, both can be realized with the Elektor RFID Reader: A PC may serve as a background system, the Elektor RFID Reader as reading device, and, for a card, the MIFARE card, which is described later.

1.2.1 Card (PICC)

The cards used and described here are all cards that are in principle compatible with the ISO/IEC 14443 standard. They consist of an IC, an antenna (which normally forms an inlay) and the card body, which encloses the whole.

We'll often refer to a contactless smartcard using the term defined in the standard: 'PICC,' or 'proximity card.' It must be noted that a PICC can generally take any form, and is thus not limited to card form. An example is the German electronic passport, which, in accordance with the ISO/IEC 14443 standard, is also a PICC.

Figure 1.4, left, shows an (old) smartcard with a transparent window on the top-right of the card. Through this, we can see the wire antenna windings, as well as the card IC in a typical module (card top-right), which is connected to both wire antennas.

Figure 1.4, right, shows a modern foil inlay, containing a printed antenna instead of a wire one. The printed antenna is cheaper to produce, but the wire one performs better, both because it is of better quality and because it provides a larger average antenna surface.

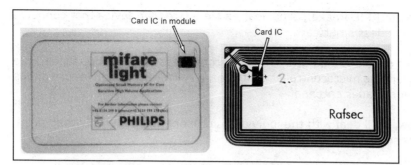

Figure 1.4.
Contactless smart-card and inlay. (Source: NXP Semiconductors / UPM Raflatac, 2009)

Wire antennas are thus used for high quality card bodies made of plastic that are required to last for long, while printed antennas are often used in, for example, paper tickets, which must be very cost-effective.

Chapter 6 will have more detail on card antenna design.

1.2.2 Reader (PCD)

The reader is naturally not solely a reader, as it communicates with the card in both directions. Hence, the expression used in the ISO/IEC 14443 standard is more accurate: Proximity Coupling Device (PCD).

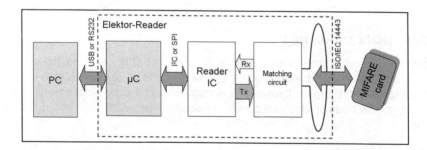

Figure 1.5.
Block diagram of a typical RFID reader.

"Reader" is the generic term for a device that can drive smartcards — in our case the contactless variety. There are many different types for many different applications. A few are optimized as battery-operated, handheld devices, while others, such as those used in underground railway turnstiles, offer a large reading range. A few are 'transparent,' that is, they simply return data upon request, while others offer special applications, including, computation. The latter often requires a special design for security considerations, so such devices will have special components, on which well-secured keys and data can be stored ("SAM" = Secure Application Module).

1 RFID FUNDAMENTALS

Our example, the Elektor RFID Reader, is basically a simple reader device that connects the PC world to a contactless smartcard.

Figure 1.5 shows a typical block diagram. The PC is connected via USB interface to the reader's microcontroller. The microcontroller controls the reader IC, which, in turn drives the interface circuit and reader antenna.

In chapters that follow, we'll deal with the individual components, especially that of the RF, or air, interface described in ISO/IEC 14443.

The concept of contactless transfer of energy and data will be examined, as well as the details of antenna design.

The firmware description will focus on the reader IC and its functions. Finally, contactless smartcard operation, from card activation to reading from (and writing to) the card, as well as PC applications, make up the final large portion of the book.

1.3 ISO/IEC 14443 application example

The ISO/IEC 14443 contactless interface has become established as the standard for many applications.

1.3.1 Public transport ticketing

Currently, the largest number of contactless smartcards are probably used as tickets for public transport. In almost all large Asian cities, the onslaught of the large crowds of people who use buses, railways and subways are accommodated only with the help of electronic tickets. The contactless smartcard combines ease-of-use with high strength and speedy processing.

The traveler need only hold their ticket close to the reader, and the turnstile opens. No other system is as user friendly.

Also, the variety of card ICs (see Chapter 4) allows the system to scale easily: The cheapest card ICs are now being used as 'cheap' paper tickets for single trips, while the more expensive ICs with larger memories and better security are used, in the same systems, as monthly passes or season tickets.

Prerequisite for the application of contactless smartcards in public transport are bus and railway networks that offer "check-in / check-out." For this, stations must be secured with automatic turnstiles that allow access to the system only to holders of valid tickets.

ISO/IEC 14443 APPLICATION EXAMPLE 1.3

Figure 1.6.
RFID ticket for subway access. (NXP Semiconductors, 2007)

1.3.2   Employee identification cards

The old time-card has now been replaced, in almost all businesses, with an electronic card. The modern employee ID card includes a PICC, which handles many tasks. Smartcards are used for everything from simple time and attendance management through to access control (the management of access privileges) to payment in cafeterias.

The advantages lie not only in robust and simple application, but also in flexibility. The same card can typically be quickly assigned to other uses. If, for example, a 'print-on-demand' service were offered to employees, this could be administered using the same card, making the installation cost low.

The old NXP Hamburg employee identification card (based on the MIFARE Classic), as shown here with the Elektor RFID Reader, has since been replaced with a MIFARE card with vastly improved safety features: the MIFARE DESFire EV1. Both cards are widely used in access control systems — the MIFARE 1K in many older systems, and the MIFARE DESFire EV1 in newer systems.

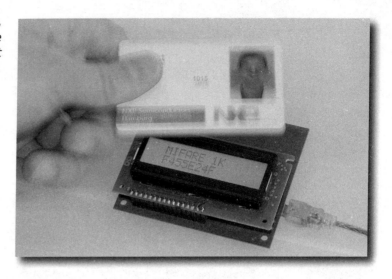

Figure 1.7.
The old employee identification card at NXP in Hamburg.

1.3.3 Electronic passports and identity cards

Germany has been issuing passports containing PICCs since November 2006, and, since the end of 2010, identity cards also containing PICCs. After authentication via the contactless interface, data is read that allows for the secure identification of persons.

The introduction of the electronic passport was a challenge, as, for the first time in terms of contactless interfaces, a worldwide open system had to be specified. It is not very useful if a country can read only their own passports — worldwide interoperability must be guaranteed. For this, you need open standards and specifications

- from the technology used for the contactless interface, to data formats and access rights;
- for compatibility and interoperability of individual components;
- for quality assurance.

Even with an already well-defined standard such as the ISO/IEC 14443 for definition of the air interface, in the early years, prior to introduction, some quite different interpretations of various technical details were allowed.

Currently, the next step is an extension of the saved data. Along with personal data (name, age, etc.), biometric data such as fingerprints or iris scans may be added.

Also, the same technology is increasingly being used for other common documents such as personal identification cards, drivers' licenses, and more.

It is clear that, for these applications, the highest security standards apply. The PICCs used are thus developed to make unauthorized data access extremely difficult. These requirements are shown in the entire chain: from PICC hardware development to the card's software development, right through to the hardware and software on the reader and system side.

We'll come back to fundamental security requirements in Chapter 4.

1.3.4 Other applications

There are many other applications for smartcards that are not described here. Examples are:

- credit cards (contactless payment);
- concert tickets;
- stadium tickets;
- football season tickets;
- parking cards;
- student ID cards;
- frequent flyer cards and other bonus point, or 'loyalty,' cards;
- coffee cards (or cards for other vending machines);
- computer access control.

1.4 Physical fundamentals

Typical smartcard applications, an operating frequency of 13.56 MHz and, above all, legal regulations for the use of radio frequencies, determined some of the technical parameters of the framework that influenced the exact ISO/IEC 14443.

Firstly, the small microcontroller in the PICC must be supplied with sufficient energy via the 13.56 MHz magnetic field. Secondly, the data must be transported as quickly as possible between PCD and the PICC.

1.4.1 Energy transmission

Typical contactless smartcards (PICCs) contain no internal power supply. They need to get all of their required energy from the magnetic field in which they operate.

For this, the PCD's transmitter coil creates an electromagnetic field with a frequency of 13.56 MHz. As long as a card is operated within this electromagnetic field, the appropriate spatial orientation assumed, the antenna coil on the card will pick up a portion of this magnetic field.

This works similarly to a transformer: When energy is transferred, the PCD forms the primary winding, while the PICC acts as the secondary.

1 RFID FUNDAMENTALS

The magnetic field induces in the antenna coil of the PICC a voltage, which is regulated by the card IC and supplies the IC with energy. This requires both a large enough magnetic field and a sufficiently strong coupling between PCD and PICC.

The transfer principle is illustrated schematically in Figure 1.8, showing the primary windings (PCD antenna), the secondary windings (PICC antenna) and the magnetic flow, represented as magnetic field lines.

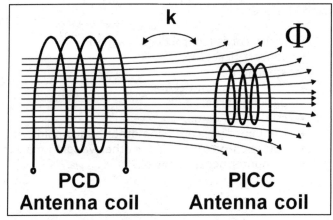

Figure 1.8. *Principle of energy transfer via antenna.*

Because, in this application, the coupling magnitude (k) is not very large, only a small portion of the field is absorbed by the card, i.e. the actual efficiency is very low. While the transfer between primary and secondary windings in 'normal' transformers may reach efficiencies of over 90%, the efficiency here is less than 10%.

The magnetic field strength is limited by the PCD power supply size as well as the legal radiation limits. The PCD antenna produces not only a magnetic field, but also an electromagnetic field, especially in the far-field, which could interfere with other devices and radio services over longer distances.

The card IC must be able to get by with very little energy.

> **Note** An ISO/IEC 14443 card reader device can be viewed as a radio transmitter: It produces a magnetic field with a frequency of 13.56 MHz as well an electromagnetic field of this frequency (at the very least). Thus, the transmission stages, filters and antenna design must meet all radio transmission equipment regulations! See also Chapter 2.

> **Note** The PCD energy must be available to the PICC during the entire transaction, i.e. during a transaction, the magnetic field may not be interrupted (with the exception of short pulses for data transfer).

1.4.2 Data transmission from PCD to PICC

ISO/IEC 14443 provides for four different data transmission speeds, applicable to both the downlink (PCD to PICC) and the uplink (PICC to PCD): 106 Kb/s, 212 Kb/s, 424 Kb/s and 848 Kb/s.

Not all readers and cards support all four data rates, but they are all required to support at least the 106 Kb/s standard data rate. Every card is activated at this data rate, and then, on demand—should both PCD and PICC support it — a switch to a higher data rate can then take place. Whether a higher data rate is necessary or appropriate depends on the application. Because higher data rates use a higher bandwidth (as well as lowering the system quality), one may achieve a higher range by sticking to the standard data rate (and higher transmission quality). We'll return to the subject of PCD antenna design.

1.4.2.1 Amplitude modulation

Data transmission from PCD to PICC is done by amplitude modulation (AM) of the magnetic field.

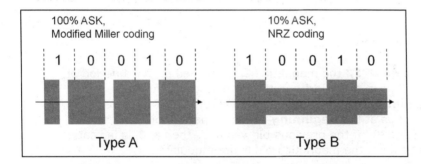

Figure 1.9.
Data transmission from PCD to PICC.

In the simplest case, the magnetic field is simply turned off for a short period: This transmitted pulse can be decoded by the PICC. This principle is illustrated in Figure 1.9. There are also other, more complicated processes used than simple '100% AM' (or on-off keying), as discussed below.

1.4.2.2 Standard data rate (106 Kb/s)

The ISO/IEC 14443 standard defines two different methods: Type A and Type B.

In **Type A**, the magnetic field is actually turned off completely for a brief period. This process uses 100% amplitude modulation and has the advantage of being simple and robust. The disadvantage is that, while the transmission is interrupted, i.e., during the transmitted pulse, the card has no energy available to it.

It is thus clear that the transmitted pulses must be as short as possible. To ensure this, the transmitted data signal is encoded using a '**Modified Miller coding**,' prior to modulation, as shown in Figure 1.10.

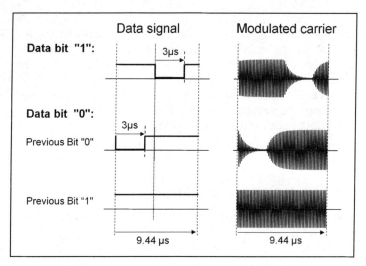

Figure 1.10.
Modified Miller coding with 100% AM for Type A.

Miller encoding uses only short pulses, as follows.

- To transmit a **digital '1,'** the magnetic field is turned off for a period of approximately 3 μs in the **middle** of the bit period.
- To transmit a **digital '0,'** the magnetic field is turned off for a period of approximately 3 μs at the **beginning** of the bit period, but only if the previous bit was a '0.' If the previous bit was a '1,' then a '0' is indicated by no pulse at all, i.e. the magnetic field is not modulated.

The time period for one bit equals $128 / f_c$ = 128 / 13.56 MHz ≈ 1 / 105.9 Kb/s ≈ 9.44 μs. There will thus, at most, be a short pulse of around 30% of the bit period, and the card can get enough energy during the remainder of the pulse.

Type B uses a transmitted data signal that is not 'specially' coded, but the magnetic field is only modulated at 10% of the amplitude (otherwise known as 'non-return-to-zero, or NRZ coding'). This requires a modulation index of at least 8% and at most 14% (m = 8–14%).

The diagram in Figure 1.9 shows that, for a digital '1,' the magnetic field remains unmodulated, while for a digital '0,' the field is modulated by 10%, for the duration of the bit.

This method has the advantage that the PICC has a continuous supply of energy, even though, in the worst case (when several '0's are sent contiguously), the PICC can get by with around 30% less energy (which is roughly what 10% AM translates to).

In practice, we find that with ordinary, simple PCD transmitters, in combination with the large coupling differences (which occur naturally), the nominal modulation index for ISO/IEC 14443 Type B (m = 8–14%) is difficult to achieve.

1.4.2.3 Higher data rates (up to 848 Kb/s)

As already mentioned, after the card is activated, the data rate from the PCD can be switched over to 212 Kb/s, 424 Kb/s or 848 Kb/s on demand.

In principle, Type A achieves higher data rates using the same method as in the standard data rate, but the transmission pulse is shortened by the corresponding ratio to the bit period, as shown in Table 1.2.

For Type A, most PCDs turn off the 13.56 MHz carrier for the short duration of the pulse, i.e. still 100% AM. Regardless, within the fixed system performance as regards to the shortness of the pulse, a correspondingly higher residual carrier is created, as shown in Figure 1.11, i.e. we can no longer speak of a 100% AM signal of the magnetic field. The limits indicated in the Figure are allowable maximum values for the residual carrier according to ISO/IEC 14443 Part 2. Permitted overshoot is not indicated here.

Table 1.2. Pulse transmission at all ISO/IEC 14443 data rates.

Data rate [Kb/s]	Bit length	Bit period [µs]	Pulse length, Type A	Pulse duration, Type A [µs]	Pulse length, Type B	Pulse duration, Type B [µs]
106	128/fc	9.44	40/fc	2.95	128/fc	9.44
212	64/fc	4.72	20/fc	1.47	64/fc	4.72
424	32/fc	2.36	10/fc	0.74	32/fc	2.36
848	16/fc	1.18	5/fc	0.37	16/fc	1.18

> **Note** The data rates indicated are rounded off: The exact data rates are derived from the calculation fc / x, where x = 128, 64, 32 or 16. So, the highest data rate is actually 847.5 Kb/s, for example.

> **Note** ISO/IEC 14443 Part 2 defines the exact pulse duration as well as minimum and maximum pulse rise and fall times. These are somewhat different for PCD and PICC, in order to create an overlap and thereby guarantee greater interoperability between all PICCs and all PCDs. The numbers presented here are simplified.

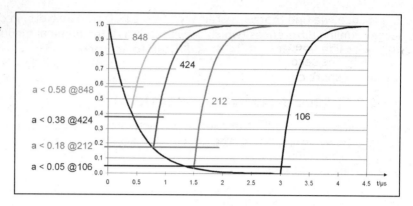

Figure 1.11. Representation of Type A PCD pulses (amplitude).

Type-B pulses, in principle, don't change at higher data rates, but simply get correspondingly shorter, as per Table 1.2.

1.4.3 Data transmission from PICC to PCD

With the help of the transformer principle used above, the data transfer from PICC to PCD can be illustrated as well.

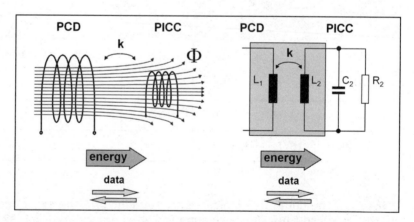

Figure 1.12. Load modulation: transformer principle.

Figure 1.12, left, shows the alignment of the antenna coils and the basic magnetic field line characteristics, while Figure 1.12, right, shows a simplified equivalent electrical circuit: The left-hand side, consisting only of L1, represents the PCD antenna coil or primary side (matching elements omitted for clarity), while the right-hand side is the simplified equivalent representation of the PICC. Basically, L2 is the PICC antenna coil, R2 is an input resistor and C2 an input capacitor on the card IC. The PICC antenna coil could be described as the secondary side of the 'transformer.'

1.4.3.1 Load modulation

The principle used for data transmission from PICC to PCD is called load modulation. Basically, the secondary-side transformer can be imagined as being switched from open circuit to short circuit. This switching between idle (Load 1) and short circuit (Load 2) can be detected on the primary side and decoded.

In this case, the load switching is not as extreme — the 'open circuit' or 'idle' means that the PICC presents an unmodulated load (Load 1), while 'short circuit' means the PICC is presenting a modulated load (Load 2). The corresponding signal (modulated / unmodulated) is received on the primary side and decoded. Figure 1.13 demonstrates this principle for the standard data rate with a Type-A signal.

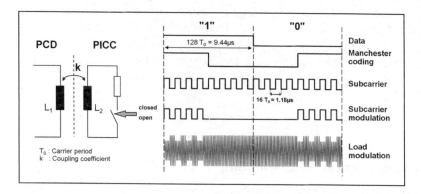

Figure 1.13.
Load modulation for Type A at 106 Kb/s.

According to ISO/IEC 14443, the PICC always uses a subcarrier of fc/16, or 847.5 kHz, to modulate the 13.56 MHz carrier signal.

1.4.3.2 Subcarrier modulation: Manchester encoding using ASK

For the transmission from PICC to PCD at the standard 106 Kb/s data rate, **Type A** uses Manchester encoding, wherein a '1' is represented by a falling edge and a '0' by a rising edge, both of which occur the middle of a bit period (see Figure 1.13, right).

This encoding is necessary for '**collision detection**' during card activation, so that, if there's a superposition of two Manchester-encoded signals, the receiver can differentiate between a '1,' a '0,' and a collision (the superposition of a '0' and a '1'). This Type A '**anti-collision**,' is used to select a single card from a stack of them.

The Manchester-encoded data signal modulates the 847.5 kHz subcarrier with 100% amplitude modulation (100% AM = 'ASK' or 'OOK'). This modulated subcarrier is applied to the load switcher, as shown in Figure 1.13, left. A vector representation of the switch in load modulation is shown in Figure 1.14.

1 RFID FUNDAMENTALS

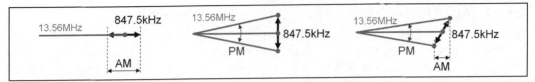

Figure 1.14. *Vector representation of load modulation switching.*

> **Note** The PICC's load switching may, in accordance with ISO/IEC 14443, be resistive or capacitive, i.e. the resulting load may be either amplitude modulated (Figure 1.14, left) or phase modulated (Figure 1.14, center), or both (Figure 1.14, right).
>
> Also, in practice, coupling effects may lead to amplitude modulation (resistive load) being received as phase modulation by the PCD.
>
> The receiver must, therefore, be able to demodulate both AM and PM. Using a simple AM detector can result in 'coverage gaps,' and is not sufficient. The PCD device used in this book uses a special type of I/Q demodulator.

1.4.3.3 Subcarrier modulation: NRZ encoding with BPSK

For Type A data transfer from PICC to PCD at higher data rates, and at all data rates for Type B, NRZ encoding is used. The NRZ-encoded data signal modulates the 847.5 kHz subcarrier using binary phase shift keying (BPSK). This modulated subcarrier is applied to the load switcher as shown in Figure 1.15 — in this example, Type A at 424 Kb/s.

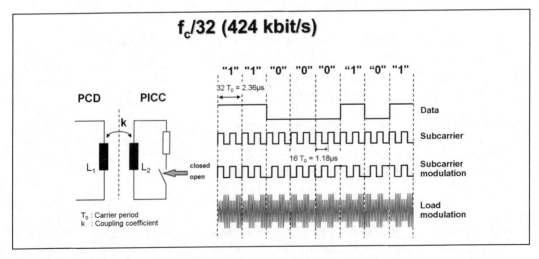

Figure 1.15. *Load modulation for Type A at 424 Kb/s.*

2
Overview of the Relevant Standards

Renke Bienert

While this book is intended as a practical work, it's difficult to get around a discussion of the relevant standards.

The world of standardization is complex, and we'll consider only the standards that are most important for the understanding of contactless interfaces. The ISO/IEC 14443 standard is the single most important standard.

2.1   ISO/IEC 14443

At the end of the 90s, the most important parts of all the so-called 'proximity' technologies were brought together into the ISO/IEC 14443 standard. At the turn of the millennium, the four parts of the standard were published in succession.

Essentially, there were two different transmission principles, one based on the MIFARE card from Philips Semiconductor, which was already sold in large quantities (Type A), and the other based on a French technology by the Innovatron company (Type B). Because a single technology could not be agreed upon, the compromise was made to allow a standard card to use either of the two technologies, while the reader had to support both.

These different features can be found mainly in Parts 2 and 3 of the standard, while Parts 1 and 4 are basically the same for both types.

Once the Type A and Type B standards were published, other companies wanted to jump in with their own technologies, and a vote was held over an extension into types C, D, E, F and G — all containing proprietary methods by each company.

However, the ISO/IEC working group did not want to unnecessarily extend and complicate the standard, and, besides, there was not even an appropriate proposal to make any of the 'new' types compatible with even the common part, ISO/IEC 14443 Part 4. These extensions did not gain a majority vote, and were rejected.

The JTC1 / SC17 / WG8 (Joint Technical Committee 1 / Sub-Committee 17 / Working Group 8) is the working group concerned with this standard. The members of this working group are representatives of a variety of national working groups. In Germany, this is supported by DIN (Deutsche Industrie Norm) and is comprised of corporate representation and interests that are concerned with contactless smartcards in Germany.

Over the years that followed, this Working Group, WG8 (and its associated Task Force 2), initiated a whole series of amendments to the published standard. These include, for example, higher data rates.

ISO/IEC standards are typically subject to revision after around five years, at which point the standards and their amendments are compiled into a new document (purely an editing job), and possible shortcomings are corrected or missing features added.

An ISO/IEC 14443 revision began in 2005. Some of the revised parts have already been published, while others are still being worked on or awaiting a vote (at time of writing).

If in doubt, it is recommended that the current status be checked. Current information regarding standardization work related to the ISO/IEC 14443 can be found at *http://www.wg8.de,* and the published documents are available from *http://www.iso.ch*.

> **Note** The ISO/IEC 14443 standard cannot be considered without its associated test methods. These are in Part 6 of ISO/IEC 10373.

> **Note** ISO/IEC 14443 defines how contactless smartcards are activated and how data is exchanged. It does not describe any application-specific card commands, encryption or authentication mechanisms and is application-independent.

2.1.1 Part 1: physical properties

Part 1 of ISO/IEC 14443 is the shortest part: it essentially defines just the card's physical format. The card antenna may take just about any form or size, but it is specifically stated that interoperability is only guaranteed if the antenna meets Class 1 specifications, which define the minimum and maximum size of the antenna coil, as shown in Figure 2.1.

Originally, the standard was purely a card-oriented standard, but more and more form factors are becoming available. Therefore, there is currently an amendment being worked on that will define other, smaller antenna types (Class 2 — Class 5).

These new classes (sizes) naturally have an influence on the electrical properties of the PICC, so the new classes also require an amendment to Part 2 (RF Signal and Power Interface), as well as an amendment to ISO/IEC 10373-6.

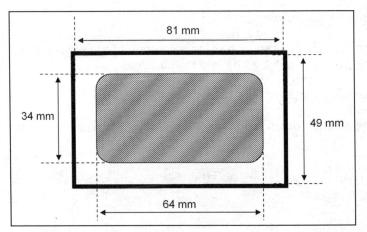

Figure 2.1.
The Class 1 antenna definition of an ID1 card.

> **Note** Initially, ISO/IEC 14443 Part 1 included specifications regarding UV-resistance, ESD characteristics, etc. Under revision, these specifications were transferred to Part 1 of ISO/IEC 7810, which covers smartcards in general, and ISO/IEC 14443 references it.

2.1.2 Part 2: RF properties and signals

ISO/IEC 14443 Part 2 is called "RF Signal and Power Interface" and defines electrical characteristics. Most of the electrical characteristics defined in Part 2 have already been described in Section 1.4 of this book.

ISO/IEC 14443 Part also defines a minimum field strength that a PCD must deliver (and from which a PICC must naturally function) of H_{min} = 1.5 A/m. It also defines the PCD's maximum field strength (and the maximum at which the PICC is required to function) of H_{max} = 7.5 A/m.

Within these two limits, a PICC must function and deliver a load modulation amplitude of at least $U_{LMA,PICC} = 22/H^{0.5}$. The PCD must have a sensitivity of at least $U_{LMA,PCD} = 18/H^{0.5}$. Both values are shown as a function of field strength in Figure 2.2.

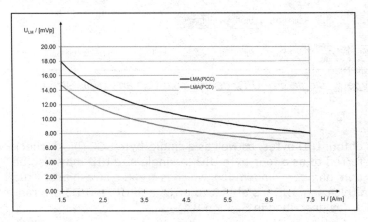

Figure 2.2.
Load modulation amplitude.

2 OVERVIEW OF THE RELEVANT STANDARDS

A few other values are defined in detail, such as:

- the overshoot permitted in transmitted pulses, as a function of the pulse edges;
- the exact relationship between pulse rising and falling edges;
- the 'reader-talks-first' principle;
- the modulation index as a function of the field strength (Type B).

as well as all of the modulation and coding techniques for Type A and Type B, as described in Section 1.4.

> **Note** For exact measurements of these values, the ISO/IEC 10373-6-defined test setups and procedures are used.

2.1.3 Part 3: Card selection and activation

Part 3 of ISO/IEC 14443 describes card activation for Types A and B. Because a contactless reader may be within range of several cards simultaneously, there must be a method by which the reader may select an individual card. Part 3 defines two such methods: one for Type A and one for Type B. Type A›s method is based on an 'arbitrary bit selection', while Type B relies on a time division method.

> **Note** The reader must support both Type A and Type B, while the card is either of Type A or Type B, for the entire system to be fully ISO/IEC 14443 compatible.

2.1.3.1 Type A: UIDs

Type A cards use UIDs (unique identifiers) to select individual cards. These UIDs come in three different lengths:

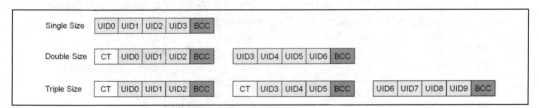

Fig. 2.3. UID lengths for Type A: single size UID, double size UID and triple size UID.

Single size UID

The single size UID consists of four UID bytes, as well as a single-byte BCC (block check character). It is thus also referred to as a four-byte UID. A single size UID may also be a dynamically-generated random number (random ID), which is created afresh at startup (power-on reset, or POR). In this case, UID0 is preset to 08_{hex} and UID1–UID3 are randomly generated at each POR (as described in Section 8.1).

A single size UID may also be a fixed integer, which may also be used for many other cards (non-unique ID, or NUID), as shown in Table 2.1.

> **Note** The BCC (block check character) is a check byte and is always calculated as the XOR of the previous four bytes.

> **Note** The UID is not a serial number in the strict sense, i.e. there is no clear definition on how a UID is converted to an integer. As you can see from the description, the single size UID does not have to be unique.

Table 2.1. Typical single size UIDs.

First byte (UID0)	Meaning
08_{hex}	The remaining 3 bytes (UID1–UID3) are generated dynamically upon POR and will not change as long as the card is supplied with energy. Because this UID usually changes every time the card is newly introduced into a reader's magnetic field, it is also called a random ID (RID).
xF_{hex}	The UID is fixed, but not necessarily unique (i.e. the same four-byte UID may occur in different cards).
$x0-x7_{hex}$	This number range is for proprietary use (MIFARE cards).
$x9-xE_{hex}$	This number range is for proprietary use (MIFARE cards).
88_{hex} ("CT")	The UID is longer than four bytes.

Double size UID

The double size UID consists of a cascade tag (CT, 88_{hex}), a seven-byte UID (UID0–UID6) and the BCC, as shown in Figure 2.3, center. It may also be referred to as a seven-byte UID.

> **Note** The CT (cascade tag) byte indicates whether or not another cascade level is necessary. This information is redundant with the SAK cascade bit (see below) and need not, in principle, be evaluated.

In a double size UID, UID0 contains a manufacturer code, which identifies the chip maker.

Triple size UID

The triple size UID is composed of two CTs, a ten-byte UID and the BCC, as shown in Figure 2.3, bottom. It may also be referred to as a ten-byte UID. In a triple size UID, UID0 contains a manufacturer code, which identifies the chip maker.

There are currently no cards with triple size UIDs.

2 OVERVIEW OF THE RELEVANT STANDARDS

> **Note** Naturally, the card IC manufacturers ensure that a single size UID's UID0 and a double size UID's UID3 are never the same as the CT.
>
> Unfortunately, there are Chinese 'clones' that don't abide by international standards, and use invalid UID0s. Because the open standard is publicly available, manufacturers aren't able to prevent such misuse. An application requiring security should therefore always make use of techniques such as encryption using secret keys.

2.1.3.2 Type A: card activation

Figure 2.4 shows two **simplified** state diagrams for each Type A card's activation procedure. The left diagram is valid, in principle, for all cards with single size UIDs, the right for the double-sized.

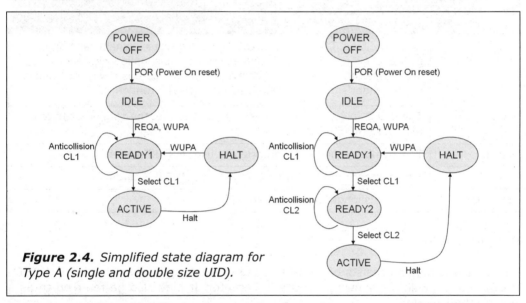

Figure 2.4. Simplified state diagram for Type A (single and double size UID).

For card activation, the following commands are used:

REQA: The REQA (Request Command, Type A = 26_{hex}) is used to poll the card. If a card is active, that is, when it has enough energy reach POR and has thus automatically entered the IDLE state, it responds with an ATQA (Answer to Request, Type A) and enters the READY1 state. This tells the reader that there is at least one card within range.

> **Note** The REQA is a 'short frame command', which is only seven bits long. No CRC is transmitted with the command, nor returned in the card's response.

Anti-collision CL1: The Anti-collision CL1 (Anti-collision, Cascade Level 1 = $93\ 00_{hex}$) requests the first four bytes of the UID. In the case of a single size UID, the card responds with its complete UID. In the case of a double or triple size UID, the card responds with the CT and the first three bytes of the UID.

> **Note** An Anti-collision command does not change the state of the card (and can thus be repeated as often as required). It may also be skipped if the UID is already known.
>
> Neither the Anti-collision command nor the card's response thereto make use of a CRC.

When several cards are in their READY1 state, they will obviously all respond with their UIDs, which the reader identifies as 'collisions.' These collisions are detected at the bit level, and resolved with procedure involving reiterating the Anti-collision command (see below).

> **Note** The entire Anti-collision procedure is required, whether there is just one card or several cards within range of the reader.

Select CL1: The Select CL1 (Select Cascade Level 1 = 93 70 xx xx xx xx xx CRC1 CRC2$_{hex}$) command selects the first four bytes of the UID. The card responds with SAK CL1 (Select Acknowledge) and enters the ACTIVE state (if it has a single size UID) or the READY2 state (in the case of a double or triple size UID).

> **Note** The Select and HLTA commands are the only ones for card activation that contain CRCs, in both the command and in the card's response thereto.

Anti-collision CL2: The Anti-collision CL2 (Anti-collision Cascade Level 2 = 95 00$_{hex}$) command requests the next four bytes of the UID. In the case of a double size UID, the card answers with the remaining four bytes of the UID, while a card with a triple size UID will respond with CT and the next three bytes of the UID.

Several cards may also be in the READY2 state, responding simultaneously, which the reader device would again detect as collisions. Again, the collisions are detected at the bit level and are resolved using the Anti-collision command repetition procedure (see below).

Select CL2: The Select CL2 (Select Cascade Level 2 = 95 70 xx xx xx xx xx CRC1 CRC2$_{hex}$) selects the next four bytes of the UID. The card responds with SAK CL2 (Select Acknowledge, Cascade Level 2) and enters the ACTIVE state (if a double size UID) or the READY3 state (if a triple size UID).

Anti-collision CL3: The Anti-collision CL3 (Anti-collision, Cascade Level 3 = 97 00$_{hex}$) requests the final four bytes of a triple size UID from the card. The card responds with the remaining four UID bytes.

> **Note** The state diagram for a triple size UID is absent from the display in Figure 2.4. For a triple size UID, the diagram is easily extended by inserting a READY3 state between the READY2 state and the ACTIVE state.

Select CL3: The Select CL3 (Select Cascade Level 3 = 97 70 xx xx xx xx xx CRC1 CRC2$_{hex}$) selects the last four bytes of the UID. The card responds with SAK CL3 (Select Acknowledge, Cascade Level 3) and enters the ACTIVE state.

HLTA: If the card is in the ACTIVE state, it can be forced into the HALT state using the HLTA (Halt, Type A = 50 00 CRC1 CRC2$_{hex}$) command. The card can be revived from the HALT state with the WUPA command.

WUPA: A card in the IDLE state or the HALT state responds to the WUPA (Wake Up, Type A = 52$_{hex}$) command with an ATQA and enters the READY1 state.

> **Note** WUPA is a 'short frame command,' and is only 7 bits long. Neither the command nor the card's response use a CRC.

> **Note** The difference between WUPA and REQA: WUPA brings the card out of either the IDLE state or the HALT state into the READY1 state, while REQA does this only for the IDLE state.

2.1.3.3 Type A: SAK encoding

In order for the reader to know whether the card is fully selected and in the ACTIVE state, the SAK (Select Acknowledge) response must be evaluated, as shown in Table 2.2.

> **Note** Only the SAK encoding defined in ISO/IEC 14443 is described here. The other permitted proprietary encodings and their applications are described in section 8.1.2.

If SAK Bit 3 = 0, then the UID is complete and the card is in the ACTIVE state.

If SAK Bit 3 = 1, then the UID is not yet complete and the card is still in the READY1 or the READY2 state, in which case an additional Anti-collision command must be sent, followed by another Select, in order to switch the card into the ACTIVE state.

Bits 3, 6 and 7 are defined in the ISO standard as shown in Table 2.2. The other remaining bits are for proprietary purposes.

The card activation principle for Type A cards is shown in Figure 2.5, left. It is generally valid for all existing UID sizes (four, seven or ten bytes). As this is a simplified illustration, collisions that occur when multiple cards are in range aren't considered. Such conflicts and their resolutions, which a reader device **must**, nonetheless, support, are described in Section 2.1.3.4, as well as in the practice chapter.

Table 2.2. SAK coding in accordance with ISO/IEC 14443 and ISO/IEC 18092.

b8	b7	b6	b5	b4	b3	b2	b1	Meaning
0	0	0	0	0	1	0	0	Cascade bit = 1: the UID is not yet complete.
		1			0			The card supports the ISO/IEC 14443-4 (T=CL) protocol.
		0			0			The card does not support the ISO/IEC 14443-4 (T=CL) protocol.
	0				0			The card does not support NFC-IP1.
	1				0			The card supports NFC-IP1

> **Note** Be aware! The bit numbering in ISO/IEC 14443 starts with Bit 1! So, a byte consists of Bits 1–8, instead of the customary Bits 0–7
>
> ISO/IEC 14443 SAK Bit 3 is therefore the 'normal' Bit 2.

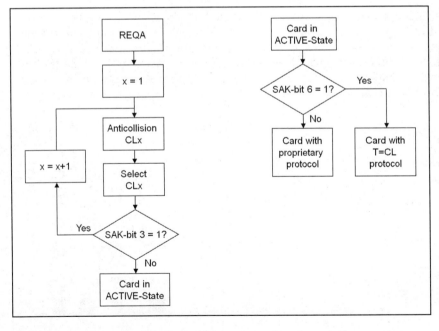

Figure 2.5. Simplified principle of card activation for Type A.

Example command sequence for a single-size UID:

Four-byte UID card activation:

REQA—Anti-collision CL1—Select CL1

2 OVERVIEW OF THE RELEVANT STANDARDS

Example command sequence for a double size UID:

Seven-byte UID card activation:

REQA—Anti-collision CL1—Select CL1—Anti-collision CL2—Select CL2

From the ACTIVE state, one may

- switch the card into the T=CL protocol layer (see next section) if the card is ISO/IEC 14443-4-compatible;
- execute the MIFARE Classic or MIFARE Ultralight commands if the card supports these; or
- bring the card to the HALT state using the HLTA command.

The encoding of the card's SAK bit (Bit 6) indicates whether the card supports the T=CL protocol (see Table 2.2 or Figure 2.5, right).

Note	The proprietary SAK bits may be used to identify the card type, as in the case of MIFARE cards (see Section 8.2).

2.1.3.4 Type A: collision-detection and conflict resolution

If several cards are in the reader's field, all of them will respond simultaneously (bit-synchronously) to REQA. Possible collisions in the ATQA will be ignored by the reader. All cards will respond to an Anti-collision command with their UIDs. Because the UID is 'unique,' Card 1's UID must differ from that of Card 2 by at least one bit. Because Type A cards employ Manchester coding (see also 1.4.3.2), a reader can distinguish between a '1,' a '0,' and a collision, as shown in Figure 2.6.

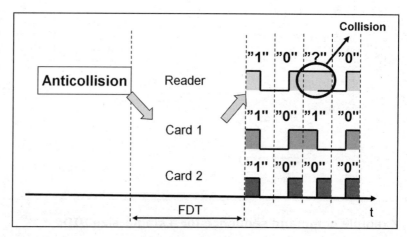

Figure 2.6. Simplified collision-detection principle, Type A.

One prerequisite is that all cards respond at **exactly** the same time: Precisely after the (in this instance) fixed frame delay time (FDT).

In Figure 2.6, one card has a UID with the sequence 1010_{bin}, the other 1000_{bin} (each as part of the complete UID). With the aid of Manchester coding, the reader picks up $10X0_{bin}$, where X represents neither a '1' nor a '0,' but an indeterminate state, or collision.

From this, the reader knows that more than one card is in range. Secondly, it knows that one card must have the sequence $10\mathbf{1}0_{bin}$ and the other the sequence $10\mathbf{0}0_{bin}$. In Figure 2.7, the anti-collision principle is demonstrated in an example. There are three cards in the field, with UIDs having sequences 1011_{bin} (Card 1), 0101_{bin} (Card 2) and 1010_{bin} (Card 3). The first Anti-collision command detects a collision in the first bit of the sequence and the command is repeated, this time with the addition of a '1' in the first digit of the sequence. With this, all cards with a '1' at this point in the sequence respond, i.e. only Cards 1 and 3.

The reader detects another collision, this time in the fourth bit of the sequence, and repeats the Anti-collision command, this time with 1011_{bin} as a parameter. Only Card 1 will respond, i.e. no further collision is detected. The reader may now select the final remaining card.

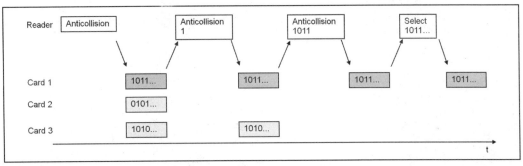

Fig. 2.7. *Type A conflict resolution principle.*

Cards 2 and 3 automatically fall back to the IDLE state when an Anti-collision command with invalid subsequent bytes are received.

If the reader device subsequently wishes to switch to selection of Card 3, Card 1 must be switched into the HALT state using HLTA, and the REQA–Anti-collision–Select sequence must be repeated.

Similarly, after that, the reader might send Card 3 into the HALT state and use another REQA–Anti-collision-Select sequence to select Card 2.

Note	Should an error occur, the card always returns to the IDLE state. Exception: If the error occurs and the card was previously been in the HALT state, the card falls back into the HALT state.
	You could picture this as the HLTA command setting a HALT flag on the card, which is cleared only at POR. If this flag is set, the card will always return to the HALT state in the event of an error.
	This applies to all error conditions during the card activation sequence (or Part 3 of ISO/IEC 14443 as it applies to Type A cards).

2 OVERVIEW OF THE RELEVANT STANDARDS

2.1.3.5 Type B card activation

Type B cards don't make use of UIDs, but, instead, random numbers for collision avoidance. Type B cards have no collision detection and resolution at the bit level.

In Type B, the reader offers several time slots (N), from which the card, with the help of a random number (R), selects. When multiple cards select the same time slot (causing the reader to receive invalid data), the process is repeated.

A (very) simplified Type B state diagram is shown in Figure 2.8.

The indices represent the relevant parameters (AFI, N and R), as explained below.

> **Note** To understand the state diagram, one has to know the meaning of the parameters AFI, PUPI, N, and R.

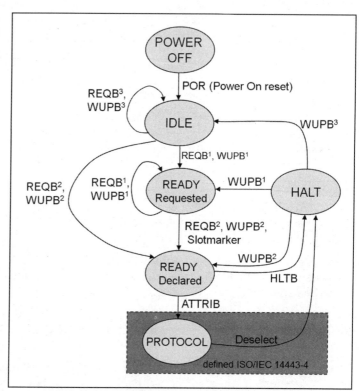

Figure 2.8. Simplified Type B state diagram.

The following four commands are used for card activation:

REQB: The REQB (Request command, Type B = 05 AFI PARAM CRC1 CRC2$_{hex}$) is used to poll the card. When a card is active (i.e. when it has enough energy to execute a power-on reset and automatically enter the IDLE state) and the parameters of the REQB command are applicable, the card enters the next state and responds with ATQB (Answer to Request, Type B).

REQB[1] or WUPB[1]

When

- the AFI is valid;
- the reader sets more than one time slot (N > 1), and
- the random number is greater than 1 (R > 1),

the card responds and enters (or remains in) the READY_REQUESTED state.

REQB[2] or WUPB[2]

When

- the AFI is valid;
- the reader sets more than one time slot (N > 1), and
- the random number is equal to 1 (R = 1),

or

- the AFI is valid and
- the reader sets exactly one time slot (N = 1),

the card responds with ATQB and enters the READY_DECLARED state.

REQB[3] or WUPB[3]

When

- the AFI is not valid

the card does not respond at all, and enters or remains in the IDLE state.

WUPB: The WUPB (Wake Up, Type B = 05 AFI PARAM CRC1 CRC2$_{hex}$) command differs from REQB only in the PARAM byte (see below). It is essentially used to bring one or more cards out of the HALT state and into their respective Ready states, or it may also be used to poll the cards.

| Note | The difference between WUPB and REQB is that a WUPB brings the cards from either the IDLE state or the HALT state into the READY_RE-QUESTED state or READY_DECLARED state, while REQB only applies only to the IDLE state. |

| Note | A WUPB with an invalid AFI returns the card to the IDLE state. |

ATTRIB: The ATTRIB (ATTRIB, Type B = 1D [four-byte PUPI] [four-byte parameter] [optional higher layer INF bytes] CRC1 CRC2$_{hex}$) command selects a card using its PUPI. The selected card responds with Answer-to-ATTRIB and enters the PROTOCOL state.

> **Note** The ATTRIB command and the Answer-to-ATTRIB response contain the parameters necessary for engaging up the protocol. In Type A, this happens with the RATS command and ATS response (see Chapter 2.1.4).

Slot MARKER: This (Slot MARKER, Type B = **n**5$_{hex}$ CRC1 CRC2$_{hex}$) command always initializes a new time slot (n + 1), to which a card whose random number (R) is equal to n + 1 will respond.

> **Note** There are Type B cards that do not support the Slot MARKER command. These would need to be used individually within the reader's field.

HLTB: If the card is in the READY_DECLARED state, it may be brought into the HALT state using the HLTB (Halt, Type B = 50 [four-byte PUPI] CRC1 CRC2$_{hex}$). From the HALT state, the card may only be revived using the WUPB command. Note that, in error situations, unlike Type A cards, Type Bs tend to remain in the state they're already in.

2.1.3.6 Type B: Card activation parameters

As mentioned previously, one needs to understand the following parameters for Type B card activation:

AFI: The AFI (Application Family Identifier, Type B) parameter is a byte contained in the ATQB and WUPB commands that enables (possible) preselection. Simply put, the reader can send a REQB or WUPB containing a specific AFI, and only those cards that belong to this AFI will respond with their ATQB.

If the reader sends an AFI of 00$_{hex}$, all cards are required to respond.

N: Using the N (Number of Anti-collision Slots, Type B) parameter, the reader may set the number of time slots for the Type B anti-collision procedure. N may be used to specify one, two, four, eight or 16 time slots.

> **Note** There are Type B cards that do not support N > 1. These would need to be used individually within the reader's field.

R: With a (random) R (Slot Number Chosen, Type B) parameter, the card can select one of the predefined time slots, in which the card must respond with ATQB.

PARAM: The PARAM (Parameter, Type B) byte contained in RQB and WUPB commands uses the encoding shown in Table 2.3.

Table 2.3. Encoding of PARAM and N in accordance with ISO/IEC 14443.

b8	b7	b6	b5	b4	b3	b2	b1	Meaning
0	0	0		0				REQB
0	0	0		1				WUPB
0	0	0			0	0	0	$N = 2^0 = 1$
0	0	0			0	0	1	$N = 2^1 = 2$
0	0	0			0	1	0	$N = 2^2 = 4$
0	0	0			0	1	1	$N = 2^3 = 8$
0	0	0			1	0	0	$N = 2^4 = 16$
0	0	0	0					The reader does not support an (optional) advanced ATQB.
0	0	0	1					The reader supports an (optional) advanced ATQB.

PUPI: The PUPI (Pseudo-unique PICC Identifier, Type B) parameter may be either dynamically generated (random) or fixed. If random, the random number will be generated during POR, and may not change during an activation.

> **Note** Warning: There may be (old) Type B cards that generate a new PUPI when they transition from the HALT into the IDLE state.

2.1.4 Part 4: communication protocol

Part 4 of ISO/IEC 14443 describes the transmission protocol that allows the reader to exchange data with the card. This protocol can be used to transmit proprietary card commands or higher-layer standardized commands.

The ISO/IEC 14443-4 protocol is also referred to as 'T=CL,' in line with the two protocol variants, T=0 and T=1, from the ISO/IEC 7816 standard for contact cards.

> **Note** The entire communication protocol for Types A and B uses a 2-byte CRC, as defined for Type A in Part 3 of ISO/IEC 14443. In our case, this CRC is automatically generated (when sent) or checked (when received) by the reader IC.

2.1.4.1 Protocol activation

Two commands are available for Type A protocol activation (see Figure 2.9). The first command, RATS, is mandatory and obviates the need for protocol parameters.

The second command, PPS, is optional and is currently used only for shifting up to higher data rates.

2 OVERVIEW OF THE RELEVANT STANDARDS

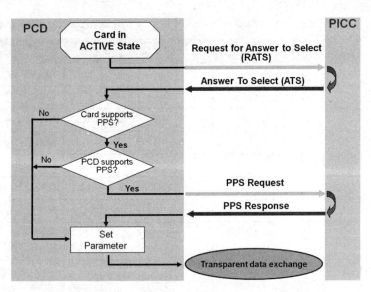

Figure 2.9. Type A protocol activation.

Note The parameters passed for Type B protocol activation are also defined in ISO/IEC 14443 Part 3.

Request for Answer To Select (RATS)

Once the card is activated and its SAK byte indicates that it supports T=CL, the reader must send a Request for Answer To Select (RATS). The card responds with an ATS.

The RATS consists of 2 bytes: the start byte, $E0_{hex}$, is followed by the parameter byte. The parameter byte contains two items of information, as illustrated in Figure 2.10.

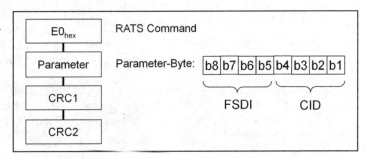

Figure 2.10. Request for Answer To Select (RATS).

The card identifier code (CID), a number between 0 and 14, is encoded in the low nibble of the parameter byte.

- A CID of 0 means there is no CID data in the exchange that follows.
- A CID of 1–14 assigns the selected CID to a card (if it supports CID).
- A CID of 15 is not permitted.

Using CIDs of between 1 and 14, one could activate multiple cards simultaneously. By assigning each new card a new CID, the protocol can make use of up to 14 cards simultaneously.

> **Note** Even though it's possible to activate multiple cards simultaneously, you should consider whether the application really warrants it. The more cards that are active, the higher the basic risk that, for example, any individual card will receive too little energy.
>
> Nevertheless, with the appropriate reader, ISO/IEC 14443 cards do work in multiple card configurations.

The high nibble of the parameter byte contains the 'Frame Size for proximity coupling Device Integer' (FSDI) value. The FSDI specifies the reader's receive buffer size (FSD) as shown in Table 2.4.

Table 2.4. Coding of FSDI according to ISO/IEC 14443.

FSDI (Hex)	0	1	2	3	4	5	6	7	8	9–F
Reader receive buffer size (FSD) (bytes)	16	24	32	40	48	64	96	128	256	RFU

Answer To Select (ATS)

Upon RATS, the card responds with ATS, as shown in Figure 2.11.

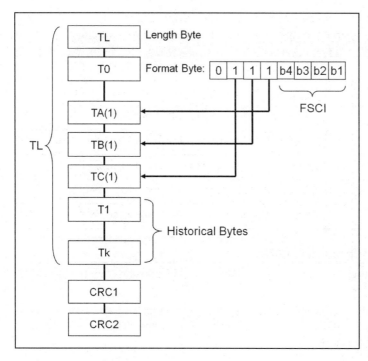

Figure 2.11.
Answer to Select (ATS).

2 OVERVIEW OF THE RELEVANT STANDARDS

The ATS consists of at least the Length Byte (TL) and the Format Byte (T0), but usually also includes the interface bytes TA(1), TB(1) and TC(1), and the Historical Bytes.

TL specifies the length of the ATS, including itself, but excludes the CRC.

> **Note** The number of historical bytes is not limited, but the entire ATS length must, of course, not exceed the size of the reader's receive buffer, or frame size (FSD).

In the low nibble of format byte **T0**, the 'Frame Size for proximity Card Integer' (FSCI) is encoded. The FSCI specifies the card's receive buffer size (FSC), as shown in Table 2.5.

Table 2.5. FSCI encoding, according to ISO/IEC 14443.

FSCI (Hex)	0	1	2	3	4	5	6	7	8	9–F
Card receive buffer (FSC)(bytes)	16	24	32	40	48	64	96	128	256	RFU

Bits 5–7 of T0 indicate whether the respective interface bytes, TA(1), TB(1) or TC(1) are present ('1') or not ('0'), as shown in Figure 2.11.

Interface byte TA (1)

Interface byte **TA(1)** indicates whether the card supports higher data rates, as shown in Table 2.6.

Table 2.6. Interface byte TA (1) encoding, according to ISO/IEC 14443.

b8	b7	b6	b5	b4	b3	b2	b1	Meaning
			0				1	DR = 2: Card supports 212 Kb/s PCD –> PICC.
			0			1		DR = 4: Card supports 424 Kb/s PCD –> PICC.
			0	1				DR = 8: Card supports 848 Kb/s PCD –> PICC.
		1	0					DS = 2: Card supports 212 Kb/s PICC –> PCD.
	1		0					DS = 2: Card supports 424 Kb/s PICC –> PCD.
1			0					DS=2: Card supports 848 Kb/s PICC –> PCD.
0			0					Card supports different send and receive data rates.
1			0					Card does not support different send and receive data rates.

DR: 'divisor receive' (seen from the card side)   DS: 'divisor send' (seen from the card side)

Example: A typical value for TA(1) is F7$_{hex}$. In this case, the card supports all possible data rates in both directions, with different data rates for uplink (PICC –> PCD) and downlink (PCD –> PICC), respectively.

Interface byte TB(1)

The TB(1) interface byte defines two time periods: the SFGT and the FWT, as shown in Figure 2.12.

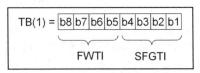

Figure 2.12.
TB(1) interface byte.

In the low nibble, the Startup Guard time Integer (SFGI) is used to set the **Startup Frame Guard Time** (**SFGT**), that is, the minimum time the reader should wait after receiving the ATS before it may send the next command. This time defines a waiting time in which the card operating system may return to full operation without being interrupted.

$$SFGT = (256 \times 16 / f_c) \times 2^{SFGI} \qquad \text{where } f_c = 13.56 \text{ MHz} \qquad \textbf{Eq. 2.1}$$

While an SFGI of 0 indicates no waiting period and an SFGI of 14 is not permitted, the smallest possible SFGT is therefore approximately 600 μs (SFGI = 1), and the largest is about 5 s (SFGI = 14).

The high nibble contains the Frame Waiting time Integer (FWI), which sets the **Frame Waiting Time** (**FWT**), that is, the maximum time that the reader may wait for a response from the card. Thus, the FWT determines the timeout period.

$$FWT = (256 \times 16 / f_c) \times 2^{FWI} \qquad \text{where } f_c = 13.56 \text{ MHz} \qquad \textbf{Eq. 2.2}$$

The shortest possible FWT is therefore about 300 μs (FWI = 0), while the longest is almost 5 s (FWI = 14). An FWI of 15 is not permitted.

Interface byte TC(1)

Only the two least significant bits are used in the TC(1) interface byte, as shown in Table 2.7.

Table 2.7. TC(1) interface byte coding according to ISO/IEC 14443.

b8	b7	b6	b5	b4	b3	b2	b1	Meaning
0	0	0	0	0	0		1	Card supports "Node ADdress" (NAD).
0	0	0	0	0	0	1		Card supports "Card IDentifier" (CID).

All the other bits must be set to '0'.

The **Node Address** (**NAD**) is defined and used in ISO/IEC 7816 and is not discussed further here.

If the card supports a 'Card Identifier,' it sets Bit 2 of the TC(1) byte, telling the reader that it will automatically accept the CID prescribed by the reader in its RATS. The card will then accept all subsequent commands only if they have the corresponding CID.

2 OVERVIEW OF THE RELEVANT STANDARDS

If the card does not support CID (i.e. if its ATS either contains no TC(1) byte, or TC(1)'s bit is '0'), then the reader may not use a CID when communicating with this card.

Historical bytes

The historical bytes are defined in ISO/IEC 7816 and make the encoding of additional information possible.

Information about the operating system or the card version number is often found in these bytes, usually simply ASCII-encoded. MIFARE uses a TLV format (tag—length—value) for an extended type code on all cards that support T=CL.

> **Note** With many cards, the content of the ATS may be changed by authorized users. This offers the advantage of coding (and identification) for specific applications, but also introduces the risk that a standard reader may not be able to activate the card due to an incompatible ATS.
>
> In addition, application-specific codes that are open to anyone may be a security concern.

Protocol and Parameter Selection (PPS) request

Once all the protocol-specific parameters have been set using RATS and ATS, the reader may use the Protocol and Parameter Selection (PPS) request to switch the card to higher data rates, as shown in Figure 2.13 — provided that the reader and card support these.

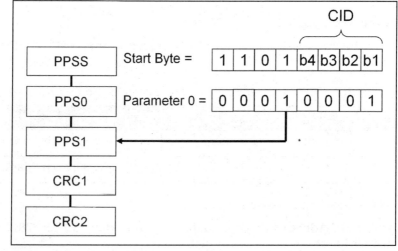

Figure 2.13. PPS Request.

The PPSS is a start byte that includes the CID previously assigned by the RATS (if the card supports a CID). Parameter Byte 0 (PPS0) that follows merely indicates that a PPS1 is next, at least in the current version of ISO/IEC 14443.

The higher data rates are then selected using Parameter Byte 1 (PPS1) as shown in Table 2.8. Typically, DSI = DRI, but if the card allows it (see Bit 8 of TA(1)), different data rates for upload and download may be selected.

Table 2.8. Parameter Byte 1 (PPS1) encoding.

b8	b7	b6	b5	b4	b3	b2	b1	Meaning
0	0	0	0			0	1	DRI = 2: PCD -> PICC = 212 Kb/s
0	0	0	0			1	0	DRI = 4: PCD -> PICC = 424 Kb/s
0	0	0	0			1	1	DRI = 8: PCD -> PICC = 848 Kb/s
0	0	0	0	0	1			DSI = 2: PICC -> PCD = 212 Kb/s
0	0	0	0	1	0			DSI = 4: PICC -> PCD = 424 Kb/s
0	0	0	0	1	1			DSI = 8: PICC -> PCD = 848 Kb/s

Note Bits 6–8 are reserved for future use (RFU) and must be '0.' A PPS request that attempts to set the default data rate (DSI = DRI = 0) is not permitted.

Protocol and Parameter Selection Response (PPS Response)

The card confirms the PPS request with a Protocol and Parameter Selection (PPS) response, as shown in Figure 2.14.

The command that follows will be of the new data rate previously specified.

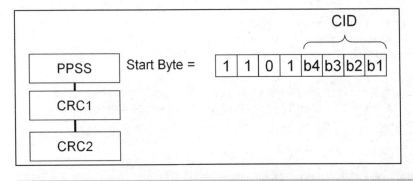

Figure 2.14. PPS response.

Note All communication for card and protocol activation, including the PPS response, takes place at the standard data rate (106 Kb/s). The first command with a higher data rate is the first command after the PPS request and corresponding response.

Note If a PPS is used, it must be sent immediately after the RATS, which must, in turn, be used immediately after Select.

2.1.4.2 T=CL protocol block structure

The T=CL protocol always uses the block structure shown in Figure 2.15. The frame sizes, FSC and FDS, dictate how large such a block may be. A block always consists of at least a Protocol Control Byte (PCB) and a CRC.

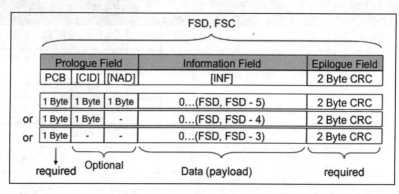

Figure 2.15. T=CL protocol block structure.

If the use of a CID was agreed upon during protocol activation, it will follow the PCB.

If the use of a NAD was agreed upon during protocol activation, it follows the CID (or PCB, if no CID was prescribed).

Depending on the block's type, it may or may not contain an information field of arbitrary length (as long as the entire block does not exceed the maximum frame size). The information field would normally contain the user data, or payload.

There are three different block types:

- I Block: transport of user data
- R Block: 'Acknowledged' and 'Not Acknowledged'
- S Block: 'Deselect' and 'Frame Waiting Time Extension'

> **Note** The T=CL block structure always uses command-and-response pairs, i.e., the reader sends a command block, and the card responds with a response block.

2.1.5 Information Block (I Block)

The I Block is used to exchange user data between card and reader. The encoding of the I Block PCBs is shown in Table 2.9.

The two most significant bits encode the block type: '00' indicates an I Block. The least-significant bit carries a one-bit **block number**, which is toggled at each interaction. This is used for error detection and correction. The first communication block begins with block number 0. Bits 3 and 4 indicate whether a NAD or CID follows.

Table 2.9. I Block PCB encoding.

b8	b7	b6	b5	b4	b3	b2	b1	Meaning
0	0	0				1		I Block
0	0	0				1	0/1	b1: 1-bit block number (it toggles, starting at '0')
0	0	0			1	1		b3: A NAD follows.
0	0	0		1		1		b4: A CID follows.
0	0	0	1			1		b5: chaining bit — more blocks will follow to complete the entire data packet.

Bit 5 is the '**chaining bit,**' which indicates whether with the current block has completed the message ('0') or additional blocks are to follow ('1'). This is needed when a message is longer than the frame size is and has to be transmitted in several parts.

When the receiver receives an I Block and Bit 5 is set, it confirms the reception of the received block with a corresponding R Block (and a matching block number). This is repeated until the receiver gets the final I Block, in which Bit 5 is cleared. Only then is the complete data packet received.

> **Note** Only I Blocks may possess a NAD. The card will respond with the same NAD that the reader sent in its I Block.
>
> When chaining, the NAD is sent only in the first block.

Example of chaining

Figure 2.16 demonstrates chaining. The first card response is broken up into three blocks. It may be a read command, wherein the amount of data to be read exceeds the frame size and must thus be broken up into several blocks. The number in parentheses represents the chaining bit. The number outside the parentheses is the index count of the block number.

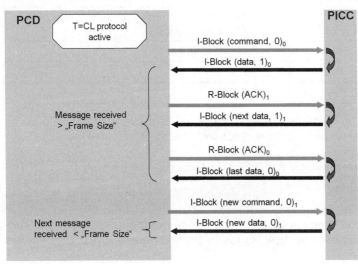

Figure 2.16. T=CL protocol chaining example.

First pair of blocks (block number 0):

The read command (first I Block with block number 0) is met with an I Block in response from the card (using the same '0' block number). The '1' chaining bit tells the reader that the answer is not yet complete.

Second pair of blocks (block number 1):

The reader sends an acknowledge as its next command (R Block with block number 1). The card responds with the second part of the answer (an I Block, also with block number 1). Once again, the chaining bit, '1,' tells the reader that the answer is still not complete.

Third pair of communications (block number 0):

The reader again sends an acknowledge as its next command (R Block, this time with block number 0). The card responds with the third part of the answer (I Block with block number 0). This time, the chaining bit, set to '0,' tells the reader that the answer is complete.

Fourth pair of communications (block number 1):

A new command now follows (with a new block number, 1), which is answered with a single I Block (no chaining).

2.1.5.1 Receive-ready Blocks (R Blocks)

As already mentioned, the R block serves only to transmit an Acknowledged (ACK) or Not Acknowledged (NAK) message and consists simply of a PCB, a CID (where specified) and the CRC. Table 2.10 lays out the encoding.

Table 2.10. R Block PCB encoding.

b8	b7	b6	b5	b4	b3	b2	b1	Meaning
1	0	1			0	1		R Block
1	0	1			0	1	0/1	1-bit block number (toggles — the first block contains 0)
1	0	1		1	0	1		A CID follows.
1	0	1	1		0	1		R(NAK)
1	0	1	0		0	1		R(ACK)

A reader must detect errors in transmission and may correct these with the help of the R Blocks, or, at worst, deselect the card (see Section 2.1.5.2).

> **Note** The card will never send an R(NAK).

Chaining

During chaining, when a valid I Block is received, the block number is toggled, and successful reception is confirmed with an R(ACK) (as in the example above). This applies to both card and reader.

For both card and reader, during chaining, when an invalid I Block is received, the block number is not toggled, and the error is also reported with an R(ACK). This means that, during chaining, should the transmitter receive an R(ACK) with a new block number, it assumes that all is well and continues the chaining by sending the next I Block. Should the transmitter receive an R(ACK) with the previous block number, on the other hand, it repeats the previous I Block.

Error detection and handling

The reader sends an R(NAK) with the **same** block number if the received data contains a transmission error (e.g. CRC error) or if the timeout (FWT) has been exceeded. Similarly, if the card receives an R(NAK) or R(ACK) (when chaining) with **the same** block number, it resends the previous block.

Presence check

If the reader sends an R(NAK) with a **new** block number, the card must respond with an R(ACK). This is used by some readers as a 'presence check,' i.e. the reader can use this R Block to verify that the card is still in range, even when no communication is currently taking place.

> **Note** The presence check is important in, for example, PC/SC readers that inform the PC operating system not only when a card has come into the field, but also when a card has left the field. While this is simple to ascertain when using cards with contacts, a contactless card requires regular polling for the presence check.

> **Note** Alternatively, the reader may do a presence check by sending an R(NAK) with an old block number, and get the card to respond with the previous I Block.
>
> A further way of performing a presence check is to send an empty I Block, which the card must respond to with an I Block.

2.1.5.2 Supervisory Blocks (S Blocks)

There are two different types of S Block: one with an information field and one without. The encoding is shown in Table 2.11.

Table 2.11. S Block PCB encoding

b8	b7	b6	b5	b4	b3	b2	b1	Meaning
1	1				0	1	0	S Block
1	0	0	0		0	1	0	S(Deselect) request and S(Deselect) response
1	0	1	1		0	1	0	S(WTX)
1	0			1	0	1	0	A CID follows.

2 OVERVIEW OF THE RELEVANT STANDARDS

S(Deselect)

The S Block without the information field is used to deselect the card and is called **S(Deselect)**.

A reader may send an S(Deselect) request. The card must respond with an S(Deselect) response, no later than approx. 4.8 ms later.

The effect is the same as for the HALT command in Part 3 of ISO/IEC 14443, bringing the card into the HALT state.

S(WTX)

If the card needs more time than the ATS-specified Frame Waiting Time (FWT) to, for example, process a command, then it may do this using an S(WTX) request, that is, a Waiting Time eXtension request. In this case, the card sends the S(WTX) request instead of a 'normal' response, and the reader responds with an S(WTX) response.

> **Note** Theoretically, an S(WTX) request may be repeated as often as is required.

In Figure 2.17, three card response cases are shown.

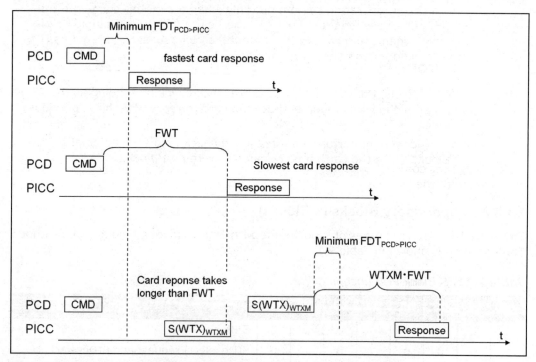

Fig. 2.17. *Frame Waiting Time and Frame Waiting Time Extension.*

In the top diagram, the card responds in the fastest possible time.

> **Note** The minimum time with which the card must comply is defined in ISO/IEC 14443 Part 3 and is called the FDT (Frame Delay Time, here the FDT between PCD Block and PICC Block) for Type A, and TR0 and TR1 for Type B.
>
> For Type A, this time depends on the selected data rates. It is about 80–90 µs. For Type B, a standard time of approximately 170 µs applies (TR0 + TR1). TR0 and TR1 may be reduced by the ATTRIB command's PCD.

In the center diagram, the card responds in the longest time possible. This time, the FWT is determined by the card. See also Section 2.1.4.1.

In the bottom diagram, the card requests an extended FWT by means of an S(WTX) request. The reader acknowledges with an S(WTX) response.

The S(WTX) includes an Information Block, in which the **WTXM** (Waiting Time eXtension Multiplier) parameter is transferred. The WTXM may be any value from 1 to 59, and sets a new, unique, extended FWT.

The normal FWT is multiplied with the WTXM and the result is the new FTW for this S(WTX) command. The new FWT applies only once, i.e. only for the current S(WTX), and the maximum FWT of approx. 5 s may not be exceeded, but the card may repeat the S(WTX) request as many times as necessary.

2.1.6 Electromagnetic disturbance (EMD)

The operation of the first contactless microcontroller cards at the end of the 90s very quickly revealed a fundamental problem. The problem was solved with the appropriate reader software, without it ever being explicitly defined. The problem arose when the controller on the card used varying amounts of energy, depending on whether its various logic blocks, such as coprocessor and memory units, were in use. This temporally variable current consumption represents not much more than a load modulation that interferes with the 'actual' intended load modulation.

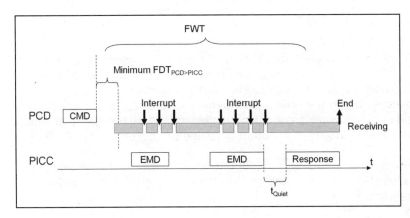

Figure 2.18. Electromagnetic disturbance (EMD).

2 OVERVIEW OF THE RELEVANT STANDARDS

This card-generated interference is known as electromagnetic disturbance (EMD) (Figure 2.18).

While the reader awaits a response from the card, it receives this interference, which it must identify as such, so that the reception can be broken off as quickly as possible. As soon as possible thereafter, the receiver must be re-enabled. This may occur several times during data transmission. The important thing is that the receiver is ready in time to receive the correct response. This is especially important under the 14443-4 protocol, because the card response is not expected at an exact point in time, but may respond during the (possibly large) time period between Minimum $FDT_{PCD>PICC}$ (approx. 80 μs) and the FWT, which may be as long as almost 5 s.

> **Note** During Type A protocol activation (anti-collision), the card response time (FDT) is fixed, but the card must, nonetheless, not present any interference, especially to existing readers, as these typically have no built-in EMD mechanisms.

The reader must thus distinguish between

- a valid card response (reception is complete);
- an invalid card response (receive error: reception is terminated with the appropriate error signal);
- no response (timeout: reception is terminated with the appropriate error signal);
- EMD (interference: reception will continue).

Actually, we must now define the difference between an erroneous card response and interference. In practice, existing data integrity safeguards are used: valid frame, parity (Type A) and CRC.

If only a few corrupt bits are received, it is almost certainly due to EMD.

For card and reader interoperability, it is necessary for the ISO/IEC 14443 standard to define a few EMD fault parameters and tolerances. At the time of writing, this is taking place in the SC17/WG8.

2.1.6.1 Rest period

The rest period is the time immediately prior to the card response, in which the card may not present 'any' EMD, so that the reader is still able to receive clean data despite, in the worst case, an EMD interruption shortly before the card's response. The rest period is determined by the speed at which the reader can identify EMD and then re-enable the receiver. It could be called a 'recovery' period, as it is determined by the speed at which the receiver can revert back to 'normal' reception following the occurrence of EMD.

The rest period is thus, on the one hand, a minimum period with which the card must comply, and, on the other, the maximum period that the reader may occupy before returning to normal reception.

> **Note** This rest period is the first time-critical requirement of the reader. It guarantees a minimum speed in communication between reader IC and microcontroller, as well as for the microcontroller itself.

2.1.6.2 Rest level

The standard defines the minimum magnitude of the card response (see Section 1.4.3.1 and Section 2.1.2), but not how low the level should be in the at-rest state. To differentiate between rest and EMD, a load modulation amplitude must be set, which the card may not exceed during the rest phase (t_{rest}). A maximum value must thus be defined for the card.

Also, the receiver's sensitivity must be limited, as it must ignore signals received below this threshold. That is not quite trivial, as the various reader antennas have very different coupling properties. In addition, it is inadvisable to reduce the reading range for cards, which are already working on less than the 14443 minimum field strength (such as, for example, all MIFARE cards).

> **Note** No currently existing MIFARE cards produce EMD above the rest level's interference threshold. This also applies to microcontroller-based MIFARE cards (e.g. MIFARE DESFire and MIFARE Plus).

2.1.6.3 Distinction between invalid card response and EMD

A distinction between an invalid card response and EMD is necessary, firstly because reception is interrupted, and, where applicable, error correction (at the protocol level) must take place. On the other hand, the receiver must continue to receive (or be switched on), so that it doesn't miss the rest of the card's response.

This definition is not as trivial as it looks at first glance, since many readers have already implemented several differentiation mechanisms. It's practical to assess these differences using the appropriate test signals, with which the readers must be tested.

2.1.6.4 The MFRC522 reader IC and EMD

The MFRC522 allows for EMD using several functions. This more practical part could just as well be in Section 6.1 or Section 9.1, but, because this has direct relevance to the functions that deal with EMD, these functions are described here.

RxNoErr

Firstly, one may simply suppress all disturbances that are shorter than 4 bits, using the RxNoErr (RxModeReg, 0x13) bit. In so doing, most card level changes will not lead to transmission interruption.

RxMultiple

On the other hand, the RxMultiple bit (RxModeReg, 0x13) may be used. If this bit is set, reception is not ended, but restarted, depending on whether valid or invalid data was received. The contents of the error register are automatically appended to the data so that one need only read all the data from the FIFO. Then, if no errors are detected (i.e. ErrorReg = 0), the MFRC522's receiver must be deactivated, e.g. by sending the IDLE command.

If an error is detected, one must ascertain whether this is as the result of an invalid card response (just a CRC error), or EMD (parity error, collision error or frame error). In the former situation, transmission is broken off and error correction is undertaken, as described in Part 4 of the 14443. In the latter, the receiver is allowed to keep waiting until valid data is received or a timeout is triggered (end of the FWT).

RxWait

The time in which the microcontroller must read the FIFO (when RxMultiple is set), is determined by RxWait (RxSelReg, 0x17). With RxWait, one sets the number of bit periods for which the receiver is turned off.

This happens either directly after switching from transmit to receive mode (when using the transceive command) or, after turning off the receiver (when using RxMultiple).

This period must not be so short that the data and error registers cannot be read in time, or so long that the rest period (t_{rest}) is exceeded.

> **Note** When the receiver is (automatically) re-enabled, the error register is automatically cleared as soon as RxWait has expired.

2.2 ISO/IEC 10373-6 test methods

To guarantee compatibility with the ISO/IEC 14443 standard, test methods that are described in the ISO/IEC 10373 standard must be used. This part has grown tremendously in recent years as, even without the new passport extensions (approx. 80 pages), there are currently (2010) over 150 pages dealing with tests, test rigs and reference equipment. The title, "Identification cards — Test methods — Part 6: Proximity cards", is somewhat misleading because it also applies to the readers of these proximity cards.

> **Note** The following description provides an insight into the methods of measurement. It is not a substitute for actually reading the standards.

2.2.1 Test equipment

For the measurements, one needs an oscilloscope (of adequate bandwidth), an impedance measurement device, a power amplifier (> 70 W) and a 13.56 MHz signal generator, as well as the equipment we shall describe below: calibration coil, test PCD setup and ReferencePICC.

2.2.1.1 Calibration coil

The calibration coil is nothing more than a single-wind rectangular coil measuring 72 mm × 42 mm with rounded corners (5 mm radius). The magnetic field strength can be calculated by measuring the voltage from such a coil:

$$\frac{H_{rms}}{A/m} = \frac{0.9 \cdot U_{pp}}{V} \qquad \text{where } U = \text{measured open-circuit voltage} \qquad Eq.\ 2.3$$

> **Note** A prerequisite for correct measurement using this coil and the corresponding equation is a homogenous magnetic field that can be integrated over the entire coil surface. This is supplied, for example, by the test PCD assembly.

2.2.1.2 Test PCD assembly

The test PCD assembly consists of a Helmholtz arrangement, as shown in Figure 2.19, left. The center coil is the transmitter coil with a radius $r = 15$ cm (see also Figure 3.24) which is matched to 50 Ω, and a power amplifier feeds the 13.56 MHz carrier to it.

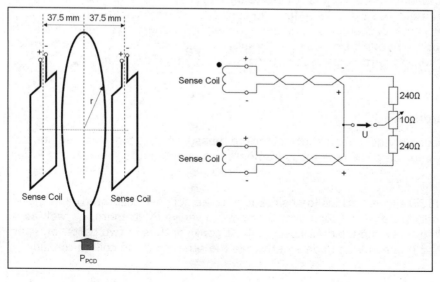

Figure 2.19. Test PCD assembly (ISO/IEC 10373-6).

2 OVERVIEW OF THE RELEVANT STANDARDS

On the left and right of the transmitter coil, each at a distance of 37.5 mm, are two 'sense coils,' with which, firstly, calibration then, later, measurements are done. The 'sense coils' are 100 mm × 70 mm in size, with rounded corners (10 mm radius).

> **Note** There are two different antenna adjustments on the transmitter coil. The (original) one is of relatively high Q factor and can thus be supplied with relatively low power. However, with such high Q factor, one cannot achieve the slope steepness for card testing at higher data rates, so there is an amternative matching with lower Q.

The sense coil connections to the bridge circuit, as shown in Figure 2.19, must be as short as possible and absolutely symmetrical, so that there is sufficient noise suppression. For this, the lead is twisted.

Prior to a measurement, the assembly is calibrated as the voltage, U, is set to zero. If $U = 0$, the assembly is symmetrical, and therefore the voltages at the two «sense coils» are exactly equal in magnitude and phase: the 13.56 MHz carrier is completely cancelled out.

For the various measurements, one card (device under test) is placed in the middle of one sense coil, and the calibration coil is placed in the middle of the other sense coil. The correct field strength is then set by means of the calibration coil, and the bridge circuit allows the measurement voltage U to be reduced. Firstly, with the voltage U, one can observe that the card is responding (that is, working) at a defined field strength. Secondly, the response (load modulation amplitude) can be measured and evaluated. The evaluation of load modulation is done with a discrete Fourier transform, which allows for the measurement of the levels of the upper and lower sidebands, independently of the phase position.

> **Note** The load of a real card is cancelled out, i.e. the card always sees the defined field strength. So, when one card taxes the assembly more than another card, both cards still behave exactly the same. However, a real reader may exhibit completely different behaviour, i.e. it could be that the second card has a higher coverage range than the first.

2.2.1.3 ReferencePICC

The test PCD assembly described above is for the measurements that must be done on the card. To evaluate the reader's measurements, we need another device: the ReferencePICC.

The ReferencePICC allows various load situations to be set (i.e. calibrated) and makes it possible, under these load conditions, to measure the field strength, as well as to generate a defined load modulation. Essentially, it consists thus of two parts: one part presents the load (main coil on underside), while the other (pick up coil on top side) is measured.

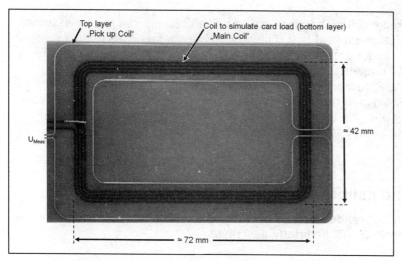

Figure 2.20. ReferencePICC (ISO/IEC 10373-6).

Load simulation

The load simulation must act as much as possible like a real card. Therefore, the main coil is the same size and shape as a typical card antenna, as can be seen in Figure 2.20 (underside). The block diagram is shown in Figure 2.21, below. Using a parallel capacitor, C, the resonance frequency is tuned to the desired value. Using R_L, the load behind the rectifier may be adjusted to simulate the load of a real card.

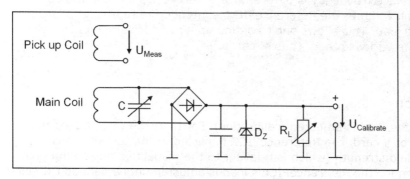

Figure 2.21. ReferencePICC block diagram (ISO/IEC 10373-6).

Measurement coil

The measurement coil on the upper side (pick-up coil) has a shape that minimizes any coupling with the main coil on the underside, while the area spanned by it is large enough that it receives enough signal from the reader's magnetic field. The reader's transmission pulses are captured at the measurement coil.

The exact size of the measuring coil is not important, because only relative signals are measured.

The complete circuit of the ReferencePICC also contains an active circuit for generating the load modulation. This is not shown here. The load is thus switched with the help of an injected subcarrier signal behind the rectifier.

2 OVERVIEW OF THE RELEVANT STANDARDS

> **Note** The required isolation between card coil and pick-up coil is only assumed when the magnetic field is sufficiently homogeneous. This does not apply to smaller reader antennas especially, for example the antenna on the Elektor RFID Reader. It is then better to use one's own pick-up coil instead of the ReferencePICC. Such a pick-up coil can certainly be made of several windings to reach a high enough voltage, and should be placed as close as possible to the reader. It is also important that it is decoupled as much as possible from the card antenna.

2.2.2 Tuning and calibration

Below, we discuss only the reader tests, so the (complex) processes for the tuning and calibration of ReferencePICC will be briefly described.

2.2.2.1 Tuning

Upon application, the ReferencePICC is first tuned to the correct resonance frequency. For this, the correct measurement setup bust be used at the impedance measuring device, and it must be set to a specific, defined load using R_L.

> **Note** The resonance frequency is different for different measurements, depending on whether the ReferencePICC is intended to represent a large or a small load. This allows one to define upper and lower limits and also to simulate various other types of cards.

2.2.2.2 Calibration

In the second step, the load is adjusted (calibration). For that, the test PCD assembly is required. Instead of a card, the ReferencePICC is placed in the test PCD assembly. The corresponding field strength (which is relevant to the respective measuring point) is set, and then, using R_L, the ReferencePICC's load is adjusted until a defined DC voltage, $U_{calibration}$, is measured. Because a change in load affects the field strength, these two steps need to be repeated a few times.

Example: We want to determine the maximum (energy) range of a reader. According to ISO/IEC 14443 there should be a field strength of 1.5 A/m. For this, we place the ReferencePICC (tuned to 13.56 MHz) in the test PCD assembly and set it to 1.5 A/m (measured at the calibration coil). Then, R_L is used to set the measured voltage, $U_{calibration}$, to 3.0 V. Adjusting R_L changes the field strength, so we must readjust the amplifier power until we again measure 1.5 A/m at the calibration coil. This, in turn, changes the $U_{calibration}$ voltage, so R_L must be readjusted. This is repeated a few times, until we get a field strength $H = 1.5$ A/m at $U_{calibration} = 3.0$ V. The ReferencePICC is now calibrated for the measurements to follow.

2.2.3 Measurements at the reader

Measurement can now be carried out using the tuned and calibrated ReferencePICC. For range measurement, for example, we place the ReferencePICC in the corresponding position at the reader antenna and measure $U_{calibration}$. In the position in which we measure the same voltage as previously set during calibration, we set the same field strength as during calibration.

2.2.3.1 Range measurement

After being calibrated in the test PCD assembly at 1.5 A/m with $U_{calibration} = 3.0$ V, the ReferencePICC is introduced slowly into the field. At this time, $U_{calibration}$ is measured. With this, we can now determine the maximum reading range: At precisely the point at which $U_{calibration} = 3.0$ V, we have precisely 1.5 A/m.

At this point, a second essential measurement is required. The transmitted pulse shapes for Type A and Type B, as well as their predetermined permitted rise and fall times and overshoots, must be checked.

A further essential measurement, also at this point, is to inject a defined load modulation (which must previously have been calibrated on the test PCD assembly), to check that the reader is able to receive this load modulation.

2.2.3.2 Measurement effort

Together with the respective adjustment and calibration, the complete testing of a reader can be quite laborious. When one considerers that the measurements also need to be done over the complete temperature range, which means that the entire test assembly must be operated in a climate chamber, under both sub-zero temperatures and at high temperatures, one would certainly expect to be able to automate several of the measurements parameters.

It is possible to replace R_L on the ReferencePICC with a voltage-controlled resistor, in order to carry out both calibration and measurement remotely: An externally-applied voltage replaces the need to turn the potentiometer.

> **Note** The automation of the tuning part of the process is a somewhat complicated, so it helps use several identical, but differently-tuned, ReferencePICCs, for example.

Aside from the ReferencePICC positioning in relation to the reader antenna, no one needs to be 'hands-on.' In a first step, one may calibrate the ReferencePICC. In a second, the ReferencePICC is positioned and the individual measurements are carried out using the stored values. If the positioning relative to the reader antenna can be done using a robot arm, then the complete measurement procedure may be carried out with relatively little effort.

> **Note** Given the fact that while there are many different card readers, but only a relatively small number of different card types, it would perhaps be better to have a measurement procedure that requires more effort on the card end (and thus less on the reader end).

2.2.4 Tests for Layer 3 and 4

The largest part of the 10373-6 consists of the functional tests for 14443 Layer 3 (card activation) and 4 (protocol). Appendices G, H and J define all possible test scenarios for both cards and readers. These are 'normative', i.e. compulsory.

Every test case contained therein describes the content of the test ("what does this case test?"), the test procedure (step-by-step instructions) as well as the extended test protocol ("what must be achieved?").

2.3 Near Field Communication (NFC)

2.3.1 Introduction

In 2002, Philips Semiconductors (now NXP Semiconductors) and Sony developed a new technology, which they called Near Field Communication (NFC). Both companies already had several years of experience in the smartcard market and several smartcards in their product portfolios.

The idea behind this was simple and logical: instead of various different contactless smartcards that must be carried around for each different application, a single mobile device, which is always with the user, could be used for the same purpose. By mobile device, naturally the first thing that comes to mind is the mobile phone. A cellphone is equipped with a contactless interface in order for it to behave like a 14443-compatible smartcard, for example.

NFC goes two steps further: for one, the cellphone battery can be used to enable an 'active' mode. In this mode, the phone may also be used as a *reader* for 14443-compatible cards. So, with an NFC-compatible cellphone, one automatically has a mobile contactless device.

On the other hand, it can also be very useful to have the functions and interfaces on a cellphone available to the smartcard world. If a phone can behave like a contactless smartcard, then the user need only load suitable card applications, via GSM, Bluetooth or other interface. Simply put, cellular users may simply download their travel tickets when necessary.

This is technically relatively easy to do, even though it naturally brings up questions about application security, protection from misuse, encryption, key management, etc. It is somewhat more difficult to describe the business model behind this, which is fast becoming ever-more complex. Who provides, at what price, which hardware or software?

The mobile manufacturer, the operator, the service provider or the smartcard system supplier? Who is responsible for which part of the application?

There is currently an NFC Forum (*http://www.nfc-forum.org*) that attempts to better specify these complex processes and also to break them down into simple use cases.

2.3.2 NFC air interface

To go over all NFC layers is beyond the scope of this book, but we will describe the bottom layer here and briefly present the three modes of operation for the air interface principle.

2.3.2.1 NFC device as card

An NFC device can be used as a normal contactless smartcard. The transfer principle, as illustrated in Figure 2.22, is exactly the same as for 'normal' contactless smartcards as described in Section 1.4. This is required, or else NFC-enabled cellphones would not be useable in existing smartcard applications.

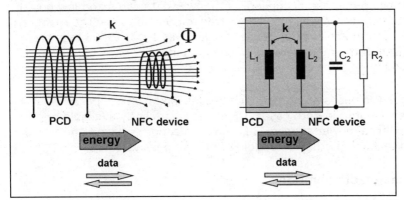

Figure 2.22.
NFC device as a smartcard.

If the NFC mobile device is required to be used in an existing smartcard infrastructure, then it must be ensured that it is in principle 14443-compatible, i.e. that it passes all of the 10373-6 tests. Again, what looks technically simple, can become very complex in practical application, even as far as the form factor of a cellphone being very different from that of a smartcard.

An NFC device can behave like a 14443 Type A card, in addition to as a Sony Felica card. In this mode, the NFC device is known as the 'NFC target.'

Note 'Felica' is a Sonly trademark. Felica cards use a proprietary Sony protocol.

2.3.2.2 NFC device as a reader

To operate as a reader, the same applies: the mobile device must, in principle, pass all applicable 10373-6 tests.

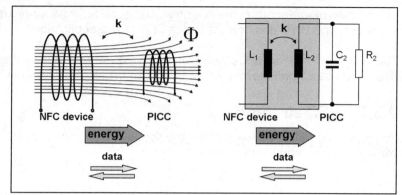

Figure 2.23. NFC device as a reader.

An NFC device can behave as a reader for Sony Felica, ISO/IEC 15693, as well as ISO/IEC 14443 Type A and B cards. In reader mode, the NFC device is referred to as an "NFC initiator."

> **Note** ISO/IEC 15693 is the standard for 'vicinity cards,' which also operate at 13.56 MHz using magnetic coupling. With lower data rates and low power usage by the cards, as well has higher transmitter power on the readers, the 15693 system achieves a greater range (over 1 m). 15693 tags are typically used in the management of goods to identify objects, but there are also ski lift systems that use them for the ski passes.

> **Note** Operating NFC devices as readers or as cards is known as "passive" mode, even though, in both cases, the devices currently available require an external power supply: This communication is based on the smartcard technology that is capable of driving passive cards.

2.3.2.3 NFC device in 'active' mode

Using NFC devices in the 'active' mode, as the name suggests, allows for active communication, in which an NFC device is present on both sides of the communication. One device begins the communication as 'initiator' and the other replies as 'target.' Both initiator and target send actively.

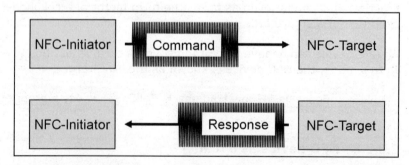

Figure 2.24. NFC device in 'Active' mode.

> **Note** The active mode is not compatible with existing smartcard systems.

3
RFID Antenna Design

Renke Bienert

In order to consider contactless smartcard (PICC) design, we'll need a basic familiarity with the fundamentals of antenna design. Firstly, it's important to explain the general operation of an antenna, and secondly we need to cover the specifics of RFID antennas on both the reader and the card end of the system. Antenna technology is neither trivial nor 'black magic'.

3.1 Theoretical Fundamentals

Although this is intended as practically-oriented book, we can't get around a discussion of a few theoretical concepts. Because contactless smartcards (PICCs) use magnetic coupling for data and energy transfer between reader (PCD) and card (PICC) antennas, antenna coils are required. These are operated at or close to their resonant frequencies. We'll only go as far into the fundamentals as is required to understand the antenna examples described in order to be able to build a functioning PCD with integrated antenna.

3.1.1 Antenna as Resonant Circuit

Figure 3.1 shows a general equivalent circuit diagram for an antenna. Electrically, each antenna is a resonant circuit with a specific 'input impedance'. This input impedance,

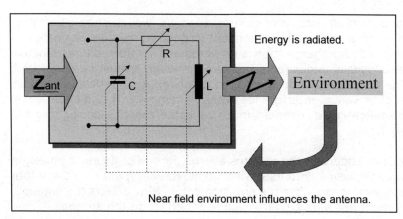

Figure 3.1.
Antenna equivalence circuit.

$\underline{Z}$, is complex, and consists of a real and an imaginary part (complex numbers are shown underlined: $\underline{Z}$, $\underline{I}$, and $\underline{U}$). The resonant frequency has the property that its input impedance is purely real, i.e., the imaginary part is zero.

In our case, the antenna is a parallel resonant circuit, for which the (simplified) electrical equivalent circuit consists of inductance and capacitance, as well as losses (represented by a resistor).

If we run this parallel circuit at its resonance, i.e. create an AC voltage whose frequency corresponds exactly to the resonant frequency, then all of the power that is input is used by the resonant circuit. This 'consumed' power is thus purely active power, and no reactive power is generated. The actual antenna power is the radiated power, excluding heat losses. Both heat losses and radiation losses are represented here as an electrical resistance.

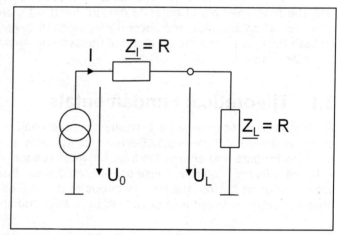

Figure 3.2. Adjustment.

We begin adjustment with the ideal, so that the load resistance (purely real) is exactly the same as the internal resistance of our driver circuit. Then, we feed the maximum power into our antenna (see Figure 3.2).

If we operate the resonant circuit outside of its resonance, i.e. we insert an alternating current whose frequency does not match its resonant frequency, then only a portion of the input power will be 'used'. Another portion is reflected and is referred to as reactive power. Reactive power is usually undesirable because, although it must be generated, it is not available to the consumer (in this case, the antenna). Also, reflected reactive power can create a lot of unwanted interference in circuit components and input leads. At a high power, a mismatch could mean so much reflected reactive power that the circuit may be destroyed.

The transmitter antenna considered here radiates a large portion of its input power, in the form of electromagnetic waves, into its environment. Basically, you could say that the antenna affects its environment, while the environment, in turn, affects the antenna. The 'near field' environment has a particularly significant effect on the antenna.

THEORETICAL FUNDAMENTALS 3.1

The distinction between near field and far field is defined by wavelength λ. All distances less than between 1/10 and ¼ λ are considered near field, while distances greater than 1 λ are considered far field. While electrical and magnetic fields must be considered separately in the near field, in the far field there is a fixed relationship between the magnetic and the electrical field.

Most of the antennas that we encounter every day are electrical antennas, i.e. they produce an electrical field in the near field. Figure 3.3 illustrates a typical example: the electrical monopole antenna. It is fed asymmetrically and will always requires a large enough and correctly shaped ground (GND). This is often a quarter wave (λ/4) ground plane antenna, and what are called 'radials' are angled obliquely downward and connected to ground.

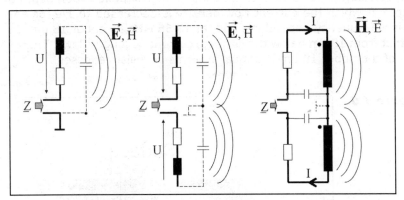

Figure 3.3.
Electrical and magnetic antennas.

Shown in the middle is the electrical dipole. It is fed symmetrically and is thus in principle independent of ground. For this reason, it is often, incidentally, used as a reference antenna.

For our RFID applications, we use magnetic antennas, as we require a magnetic coupling (Figure 3.3, right). Actually, we shouldn't speak of antennas, as we want to avoid radiating signals into the far field. A characteristic that usually makes magnetic antennas undesirable is actually of benefit to us: magnetic antennas are fairly inefficient. This means that only a small part of the input power is radiated.

In Figure 3.4, an RFID antenna's losses are shown. You may assume, for simplicity's sake, that the capacitors we use are lossless at 13.56 MHz. This would not be the case

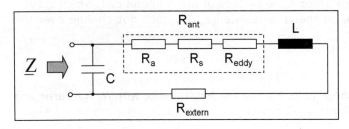

Figure 3.4.
Magnetic antenna losses.

3 RFID ANTENNA DESIGN

for the antenna coil. For clarity, the losses that make up R_{ant} are divided into three main parts:

1) R_{eddy} — eddy current losses

These losses occur when metallic surfaces are penetrated by the magnetic fields. These losses – and more particularly how to avoid them – will be discussed in more detail later (see Section 3.2.2).

2) R_s — skin-effect losses

As with any electrical conductor, resistive losses will occur in the antenna coil. Due to electromagnetic interaction in an electrical conductor, at higher frequencies the alternating current is drawn closer to the surface of the conductor. The penetration depth δ is defined as that depth at which the current flow density is decreased to 1/e, or approximately 37%. This is frequency dependent, and, at 13.56 MHz, is about 18 μm. In practice, we can say that from 18 μm below the surface, current no longer flows. Therefore, the conductivity of the upper 18 μm is critical in terms of resistive losses.

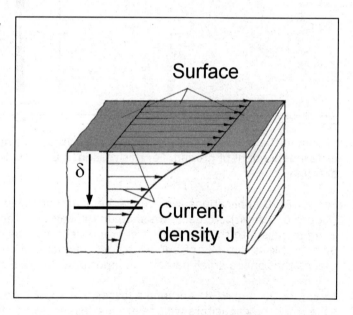

Figure 3.5. Skin effect.

It is thus especially important to protect the surface of our antenna coil, which is typically made up of circuit traces, so that the resistance, R_s, does not change over the course of time.

3) R_a – radiation losses

The radiation resistance, R_a, is negligible. For our reader antenna, with $N_1 = 6$ turns and a mean antenna area of $A \sim 29$ mm × 47 mm, the equation is:

$$R_a/_\Omega \approx 31000 \cdot \left(\frac{N_1 \cdot A}{\lambda^2}\right)^2 = 31000 \cdot \left(\frac{6 \cdot 47 \cdot 10^{-3}m \cdot 29 \cdot 10^{-3}m}{(22m)^2}\right)^2 \approx 8.9 \cdot 10^{-6}$$

Eq. 3.1

A radiation resistance of $R_a \approx 10\ \mu\Omega$ is negligibly small when compared with the other losses. This also means that scant energy is radiated: magnetic antennas have a very low efficiency.

4) Loss due to external resistance

In addition to the existing losses (R_{ant}), we build in more resistance to reduce the Q factor: the resistor $R_{external}$ in Figure 3.4 is a part of the antenna matching circuit. The associated Q factor has two main reasons:

We need to transmit data signals and we require a specific bandwidth (see Section 3.2.1).

The lower the Q factor, the lower the influence of various different cards in the detuning of the antenna.

In practice, external resistors of a few ohms are used, as described in Section 3.2.1.

3.1.2 Transformer Model

As explained in Section 1.4, magnetic coupling is used to facilitate energy and data transfer, i.e., we can think of the two antennas (reader device antenna and card antenna) as a loosely-coupled transformer.

For energy transfer, the reader antenna represents the primary side: L_1 in Figure 3.6, shown without the resonant circuit elements, C and R.

The card is the secondary side: L_2, R_2 and C_2 represent the card in the simplified electrical equivalence diagram.

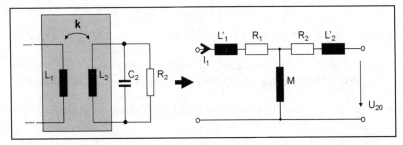

Figure 3.6.
Transformer model.

Unlike the 'normal' transformers used in power engineering, we operate our transmission system at or near its resonance. However, we can apply transformer principles to it in order to better understand and dimension our system.

3 RFID ANTENNA DESIGN

With the help of the mutual inductance, M, we can, for example, calculate the secondary voltage.

Assuming that all of the card parameters (those with index 2) are fixed, that is specified by us and not to be altered, we can also specify a few parameters that will make the maximum voltage (U_{20}) available to the card:

$$U_{20} = \omega \cdot M \cdot I_1 \qquad \text{where } \omega = 2 \cdot \pi \cdot f \qquad \text{Eq. 3.2}$$

Based on the model in Figure 3.6, the specified **parameters are M and I_1**, since the frequency is fixed.

> **Note** By assuming that all card parameters are unalterable, we appear to be ignoring the actual energy requirements of the card during this first approach, because the card parameters do actually vary. However, at the moment we are simply getting the practice right: for our reader design, we must be able to deal various different cards as they are.
>
> The derivations are also applicable in principle to varying card parameters.

Antenna current I_1:

It is easy to see that a higher power from the transmitting antenna will result in a higher voltage in the card. Thus, we try to feed as large as possible a current into the antenna coil. The components present one limitation here. Obviously, the output stage of the MF RC523 can only provide a certain maximum of current.

The European standards for current are another limitation. For example, the laws for EMC (electromagnetic compatibility), and for the use of the ISM frequency band at 13.56 MHz indirectly limit the driving current.

Mutual inductance, M:

The mutual inductance (M) is determined from:

$$M = k \cdot \sqrt{L_1 \cdot L_2} \qquad \text{where } k = \text{coupling factor} \qquad \text{Eq. 3.3}$$

This can also be understood as: the larger the coupling factor (or the closer the card to the reader), the larger the card voltage (or also, in principle, the power at resistor R_2).

Now the question arises: how do I dimension (and optimize) the reader antenna?

3.1.3 The Biot-Savart Law

The Biot-Savart law allows the calculation of spatial magnetic field strength.

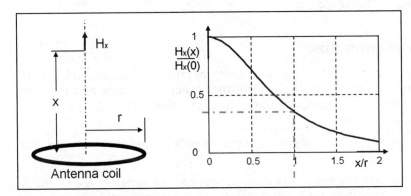

Figure 3.7. Magnetic field strength according to the Biot-Savart law.

For a circular wire loop, as shown in Figure 3.7, left, the magnetic field strength perpendicular to the wire loop may be calculated as follows:

$$H_x = I_1 \cdot \frac{N_1 \cdot r^2}{2 \cdot (r^2 + x^2)^{3/2}}$$ **Eq. 3.4**

where I_1 = current through antenna coil
N_1 = no. of turns of antenna coil
r = antenna coil radius
x = distance from antenna coil

The curve in Figure 3.7, right, graphs the field strength versus distance from the antenna coil. The field strength is normalized at zero distance, where the maximum field strength occurs, while distance is normalized to the antenna coil radius.

If the distance is much larger than the antenna radius, the following derivation can be derived:

$x \gg r$:
$$H_x(x) = \frac{I_1}{2} \cdot \frac{N_1 \cdot r^2}{x^3} \approx \frac{1}{x^3}$$ **Eq. 3.5**

The field strength decreases with the cube of the distance. In other words, for every doubling of distance, I require eight times the current. This applies both to the operation of a magnetic antenna as well as to that of a 'normal' antenna, whose radiated power can be expressed, very simplified, as the surface of a sphere. For the reading distances relevant to the RFID arena, we will not consider these situations.

In addition, something else can be noted. At zero distance, where the maximum field strength occurs, the following may be also be principally derived:

$x = 0$:
$$H_x(0) = \frac{I_1}{2} \cdot \frac{N_1}{r} \approx \frac{1}{r}$$ **Eq. 3.6**

3 RFID ANTENNA DESIGN

Field strength at zero distance is inversely proportional to antenna radius. That is, the smaller the antenna coil, the larger the maximum field strength.

The reason for this is clear: with a smaller antenna, the same energy is distributed over a smaller surface area, so the energy per unit area is larger.

3.1.4 Optimal Antenna Size

For the optimal antenna size, we determine the maximum possible coupling factor. From the law of electromagnetic induction, the transformer model equations and the Biot-Savart law, we can derive the following relationship for coupling factor k:

$$k = \alpha \cdot \frac{r^2}{2(r^2 + x^2)^{3/2}} \cdot \sqrt{\frac{N_1^2 \cdot N_2^2}{L_1 \cdot L_2}} \cdot A_2 \qquad \text{where } \mu_0 = \text{permeability constant} \qquad \textit{Eq. 3.7}$$

If we proceed according to our previous assumption that card parameters N_2, L_2 and A_2 (number of turns, inductance and card antenna coil area) are inalterable, we are left with antenna radius r, distance x, as well as number of turns N_1 and inductance L_1 as parameters, which we can vary in order to maximize k.

Thus, you can show that for large and flat coils (used here), N and L have the following relationship:

$L = L_0 \cdot N^2$, where L_0 = inductance of a single winding (single-turn inductance)

With the introduction of L_0, k may be calculated from:

$$k = \alpha \cdot \frac{r^2}{2(r^2 + x^2)^{3/2}} \cdot \frac{A_2}{\sqrt{L_{01} \cdot L_{02}}} \qquad \textit{Eq. 3.8}$$

We can thus see that the **coupling factor** depends only on r and x and is thus **a purely geometrical quantity**.

Figure 3.8 shows the graph of the coupling factor vs. antenna radius for three different reading distances. As a rough approximation, we can say that, for a given reading distance, the maximum coupling is achieved when the antenna radius is equal to the reading distance (reading distance = half the antenna diameter).

The Figure 3.8 shows the coupling factor k (Y axis) versus the antenna radius (X axis) at three different reading distances (x = 4 cm, 6 cm and 8 cm).

What is clear in the 4 cm example (top curve, the area around maximum) is that the maximum k is reached with antenna radii of between 3 and 5 cm (corresponding to a circular antenna of between 6 cm and 10 cm diameter).

With smaller antennas, k drops off very quickly, so, for example with a 3 cm antenna diameter (top curve at r = 1.5 cm) at the given 4 cm reading distance, the coupling is only half the maximum amplitude.

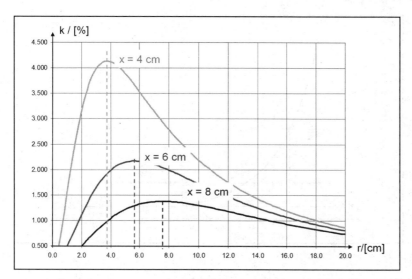

Figure 3.8.
Coupling factors for various distances.

In any case, one can also see that larger antennas don't lead to an increase in k — on the contrary, antennas that are much larger than optimum produce a lower coupling factor.

For a larger reading distance, we do need a larger antenna: the maximum coupling factor required for a 6 cm reading distance is achieved using an antenna diameter of between 10 and 14 cm (middle curve maximum in Figure 3.8).

For us there is a very simplified, but important rule of thumb:

$$r = x \rightarrow radius = distance \qquad \text{Eq. 3.9}$$

A further consideration arises when the coupling factor for two different sized antennas is graphed *vs.* reading distance. In Figure 3.9, the coupling characteristics for both a smaller 4 cm diameter and a larger 10 cm diameter antenna are graphed.

Figure 3.9 shows the coupling factor k (Y axis) *vs.* the reading distance x (x axis) for two different antenna sizes (r = 2 cm and 5 cm).

> **Note** The progression of the coupling factor corresponds to the field strength. The field strength can be altered by adjusting driver current I_1, but the coupling factor cannot. For a card's reading distance, this is significant because a reader with a small antenna but a high driver current (= high field strength) is not very useful because the card's response will not be able to be detected due to inadequate coupling. You can see that the smaller antenna has the greatest coupling at zero distance (= greater field strength).

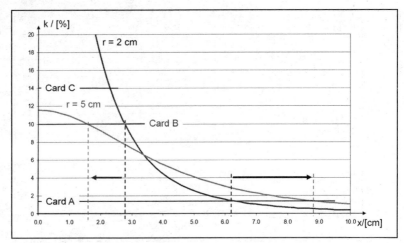

Figure 3.9. Coupling factors for different antenna sizes.

Three horizontal lines represent examples of the different energy requirements for three different cards.

For Card A, which requires a very low coupling factor (field strength), you can also see that, for the larger antenna, the range increases (in this example, from about 6.5 to almost 9 cm).

⮕ Card A: larger antenna = larger range

For Card B, which requires a larger coupling (= greater field strength), you can see that a larger reader antenna actually reduces the reading distance (here from about 3 to about 1.5 cm).

⮕ Card B: larger antenna = smaller range

Card C fares even worse in this example. It needs a much larger coupling (= greater field strength). This card, which is operable at a range of at least 2 cm using a small antenna, doesn't work at all with the larger antenna, due to insufficient coupling.

⮕ Card C: larger antenna = does not work

Note	The figures given here (for k, x, and r) are calculated values based on the previous assumptions. They serve merely to illustrate. The principle applies in practice, but the actual values will no doubt vary from the calculated values.

3.2 Reader Antennas

Using the knowledge from the previous section, we can begin designing our reader antenna. We need to bear in mind that our antenna is not only for energy transfer, but also for ISO/IEC 14443 data.

3.2.1 Antenna Quality

The higher the Q factor of the reader antenna, the greater the radiated field and thus the longer the range, at least in the idle state, i.e. with no load.

In practical application, however, two things limit the Q:

- The required bandwidth.
- The required suppression of interference.

The Q factor under no load is of less importance than the overall Q of the transmission system, which consists of both reader and card. Depending on the coupling, the card will have a larger or smaller retroactive effect on the reader. Different cards have different resonant frequencies (see Section 3.3) and different energy requirements. Thus, the determination of the antenna Q is, in principle, dependent on many (partly unknown) parameters.

3.2.1.1 Data Transmission Bandwidth

For data transfer, a certain bandwidth is required, depending on the coding and modulation methods as well as the required signal-to-noise ratio.

$$B = \frac{f}{Q}$$

where B = bandwidth
Q = Q factor
f = frequency

Eq. 3.10

A first order approximation of Q and bandwidth and the corresponding spectrum for communication according to ISO/IEC 14443 can be seen in the example in Figure 3.10.

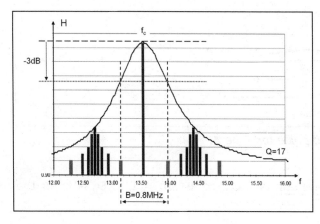

Figure 3.10.
Antenna Q factor.

3 RFID ANTENNA DESIGN

The plotted curve represents a resonant circuit with a resonant frequency of 13.56 MHz and a Q of 17. This results in a bandwidth of about 800 kHz.

The spectral lines show the carriers at 13.56 MHz and the two subcarrier sidebands (13.56 MHz ± 847.5 kHz) and sidebands for the different data rates:

- 13.56 MHz ± 847.5 kHz ± 106 Kbit/s;
- 13.56 MHz ± 847.5 kHz ± 212 Kbit/s;
- 13.56 MHz ± 847.5 kHz ± 424 Kbit/s;
- 13.56 MHz ± 847.5 kHz ± 847.5 Kbit/s.

Note This general representation is merely for illustration purposes. The spectral lines' position and amplitude are not represented exactly.

Determination of Q for Type A Transmission Pulses

However, you can more usefully determine the required Q factor by the transmitted pulses' edges. The pulse rise and fall times are also specified in ISO/IEC 14443, as shown in Figure 3.11 for a Type A transmission pulse at 106 Kbit/s.

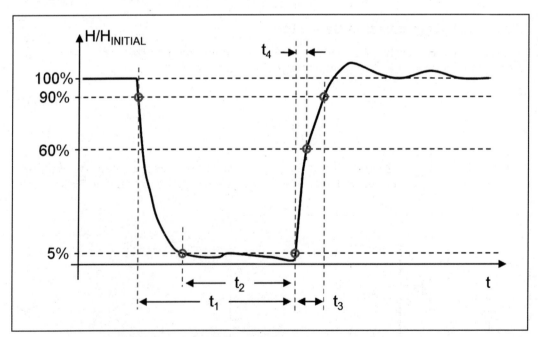

Fig. 3.11. Envelope of a transmitted pulse for Type A.

According to ISO/IEC 14443, the time unit is specified as $1/f_c = 1/13.56$ MHz for the carrier's period.

READER ANTENNAS 3.2

The reader must comply with ISO/IEC 14443 timing specifications (simplified):

$t_1 = 28 / f_c - 40.5 / f_c = 2.06 - 2.99$ µs $\approx 2 - 3$ µs

$t_2 = 7 / f_c - t_1 \qquad\qquad = 516$ ns $- t_1 \qquad > 0.5$ µs

$t_3 = 1.5 \times t_4 - 16 / f_c = 1.5 \times t_4 - 1.18$ µs < 1.2 µs

$t_4 = 0 - 6 / f_c \qquad\qquad = 0 - 442$ ns **< 0.4 µs!**

In practice, the most demanding time period with which we must comply is t_4: at the end of the pulse, the carrier must swing from 5% to 60% of the field strength in no more than 0.4 µs. In other words, if the MF RC523 can comply with t_4, the other timing requirements are usually met.

It is also important to keep the overshoot small:

For 106 Kbit/s in Type A, a maximum $H = H_{INITIAL} \pm 10\%$ is permitted.

In addition, there are local permissible maxima that may not exceed 5% of $H_{INITIAL}$ and may not be longer than 0.5 µs.

Maintaining the timing for Type A usually ensures compliance with Type B. For Type B, however, there are typically two other 'challenges': observing the overshoot restriction, and the modulation index.

Compliance with the Overshoot Restriction in Type B

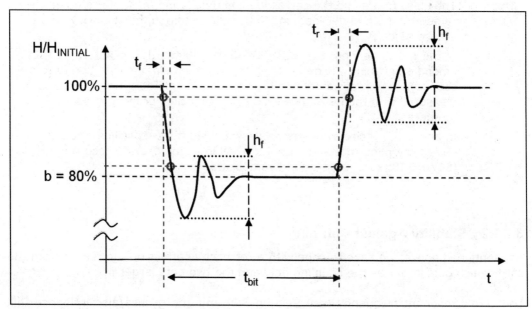

Fig. 3.12. *Envelope of a transmission pulse for Type B.*

3 RFID ANTENNA DESIGN

In Type B, the overshoots, h_r and h_f, as shown in Figure 3.12, may not exceed the following values:

$$h_f \leq \left(1 - \frac{t_f}{2 \cdot t_{f\max}}\right) \cdot 0.10 \cdot (1-b) \quad \text{where} \quad t_{\max} = 16 / f_c \text{ at 106 Kbit/s} \quad \text{Eq. 3.11}$$
$$t_{f\max} = 14 / f_c \text{ at higher bit rates}$$

$$h_r \leq \left(1 - \frac{t_r}{2 \cdot t_{r\max}}\right) \cdot 0.10 \cdot (1-b) \quad \text{where} \quad t_{r\max} = 16 / f_c \text{ at 106 Kbit/s} \quad \text{Eq. 3.12}$$
$$t_{r\max} = 14 / f_c \text{ at higher bit rates}$$

$$b = \frac{1-m}{1+m} \quad \text{where} \quad m = \text{modulation index} \quad \text{Eq. 3.13}$$

An average modulation index of about 11% results in residual carrier $b \approx 80\%$. With a time $t_f = 10 / f_c \approx 0.74$ μs, we get a maximum allowable overshoot of $h_f \approx 1.4\%$!

Compliance with the Modulation Index in Type B

The modulation index with which the reader, both in idle state and under load condition, must comply, is, according to standard $m = 8 - 14\%$. This corresponds to a residual carrier of $b \approx 75 - 85\%$.

> **Note** The pulse forms and the corresponding times (and most of the other parameters), both under load and in the idle condition, must work with a reference card.
>
> In the 150-page long ISO/IEC 10373-6 test standard, the 'Reference PICC', among other things, is defined, which, with the help of calibration (also described), guarantees a definite load and detuning. The timing must also be measured with the Reference PICC's pick-up coil in order to ensure interoperability.
>
> In addition, precise parameters are specified for the oscilloscope's required sampling rate and bandwidth. All in all, measuring according to ISO/IEC 10373-6 can be quite complicated.

3.2.1.2 Stability Against Detuning

As mentioned above, we have to cope with a retroactive effect on the reader from the card, which gets larger as the coupling between the two gets larger.

The card's resonant frequency (see Section 3.3) and energy use also influence this effect.

READER ANTENNAS 3.2

Another consideration is the reader device's environment: sometimes metal surfaces in the vicinity of the reader cannot be avoided — for example when the reader is subsequently mounted above a metal surface or someone simply approaches the reader antenna with a metal object (see Section 3.2.2).

All of these influences detune the antenna. Detuning means mismatching, lowering of field strength and, with that, lowering of range, as well as reactive power and associated interference in the reader device and the field. These are all things that should be avoided, or at least minimized as far as possible.

It may even happen that, due to detuning, the mismatch is so large that the MF RC523's driver current exceeds the maximum allowable and the IC is damaged.

As an example, the same antenna coil is presented in Figure 3.13, both with a low Q factor and with a higher one. The circuit above represents matching with a low Q factor, where the antenna is dampened with 10 Ω resistors in series. In the bottom circuit, 2 Ω resistors are used, resulting in a significantly higher Q factor.

Both antennas are matched to 40 Ω at 13.56 MHz.

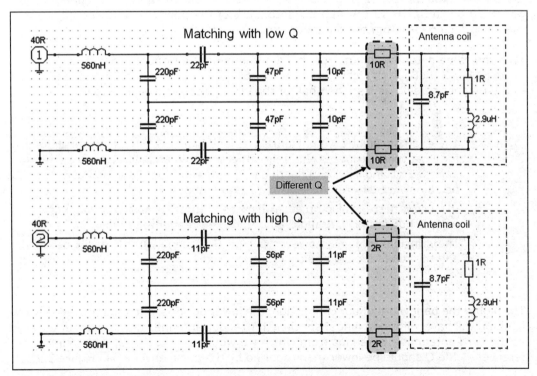

Fig. 3.13. Antenna matching with low Q and high Q factors.

The two antennas' matching simulations with and without detuning are shown in Figure 3.14. Both instances represent a purely capacitive detuning with, in this example, the same capacitance.

3 RFID ANTENNA DESIGN

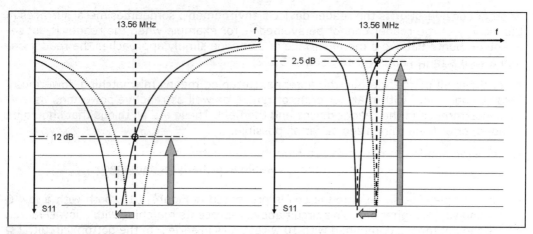

Figure 3.14. Detuning an antenna with a low Q and one with a high Q.

On the left, the wideband antenna can be seen, in which the match changes from S11 > 20 dB to S11 ≈ 12 dB. Approximately 6% of the power is reflected due to mismatching, being lost from the magnetic field.

With the same detuning, the narrow-band antenna on the right changes from S11 > 20 dB to S11 ≈ 2.5 dB! In this case, more than 55% of the power is reflected, and a card in the magnetic field has less then half of the original power available to it.

> **Note** The example illustrated in Figure 3.13 and Figure 3.14 is theoretical and is for illustrative purposes. In reality, there is no matching at the measured points (TX1 and TX2, see Section 3.2.3.1), and thus the actual details will be somewhat different to those presented above.
>
> In reality, there is also no such thing as a purely passive detuning of a few pF, but rather L, R and C all detune the reader antenna.
>
> The principle remains, however: a high Q antenna is more easily detuned and the magnetic field quickly 'falls apart'.

The higher the bandwidth of our connection, the lower our Q is, and thus the fewer problems we will have with matching. Especially for reader antennas that are about as large as the card antenna — antennas from which we would expect the largest coupling differences — the Q should be lower than required by the pulse envelope in Figure 3.2.1.1.

3.2.2 Electrically Conductive Surfaces in the Vicinity of the Antenna

Metal surfaces in the immediate vicinity of the reader antenna have several negative effects. As shown in Figure 3.15 on the left, our reader antenna's magnetic field generates eddy currents in metallic surfaces. These eddy currents in turn produce a magnetic flow opposite to that of the reader device. This — simply put — attenuates the magnetic field.

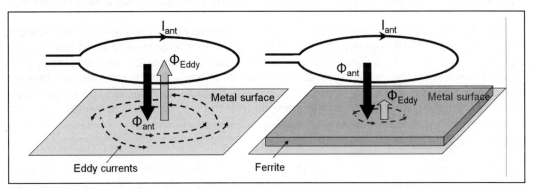

Figure 3.15. Eddy currents.

Such metallic surfaces may be present when the reader is mounted on or above a metal surface. The closer the antenna is to this surface, and the larger the surface, the larger the coupling and thus the greater the canceling effect.

> **Note** The metal surface need not have any magnetic properties — an electrically conductive surface suffices. Any varying magnetic field in the vicinity of a conductor produces an electrical voltage, and the voltage in turn produces a current — short-circuited in this instance. Every electrical current, in turn, produces a magnetic field.

In some applications, we may find that a reader device is built into a metal enclosure. A hole is cut out for the antenna, so that the magnetic field can 'get out'. In reality, the metal housing acts like a metal frame, that is, a closed winding, around the reader antenna, which reduces the field strength. And you wonder why the reading range drops to zero!

Of course, even the electronic circuit can cause such losses, if part of the circuit is placed in the middle of the antenna coil, in order to save space. A typical case: you design a reader board with all of its electronics, and then 'simply' places the antenna on the outside edge of the circuit board. We then supposedly have an antenna of the optimal size (see Section 3.1.4), and everything seems fine. Far from it, actually, if the eddy currents are not taken into account, as shown in Figure 3.16.

3 RFID ANTENNA DESIGN

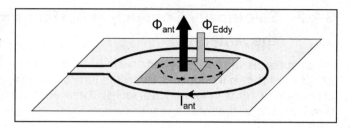

Figure 3.16.
Eddy currents due to a poorly-designed reader device.

The arrangement of the electronics in the middle of the antenna coil has another drawback:

Not only is our reader's magnetic field strongly attenuated by the eddy current effects, but a large part of the energy that should be going to the card is coupled back to the reader's own circuit. This leads to much interference, if appropriate precautions are not taken to immunize the circuit against interference. In any event, the surface in the middle of the antenna is the worst place for the electronics, as it causes the largest possible interference, not only from the antenna into the circuit, but vice-versa.

3.2.2.1 Ferrites

When magnetic or electrically conductive surfaces in the vicinity of the antenna cannot be avoided, these should be shielded by means of ferrites. Ferrites, mostly of iron oxide (Fe_2O_3), are basically, poor electrical conductors but are very good at propagating magnetic flux. No eddy currents arise in ferrite material, as the material is too poor a conductor of electrical current. The metal surface behind the ferrite material is thus shielded against the magnetic field (see Figure 3.15, right).

Figure 3.17 shows three different field strength characteristics over reading range x, for the same antenna coil.

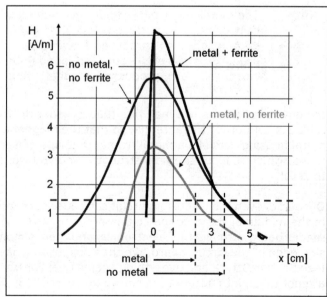

Figure 3.17.
Field strength comparison, with and without metal, with and without ferrite (NXP Semiconductors, 2009).

The symmetrical curve in the middle represents the field strength's progression without the influence of metal in the antenna's proximity. It can be clearly seen that the area above the antenna (image right) is equal to the area below (image left). The antenna in the example shown would achieve a range of almost 4 cm.

The bottom curve shows the reduction in field strength when an electrically conductive surface is introduced a few centimeters below the antenna (left). The entire field, including that above the antenna (image right) collapses. In our example, the range drops to about 2 cm.

In the top curve (both metal and ferrite), the left side shows the ferrite material's shielding effect, i.e. the ferrite introduced below the antenna 'absorbs' the magnetic flux, and the metallic surface no longer causes interference. The range is restored to almost 4 cm. It is also seen that, at zero distance, i.e. when the card is placed directly on the reader, the field strength is much higher with the ferrite than without. This requires caution that the maximum allowable values are not exceeded.

> **Note** In any case, the antenna must be suited to its environment. If ferrite is used, the adjustments must be made with the ferrite in place, as the ferrite also detunes the antenna. Ferrite increases inductance and Q. For correct field distribution, the correct ferrite material must be correctly dimensioned.

The surface of the ferrite material must not be too small, or the shielding effect will be too weak. It should also not be too large, as then the field lines will become highly concentrated in the plane of the antenna and the ferrite. In practice, favorable dimensions have emerged for medium-sized antennas, as shown in Figure 3.18, where an overlap is created by having the ferrite material be around 5 mm larger than the antenna coil.

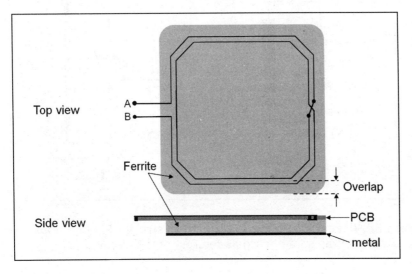

Figure 3.18. *Ferrite dimensioning.*

> **Note** Even metal without ferrite shielding detunes the antenna. So, metallic surfaces that are placed in the vicinity of the antenna as an incidental part of the device's construction, such as the batteries in the Elektor RFID Reader, must be included during antenna adjustment.

3.2.3 Balanced and Unbalanced Antennas

Since the MF RC523 has two driver outputs available, which should ideally be operated symmetrically, it is preferable to use a balanced antenna design.

In Figure 3.19, left, the principle of an unbalanced RFID antenna is shown. As can clearly be seen, the ground surface is a necessary component of the signal generation. A high power (and thus high current) feed can lead to an 'unclean' ground. A simple detuning of the antenna, for example, could lead to a mismatch and thus to reactive power in the feed line. Reactive current that flows in the ground surface causes interference in the circuit.

For the balanced feed shown in figure 3.19, right, the ground connection is not needed in principle, and, as long as the antenna is symmetrical, no current flows in or to ground. The advantage of a balanced antenna lies in its lower susceptibility to interference and reduced interference between the antenna and the rest of the electronics.

The advantage of an unbalanced antenna is the simpler connection via a standard co-axial cable. If a sensible antenna impedance of 50 Ω (or 75 Ω) is used, standard coaxial cable may be used to connect the antenna to the reader electronics.

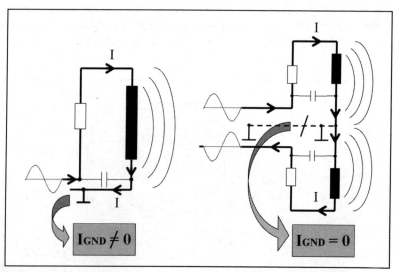

Figure 3.19. Unbalanced and balanced RFID antennas.

> **Note** The Elektor RFID Reader's antenna is a balanced, straight-fed antenna, since it is located on the same board as the rest of the circuit.

3.2.3.1 Balanced Antenna

Figure 3.20 shows the complete antenna circuit, including low-pass filter and matching circuit, as used in the Elektor RFID Reader. As can clearly be seen, the entire circuit is laid out symmetrically, i.e. all components are paired.

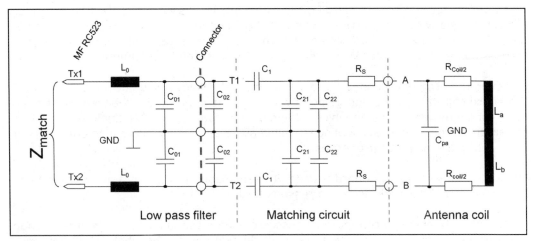

Figure 3.20. *Symmetrical RFID antenna circuit.*

The antenna coil is represented by an equivalent circuit diagram, consisting of

- the ideal coil $L_a + L_b = L_{a+b}$ (in this case the center tap connected to ground);
- a series resistor, R_{coil} (the sum of the resistive losses as shown in Section 3.2.1);
- and a parallel capacitor, C_{pa}.

We will now describe the antenna coil in more detail. The rest of the circuit follows in the practical section, in which the adjustment of a given antenna coil is described in detail.

> **Note** The inductance of a symmetrical antenna coil must always be measured between A and B, regardless of whether the middle tap is tied to ground or not. Because L_a and L_b are coupled together, you cannot separate either L_a or L_b for measurements.

Antenna Coil

Our reader's antenna coil should have as large as possible a diameter, i.e. should span a large area (if the antenna is, as is common, rectangular and not circular).

3 RFID ANTENNA DESIGN

The antenna coil's close vicinity should be as free from metallic surfaces or loops as possible. All electronic components should have the greatest possible distance from the antenna.

Outer Board Layers and Electrical Shielding

As shown in Figure 3.21, a 2-layer PCB is sufficient. In this example, the upper surface consists of two windings (between A and B), which cross each other for better symmetry with respect to the feed. The center of this coil ('center tap') which, in symmetrical feeds is connected in principle to ground, is not connected to GND here.

However, the ground is connected to a symmetrical shield, which encompasses almost the entire coil, but on the underside of the board. It must be open-ended, so that no open conductor loops exist in which eddy currents could flow. This shield is to protect the antenna from electrical coupling — while having as little as possible an effect on our RFID antenna's magnetic properties.

Even if a multilayer circuit board is used, it is advisable to use only the outer (top and bottom) layers. There are two main reasons for this:

Firstly, the tolerances in coating thicknesses between the inner layers is greater than those of the outer layers (relative to layer thickness), i.e. on the whole, the tolerances of an inner-layer coil will be higher than those of an outer-layer one. The smaller the tolerances, the better.

Secondly, smaller distances (coating thicknesses) mean a larger capacitive coupling between two inner layers (= larger C_{pa}), which reduces the antenna coil's natural resonance frequency. The higher the natural resonance, the better.

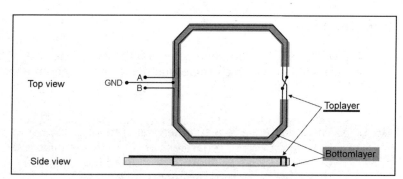

Figure 3.21. Symmetrical antenna coil with electrical shield.

Conductor Twisting

An alternative to an electrical shield is the 'twisting' of conductor tracks around each other. In Figure 3.24, an (unbalanced) antenna is shown, in which such a twisting of the two conductor loops (or tracks) is used instead of a shield. Twisting is better than an additional electrical shield, especially for larger antennas (the one shown here has a diameter of 15 cm), as the corresponding shield, along with the antenna coil's conductor tracks, result in a very high capacitance, which in turn reduces the natural resonant frequency.

Dimensioning

The number of turns should be such that an inductance of between 0.2 and 2 µH is achieved. Smaller or larger values may eventually lead to difficult adjustments with problematic capacitor sizes under certain circumstances.

The following custom equation provides a first-order estimate:

$$L_{a+b} \approx 0{,}2 \cdot \frac{[\alpha H]}{[m]} \cdot l \cdot \left(\ln(l/D) - a\right) \cdot N^{1.8} \qquad \text{Eq. 3.14}$$

where l = lead length
D = lead diameter
a = form factor: 1.07 for round antennas
     1.54 for rectangular antennas
N = number of windings
L_{a+b} = inductance between A and B

> **Note** This equation, while useful for estimates, is not accurate enough for calculation. The inductance and then the matching must be measured, regardless.

As shown in Figure 3.22, between 2 and 4 windings is typically the case. An even number of turns has the advantage that a possible center tap is located on the same side as the feed. Should you need to connect the center tap to system ground, this is easier with an even number of windings.

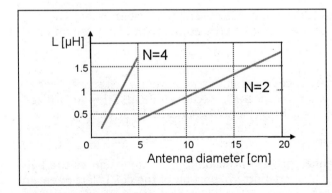

Figure 3.22.
Inductance vs. number of turns.

Both the tracks and the distance between them should be sufficiently large (at least 1 mm) in order to keep the capacitive coupling between turns as low as possible.

3.2.3.2 Unbalanced Antenna

If we want to place the antenna at a larger distance from the reader electronics, it may be more sensible to use an unbalanced feed. Even though an unbalanced feed doesn't automatically mean an unbalanced antenna, it is simpler to connect an unbalanced antenna to a simple coaxial cable.

Figure 3.23 shows an example of an unbalanced antenna circuit driven by a TX output from the reader IC.

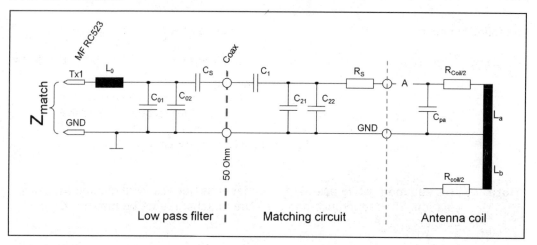

Figure 3.23. *Unbalanced antenna circuit.*

The antenna coil is represented by an equivalent circuit diagram consisting of
- the ideal coil L_a + open circuit L_b (compensation);
- a series resistor R_{coil} (sum of the resistive losses as shown in 3.2.1);
- and a parallel capacitor C_{pa}.

In the next section, this compensated antenna coil is described in more detail. The rest of the circuit follows in the practical section, in which the adjustment of a given antenna is described in detail.

> **Note** The inductance of the unbalanced, compensated antenna coil must always be measured between A and GND. Because L_a and L_b (open circuit winding) are coupled together, they cannot practically be separated for measurement.

The unbalanced, compensated antenna coil is in principle the same design as the balanced antenna described previously, but the 'bottom' portion of the coil is left open. In Figure 3.24, the antenna, as used in the ISO/IEC 10373-6 test setup, is shown as an example. As can be seen in the magnification on the right, the second coil is simply left unconnected. As you can also see, this is intended as a high-power antenna, because the external resistance R_s is envisaged as a parallel circuit consisting of 5 large resistors.

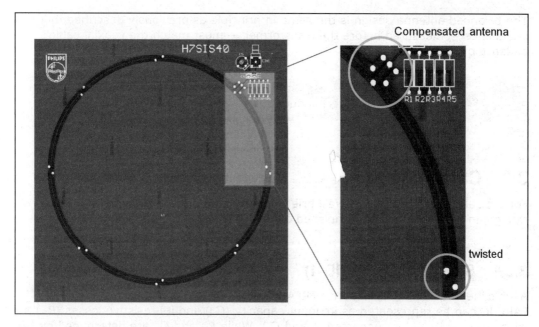

Figure 3.24. Unbalanced twisted antenna coil (compensated).

In addition, the twisting of the two coils (L_a and L_b) around each other can be seen. Alternatively, you can also use a normal balanced feed antenna design and, by using a 'balun,' feed the antenna with an unbalanced line, as shown in Figure 3.25. At its simplest, a balun may be just a corresponding RF wideband transformer as a complete component.

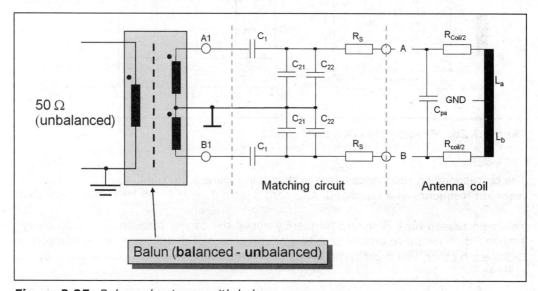

Figure 3.25. Balanced antenna with balun.

3 RFID ANTENNA DESIGN

The balanced antenna design is the same in principle as previously described, but the impedance — measured before the transformer — must match the coaxial cable's impedance of 50 Ω (or 75 Ω).

> **Note** If a 2:1 transformer is used, the impedance between A1 and B1 (without connected transformer) should be 200 Ω for a 50 Ω coaxial cable.

3.3 Card Antennas

For the sake of completeness, we'll briefly go over the design of the card antenna, although, in most cases, factory-finished cards are used.

3.3.1 Standard Cards (ID-1)

A simplified equivalent circuit for a MIFARE card is shown in Figure 3.26, left. The MIFARE IC can be represented as its input capacity (C_{in}) and its resistive losses (R_{in}), to which the card's coil is connected (L and C_p). While R_{in} and C_{in} are determined by the card IC and thus are predefined, a specific resonance can be specified in the design of the card antenna (L and C_p).

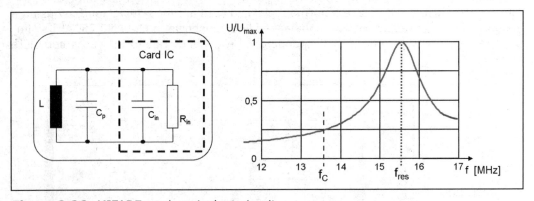

Figure 3.26. MIFARE card equivalent circuit.

The corresponding resonance curve is shown in Figure 3.26, right. As you can see, the resonant frequency of a typical MIFARE card does not lie at 13.56 MHz, but above that.

The main reason for a resonant frequency above the carrier frequency is 'stacked operation.' If you want to ensure that two or three cards will still work when stacked on top of each other, you must compensate for the detuning that will occur when they are stacked.

An example of the behavior of two stacked cards is shown in Figure 3.27. Due to the large coupling between the two cards, the common resonance frequency is reduced. At the same time, the common Q is reduced: now two ICs must be supplied with energy instead of just one.

If the resonant frequency of the individual cards is cleverly dimensioned, these two effects are compensated for, and the two cards can be used when stacked, (almost) as well as a single one, i.e. two stacked cards may well achieve the same range as a single card.

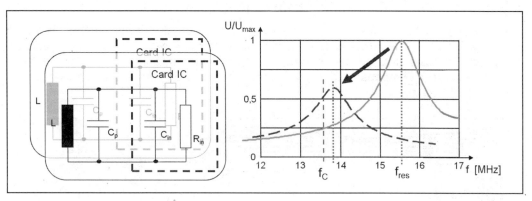

Figure 3.27. Equivalent circuit of two stacked MIFARE cards.

The disadvantage of such an adjustment is that energy is wasted (and thus the range) during single-card operation, considering that a single card's resonant frequency could be set much closer to the carrier frequency.

There are now many card antenna designs whose resonant frequency is actually closer to 13.56 MHz, and they seem to work better than other cards. However, when you place two of these cards directly next to each other, it is quickly noted that the range completely collapses, and sometimes two cards may not work together at all.

> **Note** To be precise, two individual coupled resonant circuits produce two resonant frequencies. One above and one below the individual resonant frequency. This is naturally also the case here, although the higher of the two resonance frequencies is of little interest to us, as it lies too far above the carrier.

In general, the observable trend is that the cards are, if anything, tuned to a resonant frequency that is closer to 13.56 MHz. There are two reasons for this:

The individual cards have a greater range, which makes them easier to sell.

Many new card ICs require more power than older types (because, for example, a microcontroller may be used instead of a simple state machine, or because additional crypto coprocessors also need energy), and manufacturers gladly grasp at the available 'energy reserves' that a lower resonant frequency delivers.

In practice, unfortunately, it is noted that many systems don't allow the operation of more than one card at the reader, and this forces the user to always use only one card at a time. And this, even though the technology described here supports the use of multiple cards simultaneously by all means.

3.3.2 Tokens with Smaller Inlays (Smaller than ID-1)

Another trend in the market is the increasing number of RFID tags (or 'tokens') that don't use the ID-1 standard size (standard smartcard size), but rather smaller designs. These smaller sizes have the disadvantage in that smaller antenna coils absorb less energy. This is often compensated for by the fact that the resonant frequency is set very close to 13.56 MHz.

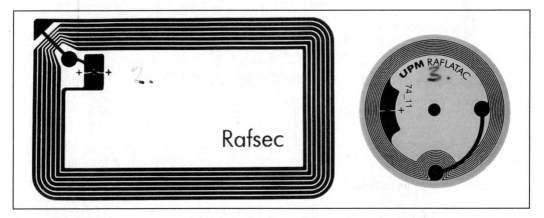

Figure 3.28. Example of two different inlays (UPM Raflatac, 2009).

Figure 3.28 shows two such inlays: On the left, an inlay in the ID-1 size for e.g. a standard card, and on the right a smaller format for a small round tag, or token.

Both are printed thin film antennas mounted directly with the semiconductor. You can see the mounted semiconductors (bridging the gaps between the contact surfaces) on the left side in each image, along with their registration marks.

Tuning the inlay to exactly 13.56 MHz should be avoided.

Firstly, any typical external influence, such as the lamination of the circuit into a plastic body or simply holding it in one's hand, detunes the frequency downward. When the resonant frequency is already at 13.56 MHz without any such influence, this external influence will have negative effects, as it will shift the resonant frequency to below that of the carrier frequency.

Secondly, if they are fairly precisely adjusted to 13.56 MHz, when used with readers that have small antennas (e.g. also ID-1) where the coupling between card and reader is very high, cards (ID-1 size) have a large retroactive effect on the reader. This can lead to a lot of undesirable interference.

> **Note** For this reason, you should also avoid reader antennas that are the same size as the card or tag antenna. If the coupling between the tag and the reader is too great, it may, for example, be very difficult to observe the required pulse forms in the transmitted signals.

3.4 Impedance Measurements with the miniVNA

For the adjustment of the PCD antenna, we need an impedance measuring device that can measure a complex impedance at 13.56 MHz in amplitude and phase. For those who don't have access to a 'big' vector network analyzer, for example, a small, relatively cheap device from the amateur radio field is suitable.

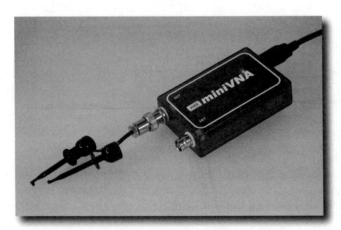

Figure 3.29.
miniVNA.

The miniVNA (see Figure 3.29) is a small USB device that can carry out simple reflection and transmission measurements within a frequency range of 100 kHz to 180 MHz. Its measurement accuracy can certainly not compete with more expensive instruments, but we will see that, with a few little tricks, we can make measurements that are precise enough for our purposes.

In Figure 3.29, the device is shown with a self-made crocodile clip adapter on the BNC input (labeled 'DUT'), that is ideal for our purposes. For the (low) frequencies that we wish to measure, this simple connection suffices.

3 RFID ANTENNA DESIGN

It should be noted here that miniVNA comes in two versions: standard and 'miniVNAPro'. The latter is recommended because it can be calibrated, allowing considerably higher accuracy for measurements. The compensation procedure as described in the following section is not needed when using miniVNAPro.

3.4.1 miniVNA User Software

The device comes with its own original software, but I will use the 'IG miniVNA' user interface from F4CLB, which you can download here: http://clbsite.free.fr/

Also, for better illustration, I use the 'ZPlots' program for Microsoft Excel, by Dan Maguire, AC6LA, downloadable here: http://www.ac6la.com/zplots.html

ZPlots allows you to work according to a Smith chart, and it allows for the computational compensation of wire lengths.

Alternatively, there is a Java tool by Dietmar Krause, DL2SBA, which has the advantage of running on many different operating systems: http://www.dl2sba.wikidot.com DL2SBA's 'vna/J' even displays a Smith chart directly after the measurement. You can therefore measure directly on the Smith chart.

3.4.2 Disadvantages and Limitations of the miniVNA

There are also two disadvantages to the miniVNA, which probably have something to do with its low price.

3.4.2.1 Lack of Sign for Imaginary Numbers

The miniVNA cannot display the sign of the imaginary part of a measurement. The software display is thus somewhat dependent on 'guesswork'. For more complex curves, this disadvantage would certainly be very troubling, but our response curves for simple antenna adjustments are so simple that we can live with this.

At best, with some guess work, ZPlots will do the job, when you can select between three different 'guess possibilities', usually leading you to the correct value.

The second disadvantage is somewhat more serious, and it needs to be carefully considered so that it can be worked around.

3.4.2.2 Lack of Calibration and Compensation

Neither the software nor the miniVNA allow for any calibration. Indeed, in many programs, calibrations are not only offered, but demanded, although these calibrate only the device itself, and not any additional cables or pin headers.

In our case it is sufficient to perform a line length compensation. Via the „Add or Subtract Transmission Line" feature, ZPlots offers a range of typical transmission lines, as well as the ability to define one's own. In **my** case, for the crocodile clip adapter shown

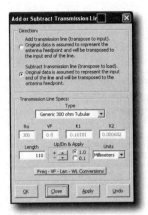

Figure 3.30.
Lead length compensation with ZPlots.

in the image above, a lead length of 11 cm (this corresponds roughly to the length of the adapter) with a transmission line setting of "Generic 300-ohm Tubular" offers good compensation, if this length is subtracted using the 'Subtract Transmission Line' feature (settings as shown in Figure 3.30).

| Note | Before every generation of a new measurement result, the compensation must be carried out again. If, after measurement, you subsequently reload the .csv file, the compensation must be carried out in once again! |

| Note | The compensation must be ascertained according to the adapter used. In some cases, correct measurements can also be made without compensation, typically, for example, when an adapter is made from 50 Ω coaxial cable. |

3.4.3 How Do I Find the Correct Compensation?

Finding the correct compensation setting is easiest with the following steps:

1) Perform calibration using 'IG miniVNA' without adapter.

2) Set the frequency range, preferably to between 10 MHz and 30 MHz.

3) In the measurement section, measure a short circuit, and save this as **Short.csv**.

4) Perform a measurement with a 47 Ω (or better, 50 Ω) resistor (see Figure 3.31) and save this as **47 _ ohm.csv**.

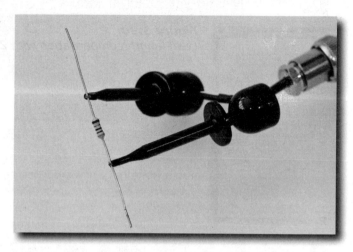

Figure 3.31.
47 Ω resistor.

5) Start ZPlots and load **short.csv**.

6) Using 'Add or Subtract Transmission Line', let the short circuit measurement converge as much as possible onto a single point (at the short circuit point in the Smith chart) (see Figure 3.32).

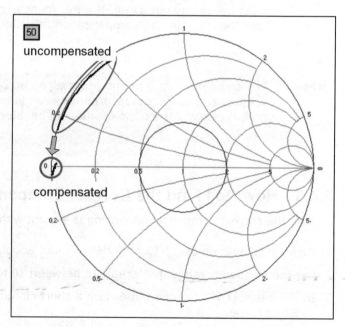

Figure 3.32.
Short circuit, representation on the Smith chart: uncompensated and compensated.

7) This compensation can then be verified by loading the **47_ohm.csv** measurement and performing the same compensation. With a well-selected compensation, this trace line will also converge almost to a point form in the middle of the Smith chart, as shown in Figure 3.33.

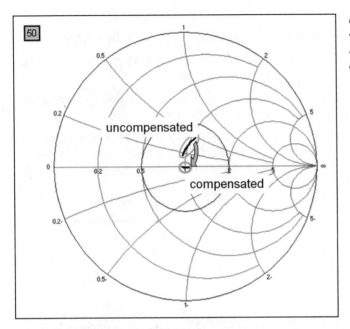

Figure 3.33.
47 Ω, representation on the Smith chart: uncompensated and compensated.

> **Note** This compensation can obviously not replace a proper calibration, but it's sufficient to offer a precise enough result for the actual antenna matching.

> **Note** Anyone unfamiliar with the Smith chart may perform the same steps using a Cartesian representation. Because an impedance curve has to be represented by two curves in the Cartesian system, the representation on a Smith chart is more clear, and therefore simpler.

3.4.4 Coil Inductance Measurement

Antenna matching begins with the measurement of the antenna coil's inductance. For this, you can use any inductance measuring instrument that is capable of representing inductance at 13.56 MHz, but, with another little trick, we can also use the miniVNA.

Although the miniVNA's accuracy will not suffice to measure inductance directly, if we know a parallel capacitance, we can calculate the inductor's resonant frequency. Unfortunately, the input capacitance of my homemade adapter is not negligible. I also don't know the capacity of my antenna coil, so this will not do to measure the resonance frequency.

However, I can measure the resonance frequency using various (known) capacitor values, and then measure the inductance.

This can be clarified in the following steps:

1) Calibrate the miniVNA and determine the compensation (see above).

2) Measure the impedance of the antenna coil. Save the result as **I+0pF.csv**.

3) Connect a known capacitor C1 = 10 pF in parallel to the antenna coil and measure the combined impedance. Save the result as **I+10pF.csv**.

4) Connect a known capacitor C1 = 47 pF in parallel to the antenna coil and measure the combined impedance. Save the result as **I+47pF.csv**.

5) In ZPlots, determine the resonance frequency of each of the three saved .csv files, as shown in Figure 3.34. For this, after loading the previous .csv files, we compensate for the measurement adapter, and then determine the measurement point from the smallest angle, theta.

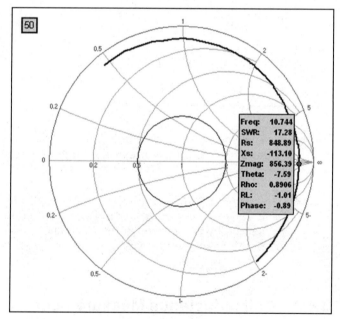

Figure 3.34. Smith chart representation of an antenna coil with 47 pF in parallel.

6) Calculate the inductance using the three frequencies (actually we only need two measurements, but the more measurements we have, the more accurate the result).

In our example, we get the following values as resonance frequencies:

$$\text{I+0pF:} \quad f_1 = 19.038 \text{ MHz;}$$
$$\text{I+10pF:} \quad f_{2a} = 15.906 \text{ MHz;}$$
$$\text{I+47pF:} \quad f_{2b} = 10.744 \text{ MHz.}$$

From the simple equation for the resonance frequency, $f = \dfrac{1}{2 \cdot \pi \cdot \sqrt{L \cdot C}}$, we derive the parallel capacitance and inductance:

Measurement 1 (I+0pF and I+10pF):

$$C_1 = C_2 \cdot \dfrac{f_2^2}{f_1^2 - f_2^2} = 10pF \cdot \dfrac{(15.906 MHz)^2}{(19.038 MHz)^2 - (15.906 MHz)^2} = 23.117 pF \qquad \text{Eq. 3.15}$$

$$L = \dfrac{1}{4 \cdot \pi^2 \cdot f_1^2 \cdot C_1} = 3.02 \mu H$$

Measurement 2 (I+0pF and I+47pF):

$$C_1 = C_2 \cdot \dfrac{f_2^2}{f_1^2 - f_2^2} = 47pF \cdot \dfrac{(10.744 MHz)^2}{(19.038 MHz)^2 - (10.744 MHz)^2} = 21.964 pF \qquad \text{Eq. 3.16}$$

$$L = \dfrac{1}{4 \cdot \pi^2 \cdot f_1^2 \cdot C_1} = 3.18 \mu H$$

Measurement 3 (I+10 pF and I+47pF):

$$C_1 = C_2 \cdot \dfrac{f_2^2}{f_1^2 - f_2^2} = 37pF \cdot \dfrac{(10.744 MHz)^2}{(15.906 MHz)^2 - (10.774 MHz)^2} = 31.047 pF \qquad \text{Eq. 3.17}$$

$$L = \dfrac{1}{4 \cdot \pi^2 \cdot f_1^2 \cdot C_1} = 3.22 \mu H$$

From these three measurements, we get the following result: $L = 3.14 \alpha H$

The 'natural resonance frequency' of less than 20 MHz (I+0pF) would be too low for a PCD antenna coil. The coil wouldn't be reliably aligned, especially not when variations in the production line are considered. In our case, this low natural resonance is due to the capacitance of the measuring adapter (about 12 pF) being included in the measurement. The natural resonance of the example antenna is actually 28 MHz. A value that can only be measured with an appropriate, calibratable instrument, and is still very low. In practice, we find that a natural resonance of above 30 MHz is desirable.

If no resonance is found when measuring without a parallel capacitor, you must either extend the frequency range, or measure only with parallel capacitors in circuit.

Further measurement using other known capacitor values reduces measurement error. Also, the more precisely the values of the parallel capacitors are known, the more accurate the result.

3 RFID ANTENNA DESIGN

> **Note** Comment on the miniVNA: unfortunately, the miniVNA does not save the sign of the imaginary component. For this, ZPlot shows three different possible representations in the window, 'Resolve Sign of X'. Sometimes a few attempts must be made to work out which result is the most likely to be correct one. While the possible error is of no (or little) concern for us, an incorrect representation is still sometimes irritating.

> **Note** Comments on miniVNA: In assessing the results, bear in mind that, while the frequency accuracy of the miniVNA appears to be very high, its amplitude accuracy is limited (although adequate for our purposes). There are sometimes displacements and bumps in the measurement curves, which must be 'ignored' since they cannot be circumvented.

> **Note** The smallest phase angle, theta, can be more easily found in ZPlot's Cartesian representation mode.

In Chapter 6, the Elektor RFID Reader antenna will be measured with the miniVNA and readjusted.

4 Security and Cryptography

Renke Bienert

When we begin considering security and cryptography we must at least cover the definition of the protection objective and take a closer look at threats. En example of a protection objective could be keeping data secret. The threat would then of course be the loss of secrecy, i.e. if secret data became publicly available.

One possible protection from this could be the use of an encryption scheme (you could also just store the data in a safe). The encryption method must then be defined.

Table 4.1. Protection objectives, threats, protection mechanisms.

		Example
What is to be protected?	Protection objective	Secrecy
What could go wrong?	Threat	Loss of secrecy
With which mechanism could I fend off the threat?	Protection mechanism	Encryption
What method do I use?	Algorithm	AES

Nonetheless, there is always some risk in terms of security, as there are no mechanisms or algorithms so good that they can guarantee 100% security.

4.1 Protection Objectives

In the world of smartcards, there are usually three primary protection goals: keeping data secure, protecting against the unauthorized manipulation of data, and one special protection objective that is always central in the discussion around RFID — protection of privacy.

4.1.1 Keeping Data Secure

Smartcards are used in many applications where confidentiality is important. These require the data to be stored on the card in such a manner that it cannot be read by un-

4 SECURITY AND CRYPTOGRAPHY

authorized persons. The card makes use of elaborate security measures to ensure this (see Section 4.2.2 for more on the relevant threats).

Also, the data must be transmitted between reader and card in a secure manner. For this, a secure encryption method may be used (see the relevant threats in Section 4.2.1.1).

It is therefore important that the security features are included not only on the card and the air interface, but for the entire transmission stretch. Even in the background system, there must be a secure realm in which secret data may be securely decrypted, processed and encrypted. Figure 4.1 illustrates three typical scenarios. All three represent an 'online' system, in which connection to a server is assumed, as might be found in a building management system.

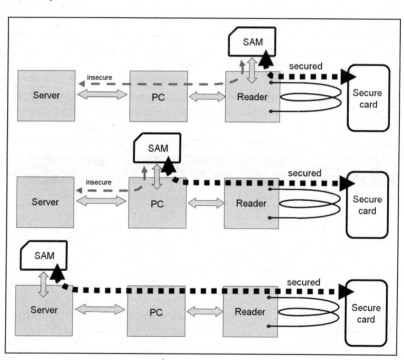

Figure 4.1. Secure communication channel.

In the top image, the secure component, known as a SAM (Secure Application Module) is part of the reader device. This secures the channel between the card and the reader, provided that the SAM is also considered 'secure'.

In the middle figure, the reader simply passes the data on transparently, and the SAM is part the PC (or host environment).

In the lower image, the PC, in turn, passes the data transparently to the server, and the entire channel between card and server is secured.

> **Note** The same representation is also applicable to other protection objectives.

4.1.2 Data Integrity

In most smartcard applications, protection against unauthorized manipulation of data is of much more importance than secrecy. This applies to an electronic ticket, for example, that stores a trip fare counter value. The fare itself is displayed openly when the user 'recharges' his or her ticket or takes a trip. Data confidentiality is not really useful here.

However, it is essential that no unauthorized person can manipulate the ticket data, and begin fare-dodging.

Here too it is important to ensure firstly that the data is stored on the card in a manner that does not allow it to be manipulated (see threats in 4.2.2).

Secondly, the transmission channel must be secure. Typically, you would use a 'signature', or a MAC (Message Authentication Code). The term 'signature' is used when using asymmetric encryption, 'MAC' when using symmetric encryption.

Both are kinds of 'digital signature' (or 'digital seal'), which ensure that the data originates from the correct sender and has not been tampered with (see Section 4.2.1.2). This is illustrated in Figure 4.2.

With a function, f_K, a MAC, or digital signature, is calculated based on the data. Both are transmitted the data, m (open or encrypted), and the digital signature. An attacker could certainly modify the data or the digital signature, but as long as the method for calculating the MAC is secure, the signature and data will no longer match after such a manipulation.

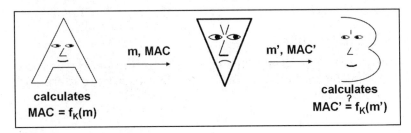

Figure 4.2.
Using a MAC, NXP Semiconductors, 2010.

The intended recipient, B (right), can also calculate the digital signature from the received data, because it knows the computation algorithm. If the calculated MAC is not the same as the received MAC, the receiver knows that either the data or the signature, or both, have been manipulated.

4.1.3 Privacy Protection

When using smartcards, especially in the field of RFID, privacy is always a central issue. In certain instances, contactless smartcards may be activated without the user's knowledge. While the reading range is physically limited, the activation and operation of the card, including the exchange of data, takes place over an (open) air interface.

For improved privacy protection, it is important that no unauthorized person can get information about the cardholder simply by activating the card. At card activation, the card returns an ISO/IED 14443-mandated ID (UID or PUPI). By surreptitiously activating the card, an attacker can read the UID. If this UID truly 'unique', the attacker will be able to make a clear association between the UID and the cardholder.

To avoid this, many cards today can be configured to return a random number instead of a fixed UID, which makes this association no longer possible.

> **Note** If the card returns an RID (random ID), there must also be a way for the authorized reader to get access to a unique and inalterable UID, as this may be needed for key diversification (see Section 4.4.4.2). The UID transfer should then ideally take place using encryption.

Also, in order to protect private data, further communication between card and reader must be encrypted if personal data is transferred. The encryption calls for authentication, to ensure that only authorized readers are able to receive this data. In addition, the actual encryption must be carried out using random numbers (see Section 4.4.1), so that an attacker cannot restore the association between the data and the cardholder.

4.2 Attacks on Smartcards

Before we take a look at the encryption itself, it may be useful to consider the possible attacks on our system. We'll consider only the most significant attacks that are particularly relevant to contactless smartcards.

4.2.1 Logical Attacks

4.2.1.1 Unauthorized Reading of Data

When data is to be kept secret is transferred via a communication channel, it is exposed to unauthorized reading.

This means that an attacker could, using his own reader, activate the card and attempt to read data actively. The card's limited range does make this a little difficult, but not impossible.

He could also attempt to eavesdrop on communication taking place between legitimate card and reader. With contactless smartcards, this presents no great problem, and, in the worst case, could work over several meters. The communication specifically from the reader to the card can typically be received even further away.

A very common method used to protect against unauthorized activation and unauthorized reading is mutual authentication and subsequent encryption of data. The attacker is then no longer able to actively read the card, as he cannot authenticate. He can still eavesdrop on valid communication, but he can't do anything with the encrypted data, as he is unable to interpret it.

Mutual authentication also means that the encrypted data is subsequently encrypted using keys based on random numbers (see Section 4.4). That way the same data, when transmitted multiple times, appears different each time and cannot be interpreted.

> **Note** If the card returns an RID (random ID), there must also be a way for the authorized reader to get access to a unique and inalterable UID, as this may be needed for key diversification (see Section 4.4.4.2). The UID transfer should then ideally take place using encryption.

Encryption alone may protect data confidentiality, but not (necessarily) data integrity.

4.2.1.2 Unauthorized Manipulation of Data

This logical attack involves the attacker attempting to manipulate data (threat to data integrity). For this, the attacker need not even be able to interpret this data. It is sometimes enough to repeat previous communication and thus create illegitimate transactions or even restore a card to a prior state, for example reinstating an expired fare ticket.

By using his own reader device, the attacker may either attempt to activate the card and modify the stored data, or he could attempt to repeat the data from a valid communication, to interfere with or modify data.

To prevent this, both legitimate sides of the communication (the 'real' reader and the 'real' card) must have the ability to assess the message's authenticity. In detail this means the receiver must be sure that:

- that the message comes from the 'real' sender;
- that the message was not tampered with;
- that the message is not a repeat of an older, valid message that had been received previously.

Firstly, mutual authentication will be required in order to ensure that sender and receiver can trust each other (see Section 4.4.1).

4 SECURITY AND CRYPTOGRAPHY

Secondly, it's necessary that there be some type of 'signature' calculated for the transmitted data. If this signature is also transferred, then the receiver can be sure that the information has not been manipulated. For the symmetrical encryption method used here, the term used is 'MAC' instead of a 'signature' (which is used only for asymmetrical methods) (see Section 4.4.3).

> **Note** To prevent messages being either swapped in their sequence or repeated, an additional encryption step is added, with which each new MAC, even for the same data, produces a different cryptogram by, for example, using a counter as a part of the 'init vector' (see Section 4.3.6).

The data may as well be transferred in the clear if there is no good reason for secrecy. As long as they are protected by a MAC against manipulation, the data integrity is ensured.

4.2.1.3 'Replay' Attack

In a 'replay' attack, the attacker resends an already-transmitted message. When a fixed record is read during a valid transaction, e.g. to open a door, this record may be passively encrypted. As long as it doesn't change, the attacker can listen in and then duplicate it using his own device (e.g. an NFC device).

To protect against such an attack, mutual authentication is used, in which (encrypted) random numbers are exchanged (see Section 4.4.1). The key with which the data is encrypted is then calculated based on these random numbers. Then, neither the authentication messages nor the actual (encrypted) data are predictable or repeatable.

4.2.1.4 'Relay' Attack

In a 'relay' attack, the attacker simply passes the information on and fakes the presence of the real card. Imagine, for example, an access control system using contactless cards. A reader authenticates the card, then reads credentials from it before opening the door. For this, the user must hold his card directly at the reader (Figure 4.3, top).

In our example attack (see Figure 4.3, bottom), the cardholder sits at a bar, having received a beer from Attacker 2, who has a reader capable of activating the cardholder's card. Attacker 2 is connected via wireless link directly to Attacker 1, who tries to gain entry with the 'Attacker Reader Device'. Authentication between the real card and the real reader still takes place, but this time making a detour via the 'Attacker Card Device', wireless link and Attacker Reader Device. Neither the real card nor the real reader have any way, in this example, of recognizing that they are not communicating directly with one another, but via a wireless link.

No amount of encryption can help here, as the real reader is still communicating with the real card.

ATTACKS ON SMARTCARDS 4.2

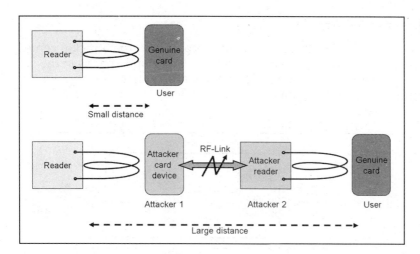

Figure 4.3.
'Relay' attack.

There are basically only two ways to recognize a relay attack or protect against one. The first is to use a second communication channel, which cannot be bypassed without it being noticed. This could, for example, be an optical channel. An electronic passport uses such protection: the machine-readable barcode must be read in order to derive the key with which you decrypt the data. This is only possible if the passport is folded open, that is, when you have optical access to the card.

Such a secondary communication channel could also be carried by the cardholder: for example, a PIN code might also be required to open the door.

The other way to recognize a relay attack is to use time measurement. Every message detour takes additional time. Even with a very high bandwidth communication channel (which the attacker would normally not have in any case), the signal requires at least the amount of time that light requires in order to traverse the greater distance. When the actual available bandwidth is limited, and electronic circuits (modulators and demodulators, amplifiers, signal conditioners) are added, the time rapidly adds up to a measurable delay.

For this, however, the reader and card must use a special, fixed command sequence that must be very fast (to allow no time for manipulation) and encrypted (to protect against data manipulation). With this command sequence, the reader can measure the time elapsed between sending the command and receiving the response, and thus recognize and ward off a possible relay attack.

The only card to offer such a command is NXP's MIFARE Plus.

4.2.1.5 'Man-in-the-Middle' Attack

This is a special form of attack, in which data is manipulated. Here, the attacker comes between the real card and the real reader and passes the information on, as in the relay attack (see Fig. 4.4). However, in contrast to the relay attack, here the attacker attempts to manipulate the transmitted data en route.

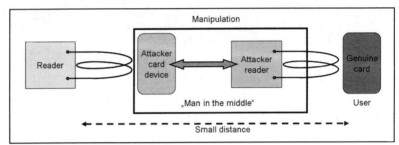

Figure 4.4. Man-in-the-middle attack.

This attack could possibly be averted by using encryption, or better, a signature (a MAC).

4.2.1.6 Denial-of-Service Attack

The denial-of-service attack attempts to disrupt the entire system so that all or part of it collapses. Imagine a group of attackers who bring down a stadium access control system. In this case, the automatic turnstiles no longer work, and several thousand people are standing outside the stadium and want to get in. In the ensuing chaos, the attackers can gain entry without valid tickets.

You can imagine several ways to disrupt such a system, so that it collapses, and only a few of them have anything to do with the cards themselves. With the appropriate jamming transmitter, the communication between reader and card can be disrupted (just as with any other radio communication). Effective protection against such an attack can take place only at the system level, i.e. the system must detect such an attack, or have a backup contingency.

4.2.2 Physical Attacks

The attacks described so far have been termed logical attacks as they don't exploit any of the card or reader's physical characteristics. Logical attacks are, in principle, independent of implementation. However, you can certainly imagine some physical attacks on smartcards or readers, directed against the implementation of the algorithms, and then come up with some ideas for effective protection.

4.2.2.1 Side Channel Attacks and Power Analysis

Side channel attacks analyze the behavior of a smartcard or other system component (e.g. reader) during, for example, a cryptographic calculation. An attempt is made to establish a correlation between the recorded data and possible keys or protected data and then infer parts of the key or the data.

With smartcards, measurement of current consumption is often used. With an insecure design, you could tell by the current consumption, for example, when a crypto coprocessor is started up, when an additional storage area is activated, or even whether ones or zeros are being loaded into individual EEPROM cells. In the simplest case, you could read the key directly, if you knew when the key was loaded. These attacks are known as SPA (simple power analysis) or DPA (differential power analysis).

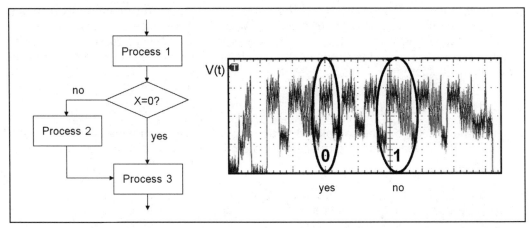

Figure 4.5. Example of an SPA (NXP Semiconductors, 2009).

In Fig 4.5, left, an example of a poor implementation is shown. Depending on the result of a comparison, Process 1 either leads to Process 2 or jumps over Process 2. This can typically be found in the usual conditional operators (IF statements):

```
x = result;                    // Process 1
if (1 == x) then
   result = result + 1;        // Process 2
do something                   // Process 3
```

The current measurement in Figure 4.5, right, shows the difference in duration of the operation, depending on whether the result was '0' or '1'. This allows the attacker in this example to infer the value of x from the waveform of the measured current.

4 SECURITY AND CRYPTOGRAPHY

> **Note** 'Differential pizza analysis' is often used as a figurative illustration: a reporter counts the number of pizzas delivered to the Pentagon in a night. Over a long period, this number remains relatively constant. Then, in one evening, a lot more are delivered, and the reporter knows that some big secret operation is taking place (suddenly many employees are working overtime). Even if this couldn't really happen, it demonstrates the DPA principle. You can make inferences about secret operations from pizza consumption.

Obviously, such measurements (and the corresponding countermeasures) are much more complex in practice than depicted here. With contactless cards, measurement of current consumption is not that easy at first glance, but attackers come up with complicated setups that take hundreds of thousands of measurements, which can later be statistically evaluated.

4.2.2.2 Reverse Engineering

Intuiting the circuit from the completed device is called reverse engineering. With smart-cards, this often means opening the cards and etching the module and the semiconductor therein.

If you etch off the individual layers of the semiconductor, you can, using a microscope with an appropriately high optical resolution, see the individual circuit components, right down to the individual transistor gates.

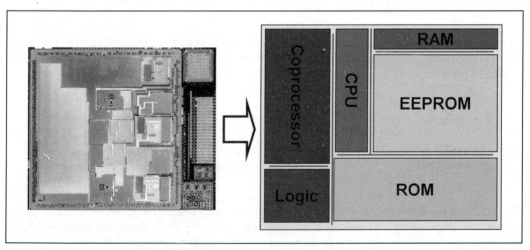

Fig. 4.6. *Example of reverse engineering (NXP Semiconductors, 2009).*

It's usually enough to be able to identify individual logic blocks, which you can then investigate further. There are now semi-automatic tools that can analyze images of semiconductor layers and deliver entire circuit components as schematic diagrams.

> **Note** The now 15-year-old MIFARE Classic from NXP Semiconductors has been analyzed like this. First, by Chinese companies who have cloned them, then in 2008 by members of the Chaos Computer Club and the Universities of Virginia and Nijmegen. This was possible because the circuit structure was openly visible, so it was, according to statements made by the researchers, easy to find the position of the shift register, in which a very simple form of encryption took place.

Today, there are many methods and processes to make secure smartcard controllers' individual circuit structures 'invisible' and fool potential attackers. For security reasons, we won't discuss these further individually. Nevertheless, new protection measures always provoke new attack methods, and security continues to be a standing battle.

4.2.2.3 Light and Laser Attacks

With various electromagnetic signals, attackers attempt a targeted attack during the processing of a transaction. Exposed semiconductor circuits are light-sensitive, so, flashes of light during the microcontroller's program flow can disrupt the program and lead to malfunctions.

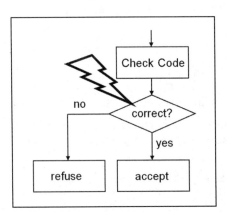

Figure 4.7.
Example of a light attack.

Figure 4.7 shows the principle of a simple light attack that influences the program counter at a specific point in time, such as when a program branch takes place, and thereby sends the program to an unforeseen section of code.

Such attacks may be carried out using simple visible light, but obviously with laser light, individual circuit elements may be targeted. At the appropriate resolution, even individual storage cells could be affected (to change from '0' to '1' or vice-versa), if no protection mechanisms are implemented.

You could also use other high-energy electromagnetic radiation (RF or X-rays) or electrostatic discharge to affect program operation.

In secure smartcard design, there are obviously many measures to protect against such attacks. These could be hardware measures, such as light sensors that, when triggered, turn off the microcontroller, or protective shields that turn off the microcontroller when it is opened. One could also take software measures to make such attacks impossible or at least extremely difficult, for example by placing sections of code at addresses that are highly unlikely to be reached unintentionally.

4.2.2.4 Temperature and Frequency

Another way of influencing the function of a microcontroller is to manipulate its operating parameters. The operating voltage could be raised, for example, or reduced during very specific program points, in order to provoke targeted malfunctions. Similar methods can be used on the operating temperature or the clock supply.

For this reason, current secure smartcards contain sensors, for example, for these important operating conditions, which monitor compliance and halt program flow as soon as any of the tolerances that could cause malfunctions are exceeded.

4.2.3 Combined Attacks

Combined attacks, in which several attack methods are used in combination, are obviously a greater danger. For example, a bus structure may be inferred using reverse engineering, after which SPA or targeted bit stream manipulation could be carried out. The number of possibilities seems limitless, and both the attack methods and defense methods are being developed continuously.

4.3 Cryptography

Method, algorithm, key, authentication, encryption, signature, MAC, DES, AES...

This book cannot and should not replace a reference book on cryptography, but will give only a brief introduction to the key cryptographic issues relevant to (contactless) smartcards.

There are essentially two different types of encryption algorithms: symmetric and asymmetric encryption.

4.3.1 Asymmetric Cryptography

In asymmetric encryption, you always need a **key pair**, which consists of two matching keys. The one is called a 'public key', while the other is a 'private key'. Both are derived using a mathematical function based on a long random number.

As shown in Figure 4.8, the public key is used to encrypt the data. The encrypted message can only be read with the matching private key.

You can also reverse the process and generate a signature using the private key, as shown in Figure 4.9. The signature can then be verified using the public key. This will ensure that this signature comes from its counterpart, the matching private key.

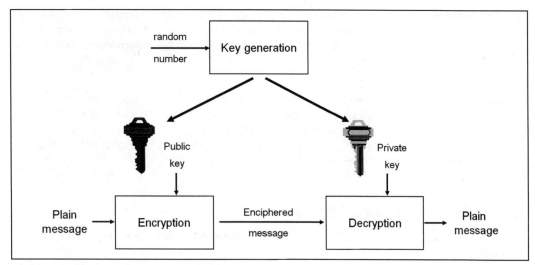

Figure 4.8. Asymmetric encryption.

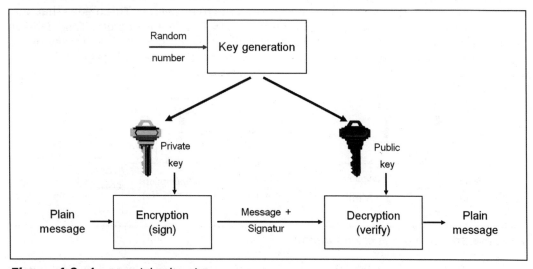

Figure 4.9. Asymmetric signature.

The core of asymmetric encryption is a mathematical one-way function, which is easier to carry out in one direction than the other, which is extremely difficult. In the world of smartcards, two methods have prevailed: firstly, the 'classic' RSA algorithm (named after Rivest, Shamir and Adleman, the three mathematicians who developed the first asymmetric method in 1977) and secondly, in recent times the method has been strengthened to be based on elliptic curves.

The advantage of asymmetric cryptography is key management: each participant in a system with many participants need only keep their own private key secret.

4 SECURITY AND CRYPTOGRAPHY

The disadvantage is the large computational cost, which makes encryption and decryption very slow. In practice, hybrid systems are often used, in which only one secret (a symmetric key) is exchanged. The symmetric key is used by a symmetric encryption method to encrypt and decrypt the actual message.

> **Note** The literature in some places lists as a caveat the fact that you only assumes that the one-way function is really simpler to carry out in one direction. There is always the theoretical possibility that (currently unknown) algorithms exist that make the reversal of the process possible with a reasonable effort.

4.3.2 Symmetric Cryptography

Symmetric cryptography is much older than asymmetric — we know of the use of such methods in ancient times.

Since then, a principle has emerged, known as 'Kerckoffs's Principle', that you can find in the six guiding principles for encryption that Auguste Kerckhoffs established in 1883. This principle says: 'the system must not be required to be secret'. This means that only the keys are kept secret, while the algorithm may not have any secrets. This leads us to standardized methods, for which the algorithm is always 'public'. The secrecy of the key is the only thing that maintains the security of the encrypted message.

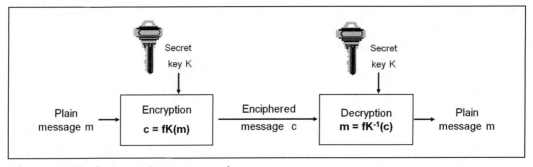

Figure 4.10. *Symmetric cryptography.*

In Figure 4.10, the principle is shown: an open message is encrypted using an (open) encryption function, fK, using secret key K. Only someone who knows this secret key K can decrypt the data using the inverse function, fK^{-1}.

The DES and AES block ciphers are the well-known standardized methods most often used in smartcard systems.

> **Note** An open process has the advantage that anyone is able to assess the 'quality' of the encryption. And thus, the security is defined purely by the key length. The use of a secret process is also referred to quite aptly as 'security by obscurity'.

The advantage of symmetric encryption is the simple implementation of the algorithm. Formerly common in hardware, today software is increasingly able to carry out symmetric encryption and decryption very quickly — much quicker than with asymmetric encryption.

The disadvantage lies in key management: the system is secure only if every potential participant has their own key, which must be kept secret. The number of secret keys increases with the square of the number of participants.

> **Note** For smartcard systems with simple symmetrical methods, this means that every card in the system uses the same key! When an attacker 'cracks' the key for a single card, he has access to the entire system. Key diversification aids in symmetrical encryption methods (see Section 4.4.4.2).

4.3.3 Block and Stream Cipher

With symmetric cryptography, there are two different methods — the stream cipher and the block cipher.

In Figure 4.11, left, a block cipher is shown that — as the name suggests — encrypts and decrypts only entire message blocks. Usually, these are blocks of 8 or 16 bytes in length. When the message to be transmitted is longer than a block, it is broken up into blocks of the corresponding block length.

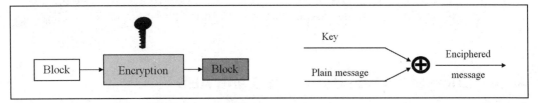

Fig. 4.11. Block and stream ciphers.

In contrast, in the stream cipher (Figure 4.11, right) each bit of the message is encrypted individually, by XORing it with the key stream.

The advantages of the stream cipher are:

- only as many bits as are contained in the open message are encrypted and transmitted, which saves time and buffer size;
- encryption and decryption is very fast (only one XOR).

The disadvantages of the stream cipher are:

- only as many bits as are contained in the open message are encrypted and transmitted. This means that, for an ACK or a NACK or a 'yes' or a 'no', which require only a single bit to encode, there is a 50% probability that an attacker could guess the message!;
- key management is extremely difficult. This is the weak point in the use of most stream ciphers.

On the other hand, for a block cipher, the advantages are:

- key management is very simple;
- even simple messages ('yes' or 'no') are securely encrypted.

The corresponding disadvantages are:

- the data must be split into blocks, and the message must always be padded to an exact block length. This means that more data is usually transmitted than contained in the message (namely when the message is smaller than an integer multiple of blocks).
- The entire block has to be received before any decryption can take place.

In practice, the block cipher has prevailed, mainly due to the difficult key management required for the stream cipher, which, with its justifiable complexity, cannot be realized securely.

> **Note** For smartcard systems with simple symmetrical methods, this means that every card in the system uses the same key! When an attacker 'cracks' the key for a single card, he has access to the entire system. Key diversification aids in symmetrical encryption methods (see Section 4.4.4.2).

4.3.4 Encryption Standards: DES and AES

The two most commonly encountered block ciphers are DES and AES.

DES (Data Encryption Standard) was developed in the 70s by the U.S. Government and published by the NIST (National Institute of Standards and Technology) in 1976. It has since been published internationally. As an 8-byte key (56 bits — parity bits don't count), it is no longer considered secure. It was demonstrated in 1999 that DES could be cracked by a computer network in 22 hours, although this was due to its small key length, not because of any latent mathematical weak points. In 2008, a special computer cracked DES within one day.

> **Note** One improvement upon (single) DES is to cascading it: triple DES (see Section 4.3.5).

DES uses a block length of 8 bytes, that is, each message that is to be encrypted must be padded so that it is a multiple of 8 bytes.

DES is a 'bit-oriented' process, making it more suited for hardware implementation than software (in contrast e.g. with AES).

The official successor to DES is AES (Advanced Encryption Standard), which was developed by Belgians Joan Daemen and Vincent Rijmen for a contest, and published in 2000. Unlike DES's process, the 'Rijndal' process used in AES is byte-oriented with different key and block lengths: keys and blocks may be 128 bits, 192 bits or 256 bits long.

As per the AES standard, the block length is limited to 128 bits (16 bytes), but all three key lengths are available. The three corresponding options are called 128-AES, 192-AES and 256-AES.

A software implementation of AES is usually much faster than of DES, and AES is also supported by experts from security authorities such as the German BSI (Deutschen Bundesamt für Sicherheit in der Informationstechnik) — the German Federal Office for Information Security.

> **Note**   The AES algorithm is freely available and may be used without licensing fees. There are many software libraries, although they are not all considered secure against side channel attacks.

4.3.5 DES Cascading

In many systems, cascading DES is used. This is called Triple DES. Here, the DES encryption is performed three times in a row (cascaded). The version with two keys, or 2-key triple DES (2K3DES), is shown in Fig. 4.12. The message is encrypted using the first half of the 16-byte key, then decrypted using the second half, and then re-encrypted using the first half again.

This arrangement of encryption-decryption-encryption has the advantage that it automatically becomes 'single DES' encryption when the first half of the key is the same as the second half. So, the same implementation can handle both DES and 3DES.

Of the key's 16 bytes, only 112 bits are used (the rest are parity bits), and experts determine the 'effective key length' to be about 80 bits.

The effective key length represents the difficulty in cracking the algorithm, and is thus basically a gauge of the algorithm's quality.

4 SECURITY AND CRYPTOGRAPHY

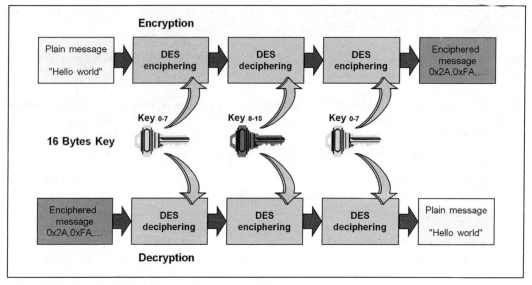

Fig. 4.12. *Triple DES with two keys (2K3DES).*

Of course, you could also use a different key for each of the three steps, as shown in Figure 4.13. This is called 3-key triple-DES (3K3DES), having a key length of 24 bytes, of which only 168 bits are used as key bits. The effective key length is considered to be around 112 bits.

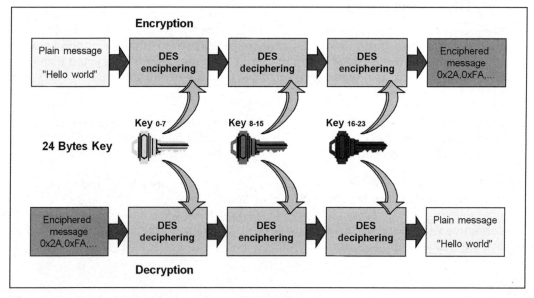

Fig. 4.13. *Triple DES with three keys (3K3DES).*

4.3.6 Operation mode

A block cipher, in principle, has a disadvantage for long messages: whenever an input block (unencrypted or 'cleartext' message) is identical, the encrypted output block is identical in turn.

4.3.6.1 Electronic Code Book (ECB)

Imagine a high-resolution image is the DES-encrypted message to be transmitted. Each 8-byte value in the original image results in an encrypted 8-byte value. When the attacker builds the encrypted blocks back up into an image, although the color and details may be fuzzy and incoherent, the contours can still be seen. Figure 4.14 shows the encryption on the left and the example on the right. You can clearly see the lettering even after encryption.

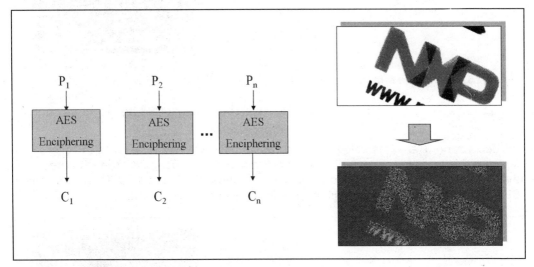

Fig. 4.14. Example of 2K3DES with ECB encryption.

This simple mode is known as ECB (Electronic Code Book). Due to this deficiency, this mode is rarely used in contactless smartcards.

4.3.6.2 Cipher Block Chaining (CBC)

In order to represent equivalent data differently in long messages and to make decryption independent of the individual block sequence, the CBC (Cipher block Chaining) mode extension was created. In CBC mode, every individual unencrypted block is XORed with the previous encrypted block, before being encrypted, as shown in Figure 4.15, left. Message encryption for sending takes place in 'CBC send mode': first XOR, then encrypt.

The first cleartext block obviously requires an 'extra' preceding block. This is called the 'initialization vector', or 'init vector' (IV). In most cases, the init vector is simply '0', but it can naturally also be generated by a counter or derived from shared secrets in order to add additional safeguards against, for example, swapping of message packets.

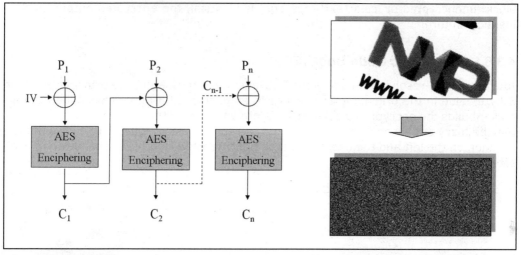

Fig. 4.15. *2K3DES example with CBC encryption (CBC send mode).*

Using this extension, as can be seen in Figure 4.15, right, there is no longer any detectable contour.

For decryption, the same calculation process must naturally take place in reverse: to decipher the message, the result of a block's decryption must be XORed with the previous encrypted block: first decrypt, then XOR, as shown in Figure 4.16. This is also called 'CBC receive mode', because it is used to receive encrypted messages.

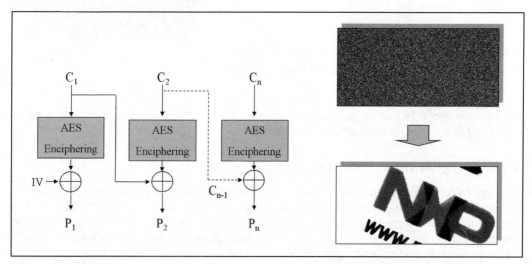

Fig. 4.16. *2K3DES example with CBC decryption (CBC receive mode).*

There are other modes of operation, but these will not be considered further here.

4.4 Application of Cryptography

The use of cryptography is not an end in itself, as already mentioned above, but exists to provide security for a system. Security requirements vary and depend largely upon the application and the value of the protected data. For contactless smartcards, the budgetary allowances for system components (e.g. cards), and required transaction speeds play a large role. As always, security costs time and money.

4.4.1 Mutual Authentication

In most situations, mutual authentication is the first step toward secure data transmission. For one, it ensures that both sides authenticate each other. The system's terminal side (reader) can be sure that it communicates with a legitimate card and the card can be sure that it communicates with a legitimate reader.

To achieve this, encrypted random numbers are exchanged, as is shown in Fig. 4.17. Usually, the reader begins authentication using the appropriate command, which normally contains additional parameters such as a key number or address, and tells the card which key it should use for authentication.

In the first step, the card will generate a random number RND B, encrypt it using a secret shared key, and respond with the encrypted random number RND B.

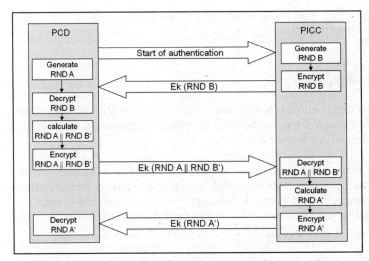

Figure 4.17.
Mutual authentication.

The reader then generates its own random number RND A, as well as decrypts the card response and reconstructs RND B.

From RND B, the reader produces B', with the help of a known algorithm. This could, for example, be a simple byte rotation. The algorithm is known to both sides: the card and the reader.

RND A and RND B' are then encrypted using the shared secret key and then sent to the card (second step).

> **Note** When we refer to a reader, we mean the reader device side of the transaction. Obviously, the encryption could take place in the reader itself, on a connected host (e.g. PC), or somewhere online in a background system, for example on a remote server. It is advisable (and more secure) to have the encryption performed by a SAM (see Figure 4.1).

The card then decrypts the message, obtaining RND A and RND B'. RND B' is then reverse-calculated into RND B and compared with the original (card-generated) RND B. At this point, if the two don't match, the card breaks the authentication process. If they are the same, the card is sure that the reader has the same secret key.

The card then derives RND A' from RND A (with perhaps the same algorithm used to calculate RND B'), encrypts RND A' and sends this back to the reader (third step).

The reader decrypts the message, derives RND A from RND A' and compares this number with its own, original RND A. At this point, should the two not match, the reader breaks off communication. If they match, the card and reader are mutually authenticated.

For this authentication method, also called '3-pass mutual authentication' as it consists of three steps, only encrypted random numbers are exchanged.

> **Note** The less predictable, i.e. the more random, the two numbers are, the more secure the authentication.

This method has several advantages:

Firstly, an attacker cannot record the interchange and 'replay' it. With the random numbers, generated on both sides, even if the same data is repeated, different data will be transferred, so a replay attack won't work here.

Secondly, you can use the transmitted random numbers to generate 'session keys'. These session keys (keys valid for one session only), which are derived from random numbers, are then used to encrypt and protect the data. Session keys are created at each new authentication, so the same data will be encrypted differently upon a re-authentication. The data transfer is also protected against a replay attack.

This process can be used with any encryption method. The length of the random numbers is usually tied to the selected method's block length, i.e., with the key length: for DES and 2K3DES, eight bytes are typically used; for 3D3DES and 128-AES, the length is usually 16 bytes.

4.4.2 Data Encryption

Data encryption generally follows after authentication, using a 'session encryption key' derived from a random number. On most cards (e.g. all MIFARE cards), the CBC mode is used. The number of transmitted encrypted data bytes is always an exact multiple of the block length ($8n$ or $16n$ bytes). Encryption alone does not guarantee the data's integrity — only its confidentiality.

4.4.3 Message Authentication Code (MAC)

To protect the data's integrity, a signature is calculated, which, for symmetric encryption, is known as a 'MAC' (Message Authentication code) (see Section 4.1.2).

Figure 4.18 shows an example of a simple MAC: all data to be protected are encrypted in sequence, using the CBC mode. The final encrypted block is appended to the transmitted data as a MAC. The receiver can now perform the same calculation, and then compare its result to the received MAC. Only when the two match has the message not been tampered with.

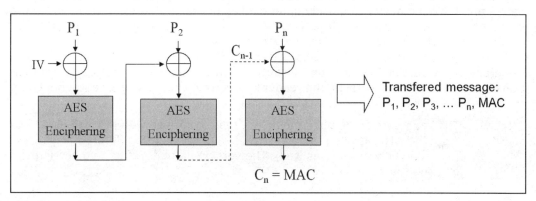

Fig. 4.18. *Example of a simple MAC.*

It does not matter whether the data used to calculate the MAC (P_1 ... P_n) is encrypted or not.

Calculation of the MAC is usually performed using the 'session MAC key', which is derived from random numbers.

In many cases, command parameters or parts of previously transmitted messages need to be protected, and thus form part of the MAC's generation. Sometimes additional secret parameters are used as input values to increase security. For example when you does not use transmitted numbers that increment at each transfer as part of the init vector, you can ensure that the same command cannot be replayed by an attacker and that the transmission sequence is not alterable.

4 SECURITY AND CRYPTOGRAPHY

A special variant is the 'Cipher Message Authentication Code' (CMAC), which is calculated according to the 'NIST Special Publication 800-38B' standard (see *http://csrc.nist.gov/publications*).

4.4.4 Key Management

Since the security relies on the secrecy of the key, proper key management is very important. The question of how a key can be stored safely is addressed in Section 4.4.5. The transfer of keys by 'downloading' new keys via extra secure communication channels is obviously also tied to this, as well as the fact that the key is 'read-never' (write only, and use).

The question of what happens when a key is compromised should be considered.

4.4.4.1 Dynamic Keys

One way to protect a system when a key leaks is to change the keys. Some systems regularly update their keys, or entire key batches. Other systems use a static key (or batch) and have the ability to quickly update a key or batch of keys in the event that a successful attack is suspected. In any event, the system must be able to securely exchange keys and work with different key versions.

Generally, a key version is stored on the card, which can be queried before authentication begins. With this key version, the reader knows which key to use for authentication.

> **Note** The key's version number should not to be confused with the 'key number'. The latter is used to allow different functions using different keys (e.g. read only or write as well). The version number refers to a version of one and the same key.

4.4.4.2 Key Diversification

A big danger for card systems with symmetric encryption is that the method requires the same key on the card and the reader. Typically, many cards are in circulation in a system, and these cards are very often quite accessible, thus easily exposed to an attacker.

Should an attacker successfully crack one card, then he has compromised the entire system because the key is used on all cards.

You could then not only read the data from the hacked card, but also make copies of the cards, for example. Imagine a ticketing system, where an attacker can make as many copies of a ticket as he wants and sell these.

UID for Key Diversification

One solution is to use the UID for key diversification. The UID is preprogrammed by the semiconductor manufacturer and cannot be copied from one card to another.

Fig. 4.19 demonstrates simple key diversification using a UID as the init vector to encrypt a 'master key' (MK). The derived key, DK, is then used for the mutual authentication (see Section 4.4.1).

The MK is saved (securely) only in the reader device, and is thus less exposed than the DK on the card. Even if the DK becomes known, this cannot be used to clone tickets, for example, because the MK is not known.

> **Note** The simple encryption as shown in Figure 4.19, with UID = IV, is but an example. Often, the CMAC algorithm is also used.

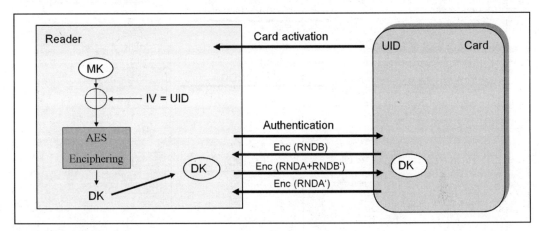

Figure 4.19. Example of simple key diversification.

RID for Data Protection

For data protection, the use of a random UID is suggested instead of a fixed one, so that undesirable parties are not able to draw inferences about or create movement profiles of the cardholder, for example. After all, the UID is, as part of the card activation, 'public', that is, anyone can read it or even actively request it, as long as they are near enough to the card.

To combat this, the ISO/IEC 14443 standard offers the option of using a random ID (RID) for card activation. At every activation, the RID is replaced with a new random number (see also Table 2.1). The new MIFARE cards (MIFARE DESFire EV1 and MIFARE Plus), for example, support this function.

Naturally, this creates a problem in that the UID is no longer available for key diversification, but must first be transmitted via a secure communication channel.

The basic sequence is shown in Figure 4.20: after the card activation using RID is done, the reader sends a command, containing encrypted random number RNDQ, to request the UID. This encryption takes place with a non-diverse key (nDK), which the card must also know.

The card decrypts RNDQ and uses it as a key to encrypt its UID. The encrypted UID is then sent back to the reader, where it is decrypted using RNDQ as a key. By this means, the UID is available for key diversification.

Because the UID's encryption is based on a key based on a random number (and is thus different every time), the UID cannot be discerned externally.

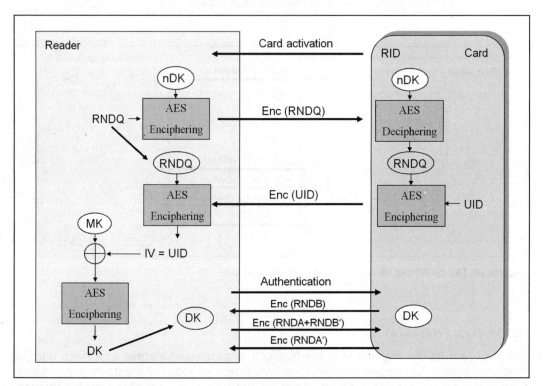

Figure 4.20. *Example of simple key diversification using an RID.*

A further security measure would be to protect the transmission of the encrypted random number with a MAC generated using a non-diverse key. The card can then ensure that only authorized readers receive a valid UID: should the MAC not match, the card returns nothing, or, even better, a random number ('junk') instead of the encrypted UID.

4.4.5 Secure Application Module (SAM)

As shown in Figure 4.1, a secure smartcard system has a SAM (Secure Application Module). This is a device that securely stores the keys and carries out the authentication, encryption and decryption processes.

The system can only be considered safe if the security standards applicable to the smartcard are used in the SAM.

A SAM could be a secure software implementation (which requires a secure hardware platform in turn). In most cases, however, a 'hardware-SAM' is utilized, which is typically also a smartcard. Smartcards are basically secure platforms, so are ideally suited for SAM use.

For security applications, many card readers have one or more contact slots, usually in the compact SIM card form, with which we're familiar from cellphones.

4.4.6 Security Assessment

Security assessment is a very complex issue. Assessing security is complicated and, in many cases, requires the user to carry out their own attacks in order to shed light on possible weaknesses.

Naturally, not everyone who builds secure systems can do this himself. Nor can they rely solely on the advertising claims of individual manufacturers. Therefore, an independent assessment is required, which ensures that:

- security is scalable and comparable;
- assessments cannot be manipulated;
- open assessment standards, that anyone can verify, are used.

An open system has emerged, with independent testing labs, which already oversee the design of smartcards. It is possible to check and assess both hardware and software. Assessments are carried out and certificates are then issued along with the application manuals. The latter should ensure that the user of the certified smartcards deploys them in such a manner extends the security assurance to the application.

This system is called the 'Common Criteria' (or CC), and various certification levels are provided (see *http://www.commoncriteriaportal.org*).

5
Introduction to Cards and Tags

Renke Bienert

This chapter provides an overview of the typical characteristics of (contactless) smartcards and describes the standard MIFARE contactless smartcards, which are used as much as possible in the practical part of this book (Chapters 8, 9).

5.1 Overview

5.1.1 Memory Cards and Microcontroller Cards

The first chip cards were purely memory cards, which essentially just stored data. Today, however, this description is very limiting, as 'simple' storage cards have different formats as well as different storage capacities compared to our smartcards. We also expect more functions from smartcards today, such as security functions, counter functions and value calculations.

Still, the term 'memory card' is quite commonly used in the smartcard arena. This term is generally used for smartcards that a system operator can purchase as a 'finished' card and use directly. These are typically inexpensive and are used in large quantities in many applications.

Reading and writing are the main features of these cards, usually associated with prior authentication and encrypted data transmission. These days, some of these 'memory cards' are in fact microcontroller cards, so, even if the user sees only a simple read and write interface externally, the smartcard is running software on a microcontroller.

For us, microcontroller cards come into play when an operating system and software is required in order to make use of the card in an application. It could be that the card contains a generic operating system and an application may be uploaded as software (for example, a Java applet).

5 INTRODUCTION TO CARDS AND TAGS

Typically, the option to upload applications makes a card more flexible, but obviously more expensive. The card's operating system may also have functions and storage space that aren't needed for the application.

5.1.2 Advantages and Disadvantages of Contactless Cards

The more complex (expensive) the card's features, the more likely it is to be provided with multiple interfaces.

The contactless interface described in this book has many advantages over the contact interface.

5.1.2.1 Robustness

The IC can easily be housed completely protected within the card body, as no direct electrical or mechanical external connections are required. The readers can also be built more robustly, as no card slots or similar are required.

5.1.2.2 Longevity

There are no open electrical contacts that could wear out or get dirty, either on the card or the reader.

5.1.2.3 Usability

It is much easier, more convenient and faster to place a card on a reader then to insert it into a corresponding opening in a reader. An application in which many people could possibly want to make use of a service may not be achievable using contact cards. Consider the automatic turnstiles in a large Asian city's subway station, and imagine that every traveler first had to insert their card into a reader slot — this would take far too long.

Two possible disadvantages must thus be taken into account, in certain instances.

5.1.2.4 Infrastructure

Contact-type cards have been on the market longer than contactless cards, as have their standards, so contact-type card readers are more widespread. Also, for a long time it was cheaper to build a contact reader than a contactless one. Low-cost RFID single-chip reader ICs (for example, the MF RC523 used here) have been around for about ten years.

Therefore, it may be more sensible to use contact cards in situations where systems are dependent on an infrastructure of existing readers, or in situations where either the advantages of contactless technology are not required or the number of readers (compared with the number of cards in use) is very high and so extremely low-cost readers are required.

A typical application that comes to mind is the SIM card in a cellphone. There is typically only one card per device. The cards are protected within the phone in any case and are not often changed, so abrasion of the contacts is not an issue.

5.1.2.5 Contact Between Card and Reader

When a contact card is in use, one can usually be fairly certain that the reader actually has direct contact with the correct card. Over-the-air communication is easier to manipulate.

Therefore, there are many attack scenarios for a contactless interface. The most dangerous attack is the so-called 'relay attack' (Section 4.2.1.4).

However, just as with contact cards, these attacks can be prevented if appropriate measures are used. For example, if the card has to be inserted into a slot, or an optical feature is used where the card's surface must be read before the reader can access the card.

The latter is the case for electronic passports, which also use the ISO/IEC 14443 contactless interface. Only when the passport is opened up — and printed information can be read — can the passport reader optically read a printed key and use the same data (as printed) to control electronic access.

In fairness, we must concede that in this situation at least part of the usability advantages are forfeited.

5.1.3 Dual-Interface Cards

There are now applications that require a few of the features of both technologies, for example if the application must use an already-existing infrastructure, and, at the same time, a second, more robust reader device will be employed. For this, there are so-called

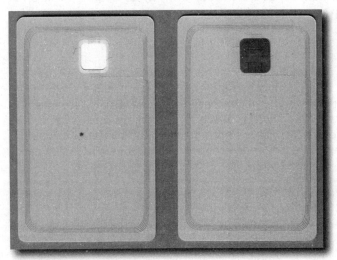

Figure 5.1.
Example of a transparent dual-interface card (PAV, 2010).

5 INTRODUCTION TO CARDS AND TAGS

'dual-interface cards' (DIF cards) which have both interfaces available. Typical examples are bank cards (which use a contact chip) that can also be used as tickets for public transport (contactless), and (contactless) access control cards (employee IDs) that can also be used to log in to PCs (requiring contact).

DIF cards are typically only suited to high-cost applications, as the chip processing in the card is more complicated than for a contact-type or contactless smartcard. All of the MIFARE cards described herein are purely contactless.

5.2 MIFARE

MIFARE™ is a registered trademark of NXP Semiconductors, B.V. (formerly Philips Semiconductors), and is made up of the words 'Mikron' and 'fare collection'. MIFARE is a brand name, not a product name.

5.2.1 MIFARE Overview

In the early 1990s, the Mikron company developed the original MIFARE card — the 'MIFARE Classic 1K', which was shipped with a reader device, in 1994.

The target application was the public transport system (fare collection), i.e. the use of a contactless smartcard as an electronic ticket for train, bus, subway or tram.

This has become a reality in many cities and regions across the world, and modern transportation systems couldn't get by without electronic travel tickets. For this reason, there now is a whole range of card ICs that are sold under the MIFARE name.

5.2.1.1 Success Story

MIFARE products very quickly established themselves on the market, and MIFARE is a success story. Aside from Philips Semiconductors (now NXP), which acquired Mikron at the end of the 1990s, Infineon (previously Siemens Semiconductor Division), also manufactures MIFARE products under license. Today there are several license holders for MIFARE technology, and much of this technology has found its way into ISO standards, such as the ISO/IEC 14443.

The MIFARE product portfolio is continuously being expanded, so that everything from simple, cheap electronic paper tickets to complex, certified high-security cards are in use. The MIFARE Classic is still sold, as always, and is probably the most widespread contactless smartcard. While the exact numbers are not disclosed by the participating companies, we can assume that there are more than a billion MIFARE Classic cards currently in use.

5.2.1.2 MIFARE Clone

Such huge success naturally attracts some imitators, such as the 'Chinese clones'. Even before the turn of the millennium, Chinese companies were attempting to tear down the

MIFARE Classic chip and recreate it through reverse engineering. There are now Chinese companies that offer clones of the MIFARE Classic, which are illegal in Europe.

5.2.1.3 MIFARE Hack

Also unsurprisingly, the widespread use of MIFARE products invited attempts to hack the proprietary MIFARE Classic encryption method. Such a hack, very similar to the reverse engineering method used previously by the Chinese, was successfully accomplished in 2008 and released publicly. The MIFARE Classic's primary flaw was the use of a proprietary, undisclosed algorithm for its outmoded security.

5.2.1.4 MIFARE Product Overview

Since 2004, the MIFARE DESFire, a MIFARE product with DES and triple-DES encryption, as well as its compatible successor, the MIFARE DESFire EV1, have also been available. Since 2002, the MIFARE Ultralight, designed for paper tickets with no encryption, and its compatible successor, the MIFARE Ultralight C (which offers triple-DES authentication), have been offered. These may be the smallest (= cheap) contactless chips with standardized encryption technology.

The direct successor to the MIFARE Classic is the newest chip from the MIFARE family: the MIFARE Plus. There was a chip with that name at the end of the 90s, but it had nothing in common with the new one, other than its name.

Table 5.1. MIFARE Product Overview

Characteristic	MIFARE Ultralight	MIFARE Ultralight C	MIFARE Classic	MIFARE Plus	MIFARE DESFire	MIFARE DESFire EV1
Interface	Type A	Type A	Type A	Type A	Type A	Type A
Memory size	64 bytes	192 bytes	320 bytes 1 KB 4 KB	2 KB 4 KB	4 KB	2 KB 4 KB 8 KB
Authentication	-	3DES	proprietary	proprietary + AES	3DES	3DES + AES
Data encryption	-	-	proprietary	proprietary + AES	3DES	3DES + AES
Special features	extremely small	small + 3DES	widespread	AES + very secure	3DES + flexible file system	3DES + AES + flexible file system

5.2.2 MIFARE Ultralight

The MIFARE Ultralight is the smallest and cheapest member of the MIFARE family. As a paper ticket, it is used in huge quantities in a few cities around the world. For example, if we used the subway in Moscow, we'd buy a MIFARE Ultralight.

The MIFARE Ultralight uses ISO/IEC 14443 card activation, but its own simplified protocol, as shown in Figure 5.2.

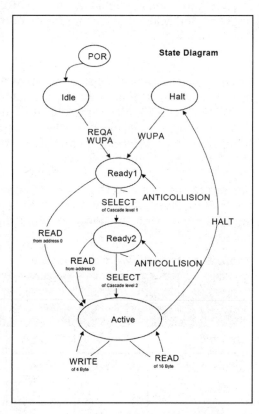

Figure 5.2.
MIFARE Ultralight state diagram (NXP Semiconductors, 2010).

The MIFARE Ultralight uses a 7-byte UID, so it must be activated using Anti-collision + Select CL1 and Anti-collision + Select CL2. Also, apart from the card activation described in ISO/IEC 14443, the MIFARE Ultralight can also be activated with a 'Read Page 0' from either the Ready1 or Ready2 state.

In the event of an error, the card returns to the Idle state, unless it was already in the Halt state, in which case it returns to the Halt state.

5.2.2.1 Instruction Set

Aside from the card activation instructions (REQA, Anti-collision CL1, Select CL1, Anti-collision CL2, Select CL2 and Halt), the MIFARE Ultralight has three others:

Read (0x30)

This command is basically compatible with the MIFARE Classic's Read command, but is only used unencrypted and reads 4 pages, or 16 bytes, at that time, beginning with the page address that is passed as a parameter (ADR).

```
Read
  0x 30 ADR  →
                ←  0x D00 D01 D02 D03 D04 D05 D06 D07 D08 D09 D10 D11 D12 D13 D14 D15
```

If a page address, such as 15, is used at which four contiguous pages are not available, addressing rolls over to the start of memory (Pages 15, 0, 1 and 2 would be read).

> **Note** "→" denotes a transmission from PCD to PICC.
> "←" denotes a transmission from PICC to PCD.
> D0 — D15 = data bytes; ADR = page address; ACK = 4-bit acknowledge (0xA)

Write (0xA2)

This command sends the page address (ADR) followed by 4 bytes of data, which are then written to the corresponding page. The card responds with a 4-bit acknowledge (ACK).

```
Write
  0x A2 ADR D0 D1 D2 D3   →
                          ←  ACK (4-bit!)
```

Compatibility Write (0xA0)

This is a two-part instruction. It writes 4 bytes to a page, but uses the same structure as the Write instruction on the MIFARE Classic.

In the first part of the instruction, the page address is sent to the card. The card responds with a 4-bit acknowledge.

In the second part, 16 bytes are transferred, of which the first four (D0 — D3) are written to the page. The remaining 12 bytes (D4 — D15) are ignored. The card responds with a 4-bit acknowledge.

```
C.Write
  0x A0 ADR                                               →
                                                          ←  ACK (4-bit!)
  0x D0 D1 D2 D3 D4 D5 D6 D7 D8 D9 D10 D11 D12 D13 D14 D15  →
                                                          ←  ACK (4-bit!)
```

> **Note** All instructions and responses (apart from REQA / ATQA and Anti-collision / UID) use a 2-byte CRC, which is automatically calculated by the reader IC and can be verified. Therefore it will be omitted here for simplicity.

5.2.2.2 Memory Organization

The MIFARE Ultralight offers a 512-bit memory, organized into 'pages' of 4 bytes each. In total, there are 16 pages × 4 bytes = 64 bytes, as shown in Figure 5.3.

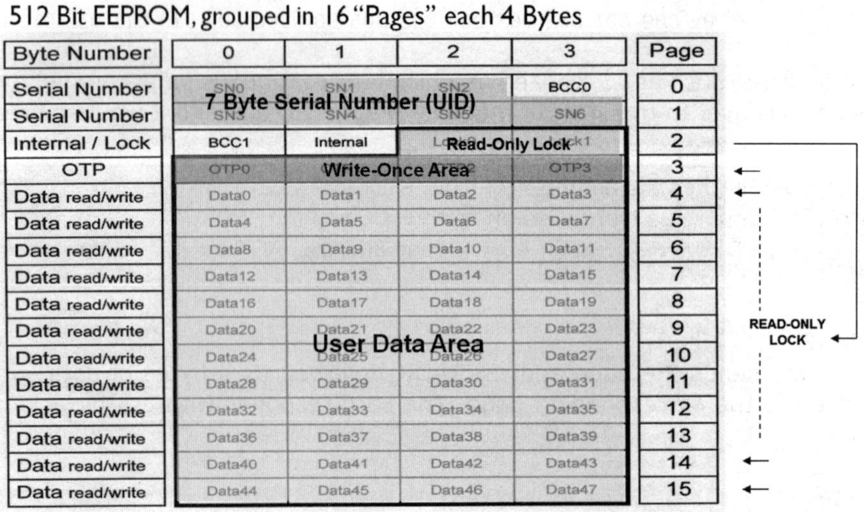

512 Bit EEPROM, grouped in 16 "Pages" each 4 Bytes

Byte Number	0	1	2	3	Page
Serial Number	SN0	SN1	SN2	BCC0	0
Serial Number	SN3	SN4	SN5	SN6	1
Internal / Lock	BCC1	Internal	Lock0	Lock1	2
OTP	OTP0	OTP1	OTP2	OTP3	3
Data read/write	Data0	Data1	Data2	Data3	4
Data read/write	Data4	Data5	Data6	Data7	5
Data read/write	Data8	Data9	Data10	Data11	6
Data read/write	Data12	Data13	Data14	Data15	7
Data read/write	Data16	Data17	Data18	Data19	8
Data read/write	Data20	Data21	Data22	Data23	9
Data read/write	Data24	Data25	Data26	Data27	10
Data read/write	Data28	Data29	Data30	Data31	11
Data read/write	Data32	Data33	Data34	Data35	12
Data read/write	Data36	Data37	Data38	Data39	13
Data read/write	Data40	Data41	Data42	Data43	14
Data read/write	Data44	Data45	Data46	Data47	15

Overlay labels: 7 Byte Serial Number (UID); Read-Only Lock; Write-Once Area; User Data Area; READ-ONLY LOCK.

Figure 5.3. MIFARE Ultralight memory organization (NXP Semiconductors, 2010).

A page is the smallest addressable unit of memory. Pages 0 and 1, as well as half of Page 2, are write-protected, and may only be read. They contain the UID and manufacturer-specific data.

Page 3 is the OTP (one-time programmable) area: all of the bits in this area are set to '0' on a new card. With the normal write command, you can set each bit to '1'. When a bit is set to '1', it cannot be reset.

Pages 4 to 15 are normal memory pages that can be easily read from and written to.

5.2.2.3 Security Functions

The only extra security feature provided by the MIFARE Ultralight is lock bit functionality, which allows individual pages to be protected from further writing, as shown in Figure 5.4.

When a lock bit (L) is set, the corresponding page blocks future write access, so from after the next POR (power-on reset = RF reset) it can only be read.

Setting a block lock bit (BL) blocks the blocking of a given memory area.

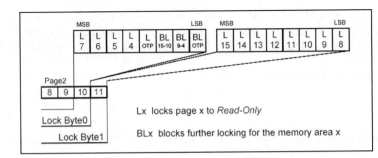

Figure 5.4.
MIFARE Ultralight Lock Bits (NXP Semiconductors, 2010).

Lock bits and block lock bits are OTP, so they cannot be restored. They become valid upon the next POR.

Examples:
- If we set all three BL bits, we cannot set any single lock bit after the next POR. Thus, all pages remain readable and writable.
- If we set L4, L5 and L6, as well as BL4 – 9, then Pages 4, 5 and 6 will be read-only from the next POR. Meanwhile, Pages 7 – 9 remain both readable and writable.

5.2.3 MIFARE Ultralight C

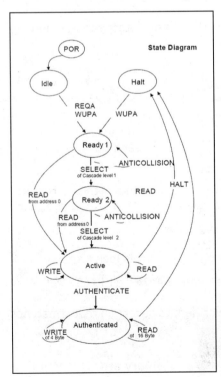

The MIFARE Ultralight C is backwards compatible with the MIFARE Ultralight, so all functional descriptions from the previous section are also valid for the MIFARE Ultralight C.

The MIFARE Ultralight C offers three times as much memory (1,536 bits = 192 bytes), a 16-bit counter and 3DES authentication.

Card activation works exactly as with the MIFARE Ultralight (see Figure 5.5). The same read and write instructions are used.

In addition to the states already described in the previous section, the MIFARE Ultralight C offers one further state: the authenticated state. This state is only reached after successful authentication (which requires knowledge of the 3DES key).

Figure 5.5.
MIFARE Ultralight C state diagram (NXP Semiconductors, 2010).

5 INTRODUCTION TO CARDS AND TAGS

Depending on the configuration, specific read and/or write accesses can be tied to successful authentication, so that only when you have successfully authenticated may you gain access to specified areas of memory.

Data transfer always takes place 'in-the-clear'.

5.2.3.1 Instruction Set

In addition to the instructions described in the previous section, the MIFARE Ultralight C has one more:

Authenticate (0x 1A 00)
This is the two-step instruction that performs authentication.

The process begins with the instruction, 0x 1A 00, to which the card responds with a status code (0xAF) followed by a 3DES-encrypted 8-byte-long random number.

```
Authenticate Step 1
0x 1A 00          →
                  ← 0x AF RBe0 RBe1 RBe2 RBe3 RBe4 RBe5 RBe6 RBe7
```

The random number must be encrypted. The MIFARE Ultralight C uses 3DES in CBC mode. The initialization vector is set to '0' at the start of authentication, and after that it is inherited from the previous cycle.

The reader must generate its own random number (RND A) and rotate random number RND B one byte to the right.

```
RND B           = dk(RBe0 — RBe7) = RB0 — RB7
RND B'          = RB1 RB2 RB3 RB4 RB5 RB6 RB7 RB0
RND A ‖ RND B'  = RA0 RA1 RA2 RA3 RA4 RA5 RA6 RA7 RB1 RB2 RB3 RB4 RB5 RB6 RB7 RB0
```

The 16 bytes that make up RND A ‖ RND B' are now encrypted:

Challenge 2 = ek(RND A ‖ RND B') = RC0 — RC15

The encrypted 16 bytes are sent to the card along with a status code. If the correct key is used, the card enters the authenticated state and responds with a status code, and the encrypted, rotated random number, RND A'.

```
Authenticate Step 2
0x AF RC00 RC01 RC02 … RC15 →
                  ← 0x 00 RAe0 RAe1 RAe2 RAe3 RAe4 RAe5 RAe6 RAe7
```

For validation, this response must be decrypted and verified, to ascertain that it matches the rotated RND A.

```
RND A' = dk(RAe0 — RAe7) = RA1 RA2 RA3 RA4 RA5 RA6 RB7 RB0 ?
```

MIFARE 5.2

> **Note** This authentication method, in which encrypted random numbers are exchanged, is known as a 'mutual 3-pass' or 'challenge-response' method. This principle is applicable to all MIFARE cards; only the type of encryption and the length of the random numbers varies.

5.2.3.2 Memory Organization

The MIFARE Ultralight C has three times the memory of the MIFARE Ultralight. The first 16 pages are exactly the same as on the MIFARE Ultralight (see Section 5.2.2.2).

Pages 16 – 39 (0x10 – 0x27) are simple extensions of the memory capacity for reading and writing.

Pages 40 – 47 (0x28 – 0x2F) have special functions, as illustrated in Figure 5.6.

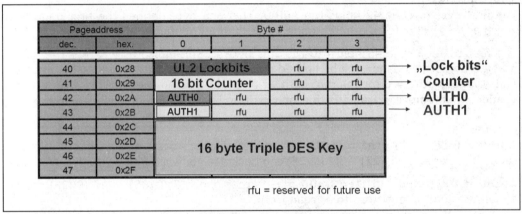

Figure 5.6. MIFARE Ultralight C (NXP Semiconductors, 2010).

5.2.3.3 Security Functions

Exactly like the MIFARE Ultralight, the MIFARE Ultralight C has lock bits to prevent writing to specific memory areas. In addition, it offers a 16-bit timer and authentication with associated access rights.

Lock Bits

The first two bytes of Page 40 are the lock bits that are valid for the extended memory area, as shown in Figure 5.7. The function of the lock bits and block lock bits is exactly the same as for the MIFARE Ultralight.

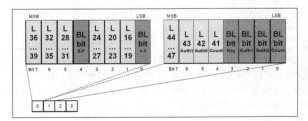

Figure 5.7.
The MIFARE Ultralight C's additional lock bits (NXP Semiconductors, 2010).

Counter

The first two bytes of Page 41 (0x29) make up a 16-bit counter. The initial value is '0' upon delivery of the IC.

The first write instruction to this page can set any arbitrary initial value.

Thereafter, the counter can only be incremented, until it reaches its maximum (0xFFFF), and then it is 'full'. There is no overflow and no reset.

Counting can be done with a normal write instruction: the least-significant nibble of the LSByte is added to. The counter can thus be increased by between 1 and 15 using a single count instruction.

> **Note** The counter can also be locked using the lock bits.

Authent Bytes

The first byte of Page 42 holds the Auth0. This one byte defines the first page from which access will only be granted upon successful authentication.

The first byte of Page 43 holds Auth1. This byte's LSbit indicates whether the access granted by authentication will be limited only to writing, or if it's valid for both reading and writing. Auth1 = 0 means that both writing and reading are possible after authentication, while Auth1 = 1 means that only writing is enabled upon authentication.

Example:
Auth0 = 0x22 and **Auth1 = 0x01** means that all pages can be read without authentication, but pages 34 (0x22) — 47 (0x2F) can be written only if successfully authenticated.

Auth0 = 0x05 and **Auth1 = 0x00** means that pages 5 (0x05) — 47 (0x2F) require authentication to be written to or read from.

5.2.4 MIFARE Classic

The MIFARE Classic is the oldest MIFARE product still in use and is manufactured and used in large quantities. Even though its encryption was hacked, its good performance and wide availability, combined with its ease of use, keep it a very attractive contender. The MIFARE Classic is older than the ISO/IEC 14443, and is thus only partially compatible with the standard. In old MIFARE Classic documentation, one can find examples of outmoded names for instructions, whose functions are ISO instruction equivalents (REQUEST IDE = REQA, REQUEST ALL = WUPA).

The MIFARE Classic is available in three memory capacities:

320 bytes (MIFARE Mini), 1 KB (MIFARE 1K) and 4 KB (MIFARE 4K)

Figure 5.8 shows a simplified state diagram, in which only two states are recognized subsequent to card activation: the ACTIVE state and the 'MIFARE Classic' state. Access to the MIFARE Classic state is controlled by MIFARE Classic authentication. As the MIFARE encryption is proprietary, one needs the corresponding encryption implementation on the reader side: this is included in the MF RC522 reader IC.

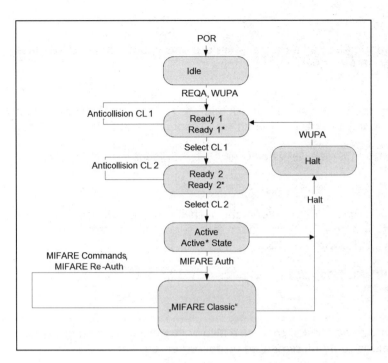

Figure 5.8.
MIFARE Classic (simplified) state diagram (using a 7-byte UID).

As with all other MIFARE Classic instructions, authentication uses a fixed command structure, consisting of a 2-byte CRC and a fixed timeout.

> **Note** This state diagram shows a card with a 7-byte UID. There are also MIFARE cards with 4-byte UIDs, for which the Anti-collision CL2, Select CL2 and 'Ready 2' and 'Ready 2*' states are absent.

Table 5.2 shows the MIFARE-specific codes for acknowledge (ACK) and not acknowledge (NAK).

Table 5.2. MIFARE 4-bit ACK and NAK.

Code (4-bit, hex)	ACK / NAK
0	Access denied for the specified address
1	CRC or parity error
2	Reserved for future use (NAK)
3	Reserved for future use (NAK)
4	Transaction underflow or overflow error (32-bit signed integer)
5	CRC or parity error
6-9	Reserved for future use (NAK)
A	Acknowledge (ACK)
B – F	Reserved for future use (NAK)

5 INTRODUCTION TO CARDS AND TAGS

5.2.4.1 Instruction Set

Following successful authentication, all data transfers are encrypted. The MIFARE Classic supports the following MIFARE Classic instructions:

MIFARE Read:
Along with this instruction, the reader sends a block address to the card. The card reads pit the specified block if the key rights (AC) match, and responds with 16 bytes of data.

> **Note** Reading and writing always takes place in 16-byte blocks.

MIFARE Write:
The reader sends the card a Write instruction along with the block address. The card acknowledges the receipt with an ACK. The reader then sends the 16 data bytes. The card confirms a successful write with an ACK.

> **Note** A reader for MIFARE Classic cards requires only a 16-byte buffer.

Value Operations
In addition to reading and writing blocks, there are also 'value block operations', which allow the card's stored values to be incremented or decremented.

This requires a 'value block', or an arbitrary data block, which contains an integer value in the valid 'value format'. Of course, the correct key rights (AC) must be used to allow a value operation.

The user must write the value block (see Section 8.4.5 for the format with an initial value) using the MIFARE Write command.

MIFARE Decrement and MIFARE Increment operations must always (!) be completed using a MIFARE Transfer, as this is when the result of the card's internal addition or subtraction is written to the target block.

There is also the MIFARE Restore, which loads a data block's content into the buffer, without modification. This instruction must also be completed with a Transfer.

The easiest way to explain the function is by means of an example.

```
Write Block 1, 640000009BFFFFFF6400000000FF00FF    // store 100 in Block 1
   Decrement Block 1, Wert 10                      // calculate: buffer = 100 - 10
   Transfer Buffer Block 2                         // write the result to Block 2
   Read Block 1: 640000009BFFFFFF6400000000FF00FF      // value: 100
   Read Block 2: 5A000000A5FFFFFF5A00000000FF00FF      // value: 90
   Restore Block 2                                 // copy Block 2 to buffer
   Transfer Block 1                                // write buffer to Block 1
   Read Block 1: 5A000000A5FFFFFF5A00000000FF00FF      // value: 90
   Read Block 2: 5A000000A5FFFFFF5A00000000FF00FF      // value: 90
```

> **Note** You can also write the buffer result back to the same block, but it's highly recommended that a backup be stored in order to 'survive' a break in communication during writing to the card. In contrast to the MIFARE DESFire, the MIFARE Classic has no backup management or 'anti-tearing'.

5.2.4.2 Memory Organization

The MIFARE Classic's memory consists of blocks and sectors. A block always consists of 16 bytes, and is a fixed entity. Addressing, reading and writing always takes place in entire blocks.

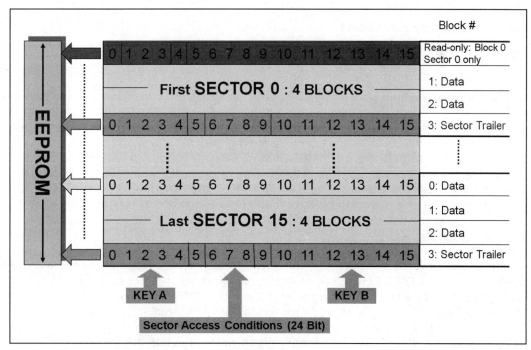

Figure 5.9. MIFARE Classic 1K memory organization.

In the MIFARE 1K, 16 sectors of 4 blocks add up to 64 blocks. The first three blocks of every sector are data blocks, which means that they can be used for reading, writing and value operations.

Exception: Sector 0, Block 0. This very first block is read-only. This block contains, among other things, the UID.

The last block in each sector is the 'Sector Trailer', which includes the Key A and Key B, as well as they key access rights (AC).

5 INTRODUCTION TO CARDS AND TAGS

> **Note** Should a key be used for its purpose (for Key A, always; for Key B, depending on the AC), then only '0000000000000000' will be read out instead of the key bytes. Note that 'write after read' (without modification) modifies the contents!

> **Note** The standard configuration of a 'virgin' MIFARE Classic card only allows the use of Key A. Key B is readable, so it cannot be used as a key. In order to use Key B, the AC must be changed to prevent this.

More detail can be found in Section 8.4 as well as in the respective data sheets, which are downloadable from *http://www.nxp.com*.

5.2.4.3 Security Functions

The MIFARE Classic encryption can no longer be considered as safe.

5.2.5 MIFARE Plus

At the end of 2008, shortly after the MIFARE Classic was hacked, the MIFARE Plus arrived on the market. The MIFARE Plus uses the same memory structure and provides a mode in which the it acts just like a MIFARE Classic. Everything else is faster, securer and more up to date. For example, the MIFARE Plus offers a fairly unique function in the smart card arena: secure and accurate verification of the FDT (Frame Delay Time). This function is known as the 'Proximity Check' and is used to detect relay attacks.

5.2.5.1 Memory Organization

Due to this backwards compatibility, the MIFARE Plus memory organization is exactly the same as the MIFARE Classic (1K or 4K). An application does not need to modify the arrangement of data on the card when switching from MIFARE Classic to MIFARE Plus, so the number of required changes to the system is minimized.

In addition to the MIFARE Classic memory organization, the MIFARE Plus has additional storage capacity where additional configuration blocks are found. These blocks allow the configuration of the (current and future) MIFARE Plus characteristics, such as the blocks in which the AES keys are stored. This additional memory is write-only.

There are two AES keys for each sector, i.e. 64 AES keys for the MIFARE Plus 2K (successor to the MIFARE Classic 1K), and 80 AES keys for the MIFARE Plus 4K. In addition, there is a range of additional keys for specific tasks.

5.2.5.2 MIFARE Plus S and MIFARE Plus X

The MIFARE Plus is available in two variants: the MIFARE Plus S and the MIFARE Plus X. Both are available with either a 2 KB or a 4 KB EEPROM.

The MIFARE Plus X ("full featured") supports all MIFARE Plus features, which is why it is export controlled (not public), so it won't be described in detail in this book.

The MIFARE Plus S supports a subset of the MIFARE Plus X's functionality, namely
- SL0, SL1 and SL3, but not SL2,
- no encryption of data – only MAC security (this is the reason that the MIFARE Plus S is not export controlled),
- no value operations in SL3, and
- only a limited set of instructions for the 'Virtual Card Architecture'.

5.2.5.3 Security Levels

As shown in Figure 5.10, the functional concept of the MIFARE Plus is based on the four security levels (SL). You can always switch to a higher security level, but never a lower one. Switching to SL2 and/or SL3 takes place with AES authentication, which uses an access key for the up-shift.

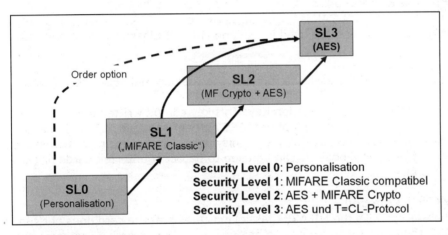

Figure 5.10. MIFARE Plus Security Level (SL) concept.

Security Level 0 (SL0)
The MIFARE Plus ships in SL0. This facilitates personalization only, and recognizes only two instructions:
- **Write Perso:** With this instruction you can write all blocks of the MIFARE Plus. A minimum number of AES keys must be written (3 for the MIFARE Plus S, 4 for the MIFARE Plus X). Also, all the keys should be written, especially the AES keys, as unwritten AES keys have a default value, known to all, and this provides no security. You can also write all other data pertaining to the application.
- **Commit Perso:** This instruction completes the personalization. The card then leaves SL0, and at the next activation awakes in either SL1 or SL3 (special version of the MIFARE Plus).

These two instructions can either be performed in the 14443-3 layer (like the MIFARE Classic instructions), or with the 14443-4 protocol using 'I Blocks'. For this, the T=CL protocol must be used, as described in Chapter 2.

> **Note** A MIFARE Plus in SL0 should never be placed 'in the field'. Personalization must take place in a secure environment, as all data in SL0 is unencrypted and transmitted insecurely.

Security Level 2 (SL2)

In SL2, the same encryption and instructions are used as in SL1, although prior AES authentication using the appropriate AES Sector Key is required. From this AES authentication, a temporary MIFARE key is generated, which will last only for the duration of the transaction. The details won't be described here, as SL2 is only applicable to the export-controlled MIFARE Plus X.

Security Level 1 (SL1)

A MIFARE Plus in SL1 behaves just like a MIFARE Classic. This means that the security is just as good (or bad) as the MIFARE Classic (with a few hardly notable exceptions).

14443-3 Layer (MIFARE Protocol)

The MIFARE Plus (2K or 4K) SL1 SAK is the same as that of a MIFARE Classic (1K or 4K). Thus, the MIFARE Plus in SL1 can be used in existing MIFARE Classic applications without alteration.

In addition to the usual MIFARE Classic instructions, there is optional AES authentication:

Authentication with the SL1 Authentication Key: This key can be written in SL0 (as can all the keys). One can thus carry out an optional AES authentication in SL1, without having to switch the card to a higher security level. This authentication has no effect on the data or any write or read privileges. It can be used to detect counterfeit cards or card emulators, for example.

14443-4 Layer (T=CL Protocol)

Even though the SAK (as with the MIFARE Classic) indicates that the card does not support the ISO/IEC 14443 protocol, it does so anyway. So, after selecting the card, one can either use the MIFARE Classic authentication (making the card a 'MIFARE Classic') or activate the card using the RATS (or PPS) protocol. With the T=CL protocol, the following AES authentication methods are available:

- **Authentication using the SL2 Switch Key:** This switches the card into SL2. This instruction is only available on the MIFARE Plus X. After this command, the card must be activated again, after which it will be in SL2.
- **Authentication using the SL3 Switch Key:** This switches the card into SL3. After this, the card must be activated again, after which it will be in SL3.
- **Authentication using the Originality Key:** This key cannot be written in SL0 mode and is secret. This instruction will only be sent by special reader devices. It is not commonly used, but helps with the detection of cloned ICs.

Security Level 3 (SL3)

In SL3, the T=CL protocol must be used. The MIFARE Plus's SAK will indicate that the card is 14443-4 compatible. Upon activating this protocol (using RATS), the card will respond with a 14443-compatible ATS. Part of this ATS will be the 'historical bytes', which are used by the MIFARE Plus to distinguish between the S and X variants. The coding is described in an application note by NXP Semiconductors, "AN10833 — MIFARE Type Identification Procedure", available for download at *http://www.nxp.com*.

Incidentally, the ATS can be changed: either you've written it while in SL0, or you can change it while in SL3, provided that you have authenticated using the corresponding key. One must be careful to comply with the ATS format, as prescribed in ISO/IEC 14443-4, or the card may become unreadable.

Basically, the MIFARE Plus uses AEC in CBC Mode for authentication and for encryption (MIFARE Plus X only), and to protect the data from manipulation (MAC).

In SL3, the MIFARE Plus makes use of complex 'secure messaging'. Everything that was insufficiently secure about the MIFARE Classic has been carefully reconsidered in order to prevent all conceivable attacks on the communication. A detailed description is beyond the scope of this book, but it can be found in the relevant data sheets and application notes from NXP Semiconductors. The principle can only be briefly outlined here: After protocol activation (RATS or PPS), AES authentication must take place. This is typically done using the appropriate Sector Key (AES Key A or AES Key B). With this first authentication (First Authenticate), the transaction is begun, and a 'Transaction Identifier' is generated by the card and sent to the reader. The read counter (R_Ctr) and write counter (W_Ctr) are set to '0'.

R_Ctr and W_Ctr are never transmitted, only incremented: R_Ctr on any read, W_Ctr on any write, decrement, increment, restore or transfer. The Transaction Identifier and R_Ctr and W_Ctr are used to create the Init Vector. This prevents instructions from being replayed during a session, or swapped in order (man-in-the-middle attack).

Because AES is generally slower than MIFARE Crypto1 (see comparison between Stream Cipher and Block Cipher in Section 4.3.3), and also because more data must be exchanged (a MAC, even when abbreviated in the MIFARE Plus, is still 8 bytes long, and adds an additional 16 bytes to a 16-byte instruction, which must be transmitted in both directions), there must be ways to speed up the entire transaction. By using some 'tricks', we can get back to execution times that are the same as the MIFARE Classic, or even shorter. These tricks are:

Multisector Authentication
If the keys for different sectors are the same, you need not re-authenticate. If all the keys are the same, you could read the entire memory using only one authentication. This saves time, especially for long transactions that span multiple sectors.

Multi-Block Read and Multi-Block Write
While the MIFARE Classic allows only one block at a time to be read or written, the MIFARE Plus allows up to three blocks at a time for writing, and up to 216 blocks at a time (the entire memory) for reading, with a single instruction. In this way, the transmission of individual instructions as well as the overhead for the associated MACs is eliminated.

> **Note** Once the number of bytes that are transmitted by an instruction exceeds the T=CL protocol's frame size, chaining takes place.

So, using just one authentication and one read instruction, the entire MIFARE Plus memory content can be read, provided that all the keys are the same.

Higher Data Rates

With the help of PPS (see Section 2.1.4.1), the communication speed can be raised from 106 Kbit/s to up to 847.5 Kbit/s. As the number of bytes is higher than the MIFARE Classic, an increased data rate brings with it a decent speed advantage. Also, operation with the higher data rate is stable and the rate does not result in reduced range, even at 847.5 Kbit/s, as long as the reader supports it.

Table 5.3. MIFARE Plus speed comparisons between SL1 and SL3.

Instruction	SL1 (@ 106)	SL3 (@ 106)	SL3 (@ 424)
Read 1 block	2 ms	4.7 ms	1.8 ms
Read 9 blocks	18 ms	18 ms	6.9 ms
Write 1 block	5.2 ms	7.3 ms	4.1 ms
Write 3 blocks	15.6 ms	10.3 ms	5.1 ms

The example times in Table 5.3 are measured on the MIFARE Plus S and show the time from the start of the instruction transfer until reception of the complete response (PCD send, PICC operation and PICC send). This does not include and data processing time in the reader. Reading 9 blocks in SL1 does not require the additional two authentications needed by the MIFARE Classic 2K, as sector boundaries are crossed.

5.2.6 MIFARE DESFire (EV1)

The MIFARE DESFire was launched in 2004, and an improved version, the MIFARE DESFire EV1, came out in 2008. The MIFARE DESFire EV1 is backwards compatible with the original MIFARE DESFire.

The MIFARE DESFire (EV1) is fully ISO/IEC 14443 compatible, that is, it uses the T=CL protocol. There are native instructions, or the ability to use ISO/IEC 7816-4 standard instructions and data structures (see also Section 8.7).

5.2.6.1 Memory Organization

In contrast to the MIFARE products described previously, the MIFARE DESFire has no rigid memory divisions, but an operating system with flexible storage allocation. Therefore, as shown in Figure 5.11, a new memory subdivision must first be created on the MIFARE DESFire EV1, much like creating a subdirectory on a hard drive. On the MIFARE DESFire EV1, these subdirectories are called 'Applications', and each is represented by an 'Application Identifier' (AID). Up to 28 'subdirectories' can be created one card.

MIFARE 5.2

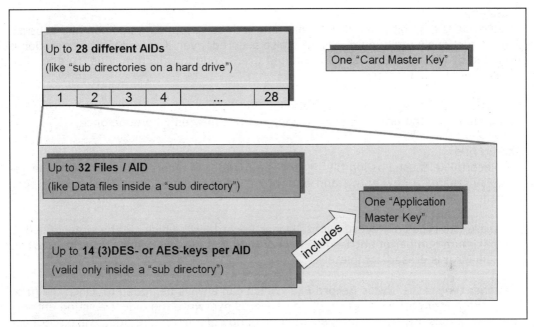

Figure 5.11. MIFARE DESFire EV1 memory organization.

Within a directory, keys and files must be created: Up to 14 keys, which may be either DES or AES keys, and up to 32 files may be created within a single AID. Different AIDs can use different encryption types (DES or AES), and access to the files may be encrypted, secured with a MAC, or completely insecure. In the latter case, not even authentication is required.

The keys (up to 14 of them) can be assigned key rights, which control access to reading and writing for specific files. There is one key, the 'Application Master Key', that may be used to allow creation and/or deletion of files, for example.

The same applies at the card level, that is, in the 'root directory': There is a 'Card Master Key' there that may be used to create and/or delete AIDs, or even reformat the entire memory.

5.2.6.2 File Types

The MIFARE DESFire (EV1) has three different file types: two types of data file, a 'Value File', and two types of 'Record File'.

5.2.6.3 Data File

A data file is used for regular storage, so it can be read from and written to. There are two types of data file: the 'Standard Data File', which has as much storage space as defined by the file size, and a 'Backup Data File', which requires twice as much space as the actual file size. The latter provides for a backup, in case a write operation is interrupted.

5 INTRODUCTION TO CARDS AND TAGS

Value File
A Value File is a 32-bit (signed) integer value with corresponding addition and subtraction instructions (credit and debit). It is also secured, even if an arithmetic operation is interrupted. There is also a special 'Limited Credit' option, which may be used to allow for a limited value increase.

This instruction allows one to refund a value that is, at most, equal to the value that was deducted during the previous transaction, and this may only be done once.

Record File
The Record File is used as log file. It stores a number of records. Each write to the Record File creates a new entry, and one may read only as many entries as have been written.

There are two types of Record File: a 'Linear Record File', which can be simply filled. If the maximum number of entries is reached, then it is full and cannot be written to further, unless it is erased completely.

The other type is the 'Cyclic Record File', which can always be added to. Once the maximum number of entries is reached, new records overwrite the oldest existing entries. These two types are also secure, so an interruption of a write operation will either have written a completely new entry, or there will be no new entry at all.

The MIFARE DESFire offers many other features and functions, which we cannot describe here individually. However, for further detail and a programming example, see Section 8.1

6 Reader Antenna Design

Renke Bienert

In the Elektor RFID Reader, we use the NXP MF RC522 reader module, hereinafter referred to as the 'RC522'. The data sheet can be downloaded from the NXP website: *www.nxp.com*.

We'll briefly consider the reader module below, and then discuss practical hardware and antenna design. A more detailed discussion, including a software description, can be found in Sections 7.2.3 and 9.1.

6.1 MF RC522 Reader Module

The RC522 is the smallest reader module, belonging to a family of ICs: Alternatively, one could use the pin-compatible RC523 or the PN512.

The RC522 supports Type A only, while the RC523 supports both A and B. The PN512 is an NFC building block – it supports both reader and card modes, so it can be used for peer-to-peer communication as per the NFC specification.

The RC522 requires a 3 V power supply (2.5 – 3.6V) (see Figure 6.1).

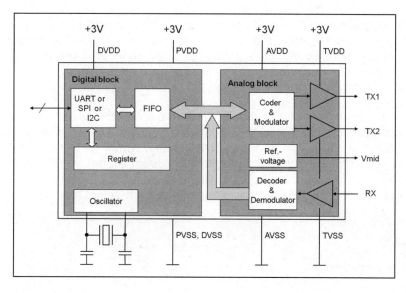

Figure 6.1.
RC522 block diagram.

6 READER ANTENNA DESIGN

6.1.1 Digital Interfaces

On the digital side, the RC522 is controlled by a microcontroller via one of the available interfaces:

6.1.1.1 UART

The UART is common serial port that supported by some devices (including PCs). Functionally equivalent to the EIA-232 (RS-232), the RC522's serial port can only be driven using 3 V levels. If the RC522 is to be connected directly to a PC, a level converter must be used.

The UART supports data rates of up to 1.2 Mbit/s

6.1.1.2 SPI

SPI (Serial Peripheral Interface) is defined by Motorola and supported by many microcontrollers. For a detailed description, consult Wikipedia.

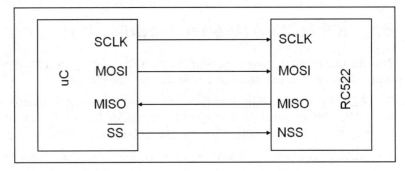

Figure 6.2.
SPI principle.

The RC522 supports SPI data rates of up to 10 Mbit/s, although one should take care as the interface has a large overhead. When reading and writing, the address must first be transmitted before data can follow. In the worst cases, when addresses are individually read or individual bytes are written, this can halve the useful data rate.

6.1.1.3 I²C

Inter-Integrated Circuit ('I squared C') was defined by Philips Semiconductors (now NXP Semiconductors). A short description can be found at Wikipedia, with a full description at *http://www.nxp.com* — just search for "I2C" and filter by 'User Manual'.

The RC522 supports 'Fast mode' at up to 340 Kbit/s and also a 'High-speed mode' of up to 3.4 MBit/s.

> **Note** The RC522 and RC523 devices do not support a parallel interface. The Elektor reader makes use of the I2C bus (see also Section 7.2.3).

6.1.2 Oscillator

To generate the 13.56 MHz carrier, a quartz crystal is required. The oscillator circuit is part of the IC, so, at its simplest, you need only connect a suitable crystal and two capacitors. The capacitor must have a resonant frequency of 2 × 13.56 MHz = 27.12 MHz, and its load capacity should be about 10 pF. With too large a load capacitance, the crystal oscillator may not always oscillate cleanly.

Alternatively, a 27.12 MHz clock signal may be supplied to the OSCIN input from an external source, but the clock frequency must be very accurate, with a very low jitter and a well-timed duty cycle, or reception problems could arise, for example.

> **Note** OSCIN is a digital input, and thus requires a corresponding digital signal if an external clock is used.

6.1.3 Analog Interfaces

The RC522 is designed to drive an RFID antenna directly. For this reason, the modulated transmission signal is available at both Tx outputs, and the Rx input must be partially coupled to the signal for reception. The next section is dedicated to a complete antenna's design.

6.1.3.1 Transmitter Outputs

Both Tx outputs are 'push-pull' outputs, which means that they alternate between 'high' and 'low' with the 13.56 MHz clock, as shown in Figure 6.3. A square wave like this produces many harmonics which must not be transmitted. Therefore, a low-pass filter is required at the Tx output, to ensure that harmonics are effectively suppressed.

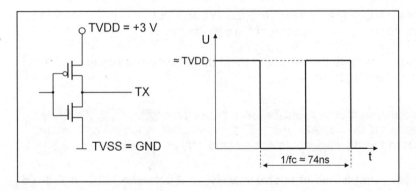

Figure 6.3.
Principle of transmitter output Tx.

Naturally, the driver stages are not quite lossless, so, under load, the voltage at the Tx output will never quite reach either VDD or 0 V, and the edges obviously can't be perfectly steep. However, the switching edges are critical and can cause problems in bad circuit designs with bad layouts.

6 READER ANTENNA DESIGN

The driver current, with which the two Tx outputs drive the antenna circuit, can simply be measured as a direct current at the TVDD supply pin. According to the datasheet, this current may not exceed 100 mA. The load connected to Tx1 and Tx2 must thus be dimensioned so that the current is as high as possible (in order to get the best output power), but it may not exceed the maximum current of 100 mA.

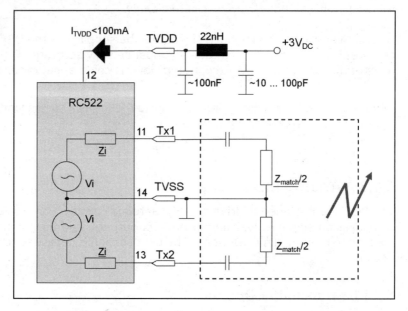

Figure 6.4. Principle of antenna matching.

For a better understanding of the antenna matching, see Figure 6.4. It shows the driver stages as voltage source V_i, a complex internal resistance, $\underline{Z_i}$, and load Z_{match}, which, DC decoupled, symmetrically load Tx1 and Tx2.

The actual internal resistance of the driver outputs, Tx1 and Tx2, is very small — in the region of a few ohms, so no power adjustments are made there.

Our load, Z_{match}, is thus chosen so that at the highest possible output power, driver current ITVDD does not exceed 100 mA.

> **Note** A power adjustment at Tx1 and Tx2 would mean that $\underline{Z_i} = Z_{match}/2$. This would mean that the same power fed to the antenna would have to be offset as heat in the IC. This would not be very efficient.

Practically, it can be seen that a load of approximately 40 Ω between Tx1 and Tx2 represents an optimum (Z_{match} = 40 Ω). For battery-operated devices, which need to be optimized for minimum power consumption, a fairly decent range can still be achieved using a smaller load (= larger impedance), without consuming the maximum current. An impedance of 50 – 60 Ω between Tx1 and Tx2 has been found to be a good compromise between maximum output power and power saving.

> **Note** One may also use only one Tx output, if, for example, an asymmetric antenna is used, and not depend on the maximum output power. With only one Tx output, however, only half the power will be achieved.

> **Note** An example of antenna matching with the correct wiring of the Tx outputs, as used in the Elektor Reader, is shown in Section 6.2.

Below, the most important registers for controlling the RC522 transmit outputs are listed:

TxModeReg (0x12)

With the TxModeReg, we do essentially two things. Firstly, turn automatic CRC generation on or off. All instructions, apart from the three Type A card activation instructions, REQA, WUPA and Anti-collision, have a CRC, which the RC522 generates automatically. Secondly, selection of one of the four different bit rates can be done via this register.

Table 6.1. TxModeReg (0x12)

Hexa-decimal	Binary	Meaning
00	0000 0000	Default. 106 Kbit/s, no CRC -> REQA, WUPA, anti-collision.
80	1000 0000	106 Kbit/s, CRC active -> all other instructions at 106 Kbit/s
90	1001 0000	212 Kbit/s, CRC active -> all other instructions at 212 Kbit/s
A0	1010 0000	424 Kbit/s, CRC active -> all other instructions at 424 Kbit/s
B0	1011 0000	848 Kbit/s, CRC active -> all other instructions at 848 Kbit/s

TxControlReg (0x14)

The Tx outputs are turned on and off using the TxControlReg. In addition, the respective output signals' polarity can be set, so the idle state can be defined as 'high' or 'low' on the respective Tx output.

Table 6.2. TxControlReg (0x14)

Hexa-decimal	Binary	Meaning
80	1000 0000	Default. Both outputs are off. -> RF Off
83	1000 0011	Both outputs are active and operate in a push-pull manner. -> RF On
03	0000 0011	Both outputs are active and working in unison (not used for our antenna).

For a balanced antenna, we drive the output in a push-pull manner: We need only turn on the RF by writing 83_{hex} to the TxControlReg.

6 READER ANTENNA DESIGN

TxASKReg (0x15)

With this register, we can force 100% amplitude modulation for Type A, among other things. Alternatively, the 100% AM can be produced using the corresponding value in ModGsPReg (0x29).

Table 6.3. TxASKReg (0x15)

Hexa-decimal	Binary	Meaning
00	0000 0000	Default. Modulation depth is determined by ModGsPReg.
40	0100 0000	100% amplitude modulation.

ModWidthReg (0x24)

ModWidthReg sets the modulation duration for Type A transmit pulses. The default value applies to the standard bit rate.

Table 6.4. ModWidthReg (0x24)

Hexa-decimal	Binary	Meaning
03	0010 0110	Default. Transmit pulse is about 3 µs long (at 106 Kbit/s).
15	0001 0101	Transmit pulse is about 1.5 µs long (at 212 Kbit/s).
08	0000 1000	Transmit pulse is about 0.75 µs long (at 424 Kbit/s).
01	0000 0001	Transmit pulse is about 0.375 µs long (at 848 Kbit/s).

GsNReg (0x27)

This register determines the output power at which individual n-drivers are switched. The upper nibble (Bits 4 – 7) determines the power during 'idle' state, that is, under no modulation. Usually, this is set to maximum (0xF = all n-drivers are active).

Table 6.5. GsNReg (0x27)

Hexa-decimal	Binary	Meaning
88	1000 1000	Default.
F0	1111 0000	Maximum output power, Type A.

The lower nibble (Bits 0 – 3), determine the modulation pulse output power. With Type A, one may either force 100% AM (TxASKReg), or set this value to 0. For Type B, the modulation index must be set using Bits 0 –3 of GsNReg. Because the antenna matching has an effect on the modulation index, it is best to try out each possible value (0 to 0xF) and measure the modulation index. The value that best achieves an m of between about 8 and 14% is then selected.

CWGsPReg (0x28)

Similar to the upper nibble of the GsNReg, Bits 0 — 5 of CWGsPReg can switch the individual p-drivers that govern the power of the unmodulated carrier. Normally, it should be set to maximum.

Table 6.6. GsNReg (0x27)

Hexa-decimal	Binary	Meaning
20	0010 0000	Default.
3F	0011 1111	Maximum output power.

ModGsPReg (0x29)

Similar to the lower nibble of the GsNReg, bits 0 — 5 can switch the individual p-drivers that govern the power of the modulated carrier. This is not relevant to Type A when 100% AM has been selected (TxASKReg). This register would normally be set to the maximum value.

Table 6.7. ModGsPReg (0x29)

Hexa-decimal	Binary	Meaning
20	0010 0000	Default.
3F	0011 1111	Maximum output power during modulation.

6.1.3.2 Receive Input

The RC522 has basically all the necessary filtering and decoding circuits required to receive a valid card response. However, it needs a clean input with a large enough usable signal: The input for our RFID reader is always the 13.56 MHz carrier along with the actual data signal (the subcarrier modulated by the card). The self-generated carrier could be considered as noise here, as it provides no useful information to the receiver.

> **Note** To define a 'sensitivity' for the RC522 here would be pointless. Normally, a receiver's sensitivity is defined as the smallest input signal at which the receiver is still able to produce a sufficiently large baseband signal (or a sufficiently low bit error rate for digital signals).
>
> Here the input signal is always dominated by a 13.56 MHz carrier whose signal strength has nothing to do with receiver sensitivity and also is not tied to the signal strength of the subcarrier.

In principle, for proper operation, only a portion of the antenna signal must be fed to the receiver. To better understand this, consider the Rx circuit in Figure 6.5, left:

6 READER ANTENNA DESIGN

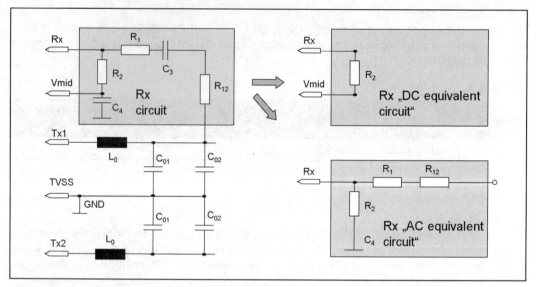

Figure 6.5. *Principle of the receiver circuit.*

This circuit can be conceptually divided into two parts: a 'DC equivalent circuit' and an 'AC equivalent circuit', as shown in Figure 6.5, right.

DC Equivalent Circuit
In the DC equivalent circuit diagram, capacitors C_4 and C_3 present an infinitely high resistance. What remains in Figure 6.5, top right, is nothing more than a resistor that feeds the (DC) voltage from the Vmid output to the Rx input. Thus, the average voltage level at the Rx input is 'clamped' to Vmid ≈ VDD/2.

AC Equivalent Circuit
For the AC equivalent circuit, the capacitors are simply considered as 'short circuits'. What remains in Figure 6.5, bottom right, is a simple voltage divider, consisting of $(R_1 + R_{12})$ and R_2, which pass a part of the antenna signal to the Rx input.

So, the alternating voltage level (of the 13.56 MHz signal) can be set with the values of R_1, R_{12} and R_2.

For correct sizing, we need only check that the specified allowable values from the datasheet are not exceeded, and correct if necessary.

As shown in Figure 6.6, the AC part of the input voltage can be measured simply at the Rx input: It should normally be about 3 V_{pp}, and it must never exceed 5 V_{pp}.

| Note | An excessive voltage at the Rx input won't necessarily destroy the RC522, but it leads to 'clipping' effects and thus erroneous reception. With too small a voltage at the input, sensitivity is forfeited, which means an inability to make use of the optimal reception range. |

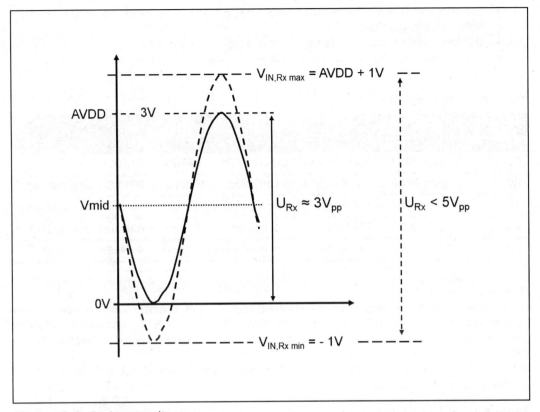

Figure 6.6. *Rx input voltage.*

> **Note** The voltage at the Rx input varies greatly with the load and detuning of the antenna!

In the following, we will briefly consider the RC522 registers that are significant to reception.

RxModeReg (0x13)
Similar to the TxModeReg, RxModeReg governs the bit rate — this time for the receive path. Also, automatic CRC checking can be turned on or off for the receive path. All card responses, apart from ATQA and UID, have a CRC.

In addition, this register offers a 'digital filter': RXNoErr (Bit 3). With this bit, one can suppress all data bursts shorter than 4 bits. This allows simple interference of all types to be effectively suppressed.

If additional filtering is required, for example, if the card generates interference prior to its responses, causing 'junk' to be received (EMD, see Section 2.1.6.4), RxMultiple (Bit 2) may help. When RxMultiple is enabled,

- the receiver turns off after a received packet and turns back on only after a delay period (RxWait),
- the last byte of the FIFO is sent to the error register, and
- the microcontroller must deactivate the receiver when the valid card response is received.

Table 6.8. RxModeReg (0x13)

Hexa-decimal	Binary	Meaning
00	0000 0000	Default. 106 Kbit/s, no CRC -> ATQA and UID.
80	1000 0000	106 Kbit/s, CRC activated, RxMultiple + RxNoErr deactivated.
90	1001 0000	212 Kbit/s, CRC activated, RxMultiple + RxNoErr deactivated.
A0	1010 0000	424 Kbit/s, CRC activated, RxMultiple + RxNoErr deactivated.
B0	1011 0000	848 Kbit/s, CRC activated, RxMultiple + RxNoErr deactivated.

RxSelReg (0x17)

In addition to receiver input configuration (Bits 6,7), this register can be used to set RxWait (Bits 0 — 5). RxWait is the amount of time that the RC522 waits after switching from transmit to receive (Transceive command) before the receiver is activated. When RxMultiple is active, the RC522 waits for the same amount of time after a (possibly invalid) frame is received, before the receiver is automatically activated. RxWait is defined as the number of data bits.

Table 6.9. RxSelReg (0x17)

Hexa-decimal	Binary	Meaning
84	1000 0100	Default. RxWait = 4 data bits → ≈ 38 µs at 106 Kbit/s.

RxThresholdReg (0x18)

RxThresholdReg defines two thresholds: The upper nibble defines MinLevel (Bits 4 — 7), while lower defines CollLevel (Bits 0 — 3). These values can actually be imagined as thresholds:
- 0 (lowest threshold) means everything gets through, while
- F (highest threshold) means nothing gets through.

> **Note** The threshold input is the digital output of the demodulated subcarrier (prior to baseband decoding). The output gain (RxGain, RFCfgReg) has a direct influence on the threshold detection's properties.

MinLevel (always in conjunction with RxGain) basically defines the receiver 'sensitivity'. All signals smaller than MinLevel are ignored. In other words, the input level must be greater than MinLevel in order for it to be recognized as a signal.

The same applies to CollLevel, except that it defines which signals will be recognized as collisions. CollLevel is only relevant to Type A at 106 Kbit/s.

Table 6.10. RxThresholdReg (0x18)

Hexa-decimal	Binary	Meaning
84	1000 0100	Default. MinLevel = 8 data bits, CollLevel = 4.
55	0101 0101	Good medium setting for Type A (ASK) at 106 Kbit/s.
75	0111 0101	Good medium setting for all higher bit rates and Type B (BPSK).

DemodReg (0x19)
Apart from the use of I and Q channels (which, aside from test purposes, should be left at their defaults), two time constants for the BPSK demodulator are defined in this register:
- TauSync (Bits 0,1) defines a time constant, which determines how fast the BPSK demodulator's PLL locks onto the subcarrier burst preceding each BPSK response from the card.
- TauRcv (Bits 2,3) defines the time constant used during demodulation of the BPSK card response.

The default values produce very good results, so they should not be changed except for test purposes.

RFCfgReg (0x26)
We set RxGain (Bits 4 – 6) in the RFCfgReg. RxGain is the amplification of the demodulated subcarrier prior to decoding. This value should always be considered in conjunction with RxThreshold.

Practice has shown that one should avoid RxGain values greater than 5. Obviously, with larger gain values, the RC522 tends to oscillate. Values that are too small will reduce the 'sensitivity'.

Table 6.11. RFCfgReg (0x26)

Hexa-decimal	Binary	Meaning
48	0100 1000	Default. RxGain = 4.

> **Note** Section 6.2 describes how to find the best possible combination of RxThresholdReg and RxGain.

6.1.4 Test Signals
The RC522 offers some useful test signals that are of particular interest during the design of a reader. These are available at the three test pins. These, together with the corresponding registers, are described briefly below.

6 READER ANTENNA DESIGN

6.1.4.1 MFOUT

With the MFOUT pin (Pin 8), the digital transmit data can be output, either Miller coded or uncoded. This signal is perfect as an oscilloscope trigger, should one wish to observe the card response directly in the magnetic field. For this, we use the lower nibble of TxSelectReg (0x16):

Table 6.12. *RxModeReg (0x13)*

Hexa-decimal	Binary	Meaning
10	0001 0000	Default. No signal on MFOUT (high impedance).
14	0001 0100	Miller-coded transmission data signal at MFOUT.
15	0001 0101	Uncoded data transmission signal at MFOUT.
17	0001 0111	Decoded received signal at MFOUT.

Both transmit signals, as shown in Figure 6.7, are recorded using an oscilloscope: uncoded above, Miller coded in the center. When the Miller-coded transmit signal is captured by Channel 1, Channel 2 can be used to show the modulated 13.56 MHz carrier, as shown in Figure 6.7, bottom. For this, the short-circuited probe connected to Channel 2 is kept in the magnetic field, so that the probe, along with the ground lead, make up a conductor loop.

> **Note** To use MFOUT, SVDD (Pin 9) must be supplied with power.

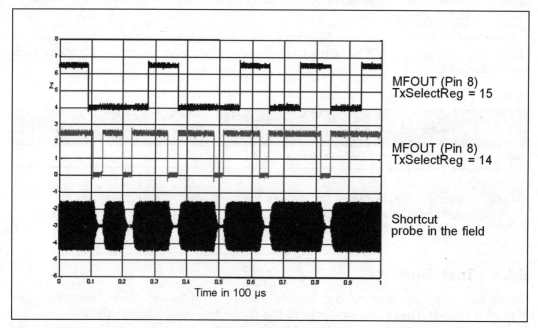

Figure 6.7. *MFOUT: Transmitted signal.*

The same can be done with the received signals, as shown in Figure 6.8. On Channel 1 (Figure 6.8, bottom), we can see the demodulated card response. At the same time, we hold the short-circuited probe on Channel 2 close to the card and can normally see the load modulation (Figure 6.8, top), if it's amplitude modulated.

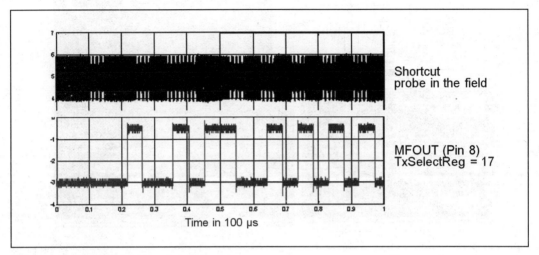

Figure 6.8. MFOUT: Received signal.

6.1.4.2 AUX1 and AUX2

Also very helpful are the other two test pins, AUX1 (Pin 19) and AUX2 (Pin 20), especially when one encounters difficulties with unexpected interference, for example. The two AUX pins are power outputs, so pull-down resistors must first be used before anything can be measured.

Fortunately, the Elektor RFID Reader's test pins are made available on exposed vias, so a short wire can be soldered to them and the test pins can be accessed from the bottom side. We simply removed a piece of the ground surface, exposing a portion of its copper, and just soldered 1 kΩ resistors between AUX1 and AUX2 and ground, as shown in Figure 6.9, left. It's easiest to use 0805 SMD components.

Three short, thin wires were then soldered to the test pins from above, making it easy to connect them to an oscilloscope.

AnalogTestReg (0x38)
The upper nibble (Bits 4 — 7) of AnalogTestReg determine the signal at the AUX1 output, while the lower nibble (Bits 0 – 3) does the same for AUX2. There are a number of test signals that can be switched to the AUX outputs, but the two most commonly used

6 READER ANTENNA DESIGN

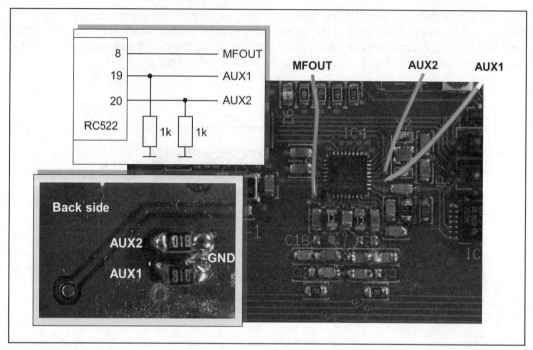

Figure 6.9. Wiring the test pins.

ones are the I and Q channels from the received signal (ADC_I and ADC_Q). They show the demodulated subcarrier. Should interference slip into the RC522's analog path via the antenna or even the digital path, this can be seen on ADC_I and ADC_Q.

Table 6.13. AnalogTestReg (0x38)

Hexa-decimal	Binary	Meaning
00	0000 0000	Default. No signal on AUX1 and AUX2 (high impedance).
56	0101 0110	I channel (ADC_I) to AUX1, Q channel (ADC_Q) to AUX2.

Figure 6.10 shows an example of the I channel measured at the AUX1 output.

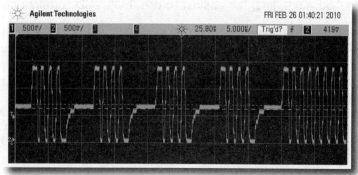

Figure 6.10. I-channel on AUX1.

6.1.5 Miscellaneous

6.1.5.1 Power Supply and GND

The RC522 has four pins for supply voltage: PVDD, DVDD, AVDD and TVDD. In principle, all of them can be connected to the same voltage, but in certain circumstances it is recommended that the lines are decoupled from each other, as shown in Figure 6.4 or Figure 6.11. Large blocking capacitors should at least be provided for TVDD, as well as the Pi filter circuit shown. If it later turns out that a filter is not needed, and that a simple blocking capacitor suffices, then the series inductor can simply be bridged with a jumper (0 Ω).

To use MFOUT, SVDD must also be supplied with power.

> **Note** Measurements on the Elektor reader with optimized antenna have shown that a larger blocking capacitor of a few µF is necessary to 'clean up' the circuit. Without this blocking capacitor, interference can be observed on the power supply and other signals, which massively disrupt reception and lead to situations where card responses can no longer be reliably decoded from marginal signals.

6.1.5.2 Tolerances

With all circuit designs, always ensure that all tolerances are observed in the analog area, especially the RF area. The RC522 itself is subject to these tolerances.

On some individual ICs, the offset of one or both of the respective amplifiers on the I and/or Q channels lie right at the A/D converter's digital threshold: With these ICs, the A/D converter output's least-significant bit is toggled, producing noise. This can be measured on ADC_I and ADC_Q. There is no way to prevent this, so it must be tolerated.

> **Note** In any case, all tolerances must be observed as far as possible, so that the reader will work properly in practice. To turn a well-designed, working prototype into a mass-produced device takes some effort, and does not fall within the scope of this book.

6.2 Antenna Design

Now that we've described the principles and basic functions of the RC522, let's turn to the practical antenna design. Using the Elektor RFID Reader as an example, we'll go over a complete antenna design.

The complete antenna circuit is shown in Figure 6.11.

6 READER ANTENNA DESIGN

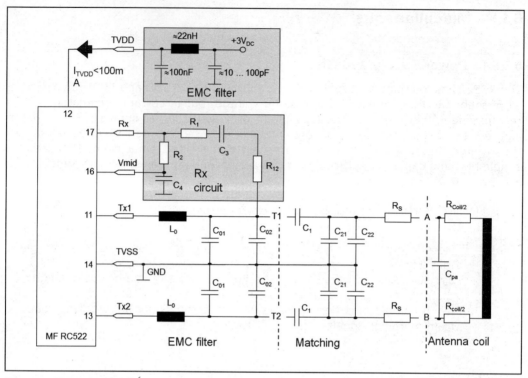

Figure 6.11. Antenna circuit with RC522.

We'll dimension the EMC filter, measure the antenna coil, then calculate, simulate and tune the matching circuit. After that, we'll verify the antenna's transmission function, the power consumption and the correct transmission pulse forms.

Once the transmission path is working properly, we'll attend to the correct dimensioning of the receiver path.

Finally, we'll examine the complete functioning of the entire antenna, and detect and eliminate interference if necessary.

6.2.1 Coil Design

The Elektor RFID Reader's mechanical characteristics basically determine the design of the coil. The board size is determined by the enclosure. Unfortunately, due to the size of the display, the antenna is inconveniently located: the coil is directly above the battery compartment. The interference caused by this must be taken into account — we must certainly consider the battery (metal behind the antenna) in the antenna design.

| Note | The procedure for a new, independent coil design is described in Section 3.2.3.1. |

6.2.1.1 Measuring the Coil Parameters

Firstly, with an impedance analyzer, we measure the coil between A and B as shown in Figure 6.12. For this, we remove the series resistors, in order to measure the coil on its own. We make two measurements: one in an open environment, and one above the batteries.

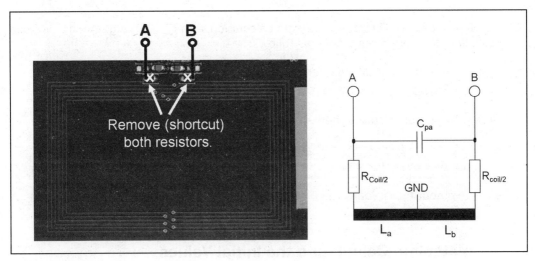

Figure 6.12. Antenna coil measurement.

The measurement can be taken as described in Section 3.4. We measure the inductivity, the parallel capacitance and the series resistance.

> **Note** The measurement appears to be more accurate than it actually is. An imprecise measurement suffices for us, as the measured values are needed only as starting points and the tuning will be done with the Analyzer.

We got the following initial values.

	Without battery	With battery	Comments
$L = L_a + L_b$	2.97 μH	2.87 μH	This value is important to begin with.
R_{coil}	0.1 Ω	0.1 Ω	This directly-measured value is very inaccurate. An approximation suffices as a starting point, however.
C_{pa}	9.3 pF	9.3 pF	This directly-measured value is very inaccurate. An approximation suffices as a starting point, however.

6.2.1.2 Determine the Q Factor and the Series Resistance

One can calculate the Q factor simply:

$$Q = \frac{\omega L}{R_{total}} = \frac{\omega L}{R_{coil} + 2 \cdot R_s} \qquad \text{Eq. 6.1}$$

We specify the desired Q factor and begin calculation with $Q < 20$, e.g. $Q = 12$. We can then calculate the series resistance as follows:

$$R_s = \frac{1}{2}\left(\frac{\omega L}{Q} - R_{coil}\right) = \frac{1}{2}\left(\frac{2\pi \cdot 13.56 MHz \cdot 2.87 \mu H}{12} - 0.1\Omega\right) = 10.24\Omega \approx 10\Omega$$

The first value in our circuit is then already provisionally determined.

> **Note** Although a Q factor of 12 seems low, the advantage is that it's easy to achieve it with a cheap E-series resistor, and it's better in any case to select a lower Q in order to ensure proper functioning both with and without metal (batteries) in proximity.

6.2.2 Matching: Calculating the Initial Values

The initial value calculations described here can easily be plugged into the ready-made Excel spreadsheet available from NXP Semiconductors (www.nxp.com). On the sheet, you input the measured values and the required parameters (Q and Z_{match}), and the values for the matching network are returned.

6.2.2.1 Parallel Equivalent Circuit

To calculate the initial values, we first replace the antenna coil with a parallel equivalent circuit, as shown in Figure 6.13.

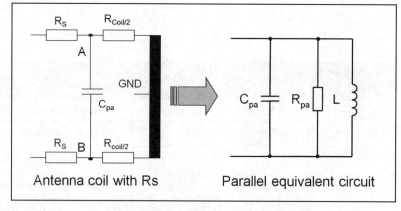

Figure 6.13. Antenna coil parallel equivalent circuit with series resistance.

L and C remain unchanged, while the parallel resistance is calculated as follows:

$$R_{parallel} = Q^2 \cdot R_{serial} \qquad \textit{Eq. 6.2}$$

So, in our case:

$$R_{pa} = Q^2 \cdot \left(2 \cdot R_s + R_{coil}\right) \cdot R_{total} = 12^2 \cdot \left(2 \cdot 10\Omega + 0.1\Omega\right) \approx 2.934\Omega$$

This equivalent circuit allows the matching circuit to be more easily calculated.

6.2.2.2 Partitioning and Simplifying the Circuit Diagram

Next, we separate the circuit blocks to make calculation easier, as shown in Figure 6.14. Then we can consider the left and right sides separately. On the left side is the low-pass filter for suppressing the harmonics, which also serves as an impedance transformer between Z_{tr} and Z_{match}.

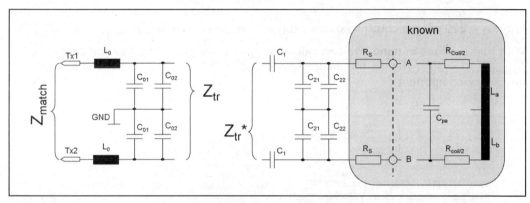

Figure 6.14. Partitioning the circuit diagram for simplicity.

All that's left on the right side is the actual matching circuit, for which we need to calculate the parallel and series capacitance.

In order to reconnect the two sides, we must ensure that the impedance of the antenna circuit (right) $Z_{tr}*$ matches complex conjugate Z_{tr} (left).

For the matching, the following applies:

$$Z_{tr} = R_{tr} + jX_{tr} \qquad \textit{Eq. 6.3}$$

$$Z_{tr}^* = R_{tr} - jX_{tr} \qquad \textit{Eq. 6.4}$$

6.2.2.3 Low-Pass Filter

Next, for the left side of the circuit, as seen in Figure 6.14, we need to calculate the values for L_0 and C_0.

6 READER ANTENNA DESIGN

> **Note** C_0 is basically implemented 'in double' as C_{01} and C_{02}, so that the required values can be reached precisely enough using a combination of two components. The same principle applies to C_1 and C_2. In practice we find that C_0, and usually also C_1, can be matched well enough using just a single component.

We begin by specifying L_0: a value of $L = 560$ nH has proven to be very useful in practice.

We can easily calculate the value of capacitor C_0 if we know the cutoff frequency of the low-pass filter. In principle, we could make this determination ourselves, but it can be seen that placing the cutoff frequency at the upper sideband is convenient: $f_{cutoff} = f_c + 847.5$ kHz.

C_0 can now be calculated:

$$C_0 = \frac{1}{\left(2\cdot\pi\cdot f_{cutoff}\right)^2 \cdot L_0} = \frac{1}{4\cdot\pi^2\cdot\left(13.56MHz + 847.5kHz\right)^2\cdot 560mH} \approx 220 pF \qquad \text{Eq. 6.5}$$

With $C_0 = 220$ pF, the cutoff frequency for our low-pass filter will be $f_{cutoff} = 14.339$ MHz.

With these values in hand, we can calculate $Z_{tr} = R_{tr} + jX_{tr}$, if we assume that $Z_{match} = R_{match} + J0 = 35\,\Omega$ (hence the desired matching at Tx1 and Tx2 is real and contains no imaginary component):

$$R_{tr} = \frac{R_{match}}{\left(1-\omega^2\cdot L_0\cdot C_0\right)^2 + \left(\omega - \frac{R_{match}}{2}\cdot C_0\right)^2} \qquad \text{Eq. 6.6}$$

$$X_{tr} = 2\cdot\omega\cdot\frac{L_0\cdot\left(1-\omega^2\cdot L_0\cdot C_0\right) - \frac{R_{match}^2}{4}\cdot C_0}{\left(1-\omega^2\cdot L_0\cdot C_0\right)^2 + \left(\omega\cdot\frac{R_{match}}{2}\cdot C_0\right)^2} \qquad \text{Eq. 6.7}$$

For our filter, the result is: $Z_{tr} = 295\,\Omega - j12\,\Omega$

6.2.2.4 Matching Network

With the low-pass filter parameters, we can calculate the missing values for C_1 and C_2:

$$C_1 \approx \frac{1}{\omega\cdot\left(\sqrt{\frac{R_{tr}\cdot R_{pa}}{4}} + \frac{X_{tr}}{2}\right)} \qquad \text{Eq. 6.8}$$

$$C_2 \approx \frac{1}{\omega^2\cdot\frac{L}{2}} - \frac{1}{\omega\cdot\sqrt{\frac{R_{tr}\cdot R_{pa}}{4}}} - 2\cdot C_{pa} \qquad \text{Eq. 6.9}$$

These rough equations give us the following values: $C_1 = 25$ pF; $C_2 = 52$ pF

We could just as easily get these values from NXP's Excel spreadsheet, as shown in Figure 6.15.

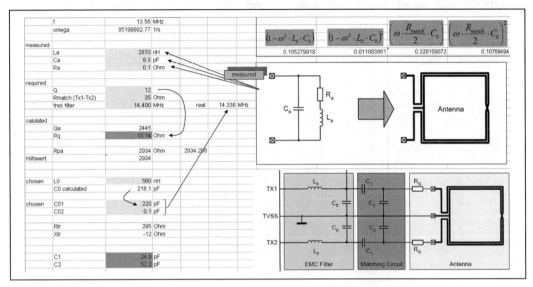

Figure 6.15. Matching calculations using Excel spreadsheet (NXP Semiconductors, 2010).

6.2.3 Matching: Simulation and Measurement

Using these approximate values, it's best to do a simulation before we begin soldering the actual components. Well-suited for the task is 'RFSIM99', a small program with which one can easily simulate linear networks.

The calculated approximate values are not particularly accurate, but a good fit can quickly be reached using the simulation. The final circuit is shown in Figure 6.16, with the corresponding results in Figure 6.17.

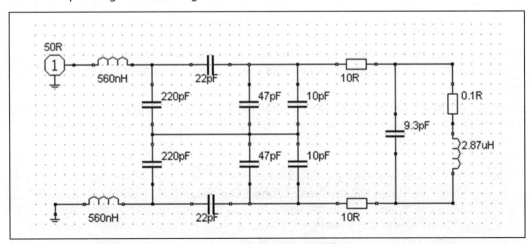

Figure 6.16. Simulation of matching with RFSIM99.

We're simulating a balanced circuit with unbalanced tools, so the actual ground connection on the real circuit (GND) is not tied to ground in the simulated circuit. We simulate exactly the way we measure: between Tx1 and Tx2.

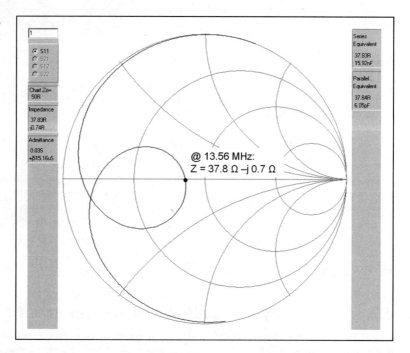

Figure 6.17. Simulation results.

The values determined by simulation are also seen in the actual matching. This may not always be, as the previous measurements of L, R_{coil} and C_{pa} are usually imprecise, and the actual adjustment may need to be reached through a process of iteration. It is thus best to start with capacitors whose values are slightly smaller than those calculated / simulated.

This has two advantages:

Firstly, unforeseen stray capacitances in the real circuit will usually raise capacitance values (meaning that the actual components are often a little smaller than the simulated values).

Secondly, it's much easier to increase a small capacitance by simply soldering a second small capacitor 'piggyback' on top of an existing one, than to try and decrease a large capacitance.

> **Note** One could simulate only one half of the circuit — the upper half, for example. One would then actually connect the ground and halve the values for R_{coil} and L, and double the value for C_{pa}. This applies in the simulation, but not in reality: the windings of coil L are coupled together, and thus neither L_a nor L_b can be measured on their own.

ANTENNA DESIGN 6.2

Figure 6.18 shows the results, measured with the miniVNA and represented in ZPlots. As we can see, the matching varies only slightly due to the low Q, even when a card is introduced into the antenna's vicinity.

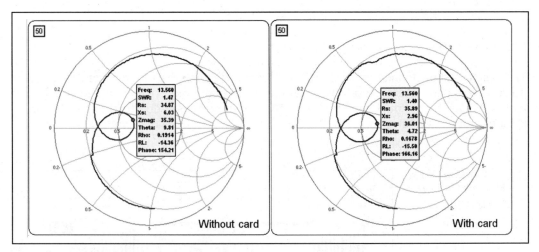

Figure 6.18. Measurement results from the miniVNA in ZPlots.

The new values for the antenna matching are now determined:

L_0 = 560 nH	(L2$_{Elektor}$ = L3$_{Elektor}$)	-> no change
C_{01} = 220 pF	(C19$_{Elektor}$ = C17$_{Elektor}$)	-> no change
C_{02} = not used	(C20$_{Elektor}$ = C18$_{Elektor}$)	-> no change
C_1 = 22 pF	(C23$_{Elektor}$ = C21$_{Elektor}$)	-> change!
	(C24$_{Elektor}$ = C22$_{Elektor}$)	-> no change
C_{21} = 47 pF	(C27$_{Elektor}$ = C25$_{Elektor}$)	-> change!
C_1 = 22 pF	(C28$_{Elektor}$ = C26$_{Elektor}$)	-> change!
R_s = 10 Ω	(R9$_{Elektor}$ = R8$_{Elektor}$)	-> change!

6.2.4 Measurements on the Transmitted Pulse

Once the antenna is optimally matched, we can start using the reader with the new adjustments. As a precaution, we first check the current consumption in the RC522's transmitter output stage, i.e. TVDD. This requires cutting the circuit traces on the Elektor Reader:

Current TVDD = 50 mA without card.

That is absolutely fine, and there is still plenty of 'play' to allow for antenna detuning when a card is in the magnetic field or due to metal surfaces close to the antenna.

> **Note** It is very interesting to induce such detuning and simultaneously measure the current at TVDD. It should never exceed the limit of 100 mA specified in the datasheet.

6 READER ANTENNA DESIGN

Next, we check the transmit pulse, which, with the low Q, should be quite clean. For this, we switch the encoded data signal to the MFOUT pin and watch the carrier signal on the oscilloscope's second channel with a short-circuited conductor loop connected.

Ch1 = MFout (trigger signal) -> reader.RC522.WriteSFR(0x16, 0x14)

Ch2 = measuring coil at 5 cm distance

The result is shown in Figure 6.19. It's important to check the rise time of the transmit pulse at this point. According to the standard, the edge must rise from below 5% to at least 60% of the residual carrier within 500 ns. In our case, due to the low Q, it does so substantially faster.

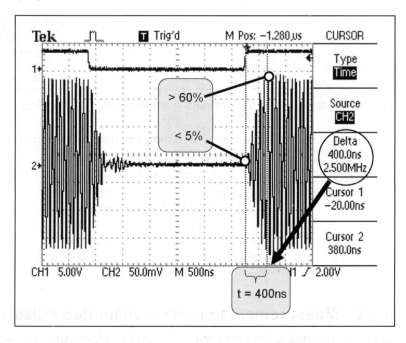

Figure 6.19. Measuring of the transmitted pulse.

We see slight overshoot at the end of the rising and falling edges: these should be small enough to fall within the ISO/IEC 14443 standard.

> **Note** This measurement is **not** standards-compliant! The standard dictates that all defined rise and fall times must be measured with a calibrated ReferencePICC. For us, the associated costs exceed the benefits, so we confine these checks to the specified minimum measurements. In this way, the defined retroactive effect of the card on the reader antenna, which can only be produced using the calibrated ReferencePICC, is completely disregarded.

6.2.5 Measurement and Adjustment of the Receive Path

Finally, proper adjustment of the receive path is done. Since the Elektor RFID Reader is already assembled, we need only do a check. The requirements for this are described in Section 6.1.3.2.

We basically just need to measure the input voltage at the Rx pin. We do this with different loads, so the voltage is measured

- with no load (idle);
- with different cards at different distances from the antenna, and
- with metal surfaces at a greater distance (< 5 cm) from the antenna.

The voltage should never exceed 5 Vpp, or likewise be less than 2 Vpp. In our example, nothing needs to be changed. The standard assembly is good enough.

> **Note** For this measurement, we need a probe tip that is suitable for RF, so that frequencies higher than 10 MHz can be measured: you must have a very high impedance and a low input capacitance, otherwise the measurement will be inaccurate.

6.2.6 Eliminating Interference

Unfortunately, further experimenting with different cards showed that our reader, with the now well-matched antenna, still had difficulty with certain cards in certain positions. This happened even with a second newly-attached reader.

There must have been something wrong, and we had to look at the test signals, especially for the cases in which the failures occurred.

For this, we fitted two 1 kΩ resistors and connected a supply voltage to SVDD, as described in Section 6.1.4, and set out making measurements. What followed was a typical troubleshooting session — a laborious process that took several hours before we found the real cause and eliminated it.

A broadband noise signal was observed superimposed on all signals, including supply voltage, when certain cards were held very close to the reader antenna. This noise signal was several tens of millivolts in amplitude (depending on card position), and had a wide bandwidth of about 200 kHz. The source of the noise couldn't be found by looking at the spectrum or by using different practical filter configurations, either.

All supply voltages were separated, in order to decouple them from one another. For this, the circuit traces, which are accessible from on the rear, fortunately, were separated, connected to thin wires and then connected individually to a supply voltage. Figure 6.20 shows the restored state at the end of all the measurements. The (repaired) separations between AVDD and TVDD as well as between TVDD and VDD can be seen. Likewise, we can see the supply going to SVDD as well as the two resistors wired to the AUX outputs.

6 READER ANTENNA DESIGN

With the separated supply lines, the voltages may be filtered separately. External power was even supplied, so that the influences of the voltage regulator and the microcontroller could be eliminated. All of this turned out to be useless.

Figure 6.20. The individual supply voltages.

Then, one after the other, registers GsNReg, CWGsPReg, RxThresholdReg and RFCfgReg were modified. That way, the dependencies could be seen, and we noticed that the output power had a decisive influence on the noise signal: as soon as the output current was reduced from around 60 mA (with card) to 52 mA (with card), using GsNReg, the noise disappeared. Apparently, there was an inefficient coupling between the output driver and the rest of the circuit, possibly even partially amplified within the RC522.

To avoid sacrificing power, an attempt was made to better filter TVDD, and this, finally, was successful: the noise was eliminated with a 10 µF ceramic capacitor (not electrolytic!) connected directly to the TVDD pin.

Once the circuit was restored to its 'original' state, C11 (100 nF) was replaced with a 10 µF capacitor, the circuit worked as intended. Even more cards were able to be detected and activated. All of the cards worked from zero distance to the maximum range.

As is shown in Figure 6.21, it was actually easier to solder the new C11 (10 µF) on top of the old one (100 nF), 'piggyback' style, without removing the original one.

ANTENNA DESIGN 6.2

Figure 6.21. C11 'piggyback'.

6.2.7 Range Checking

Finally, a few words on range checking. The easiest is to capture the test signals, for example the MFOUT signal:

As described in Section 6.7, the encoded signal is switched to the oscilloscope's Channel 1. Channel 2 is short-circuited with a conductor loop: just use the probe's normal ground for simplicity. This short-circuited conductor loop is held close to the card, as the card is introduced slowly into the magnetic field.

Now, when the card is continuously 'polled', that is, when a REQA is sent repeatedly, the card response will be seen in the field, as soon as the card begins to work, that is, as soon as it responds. The card's response is seen on Channel 2, as long as a high enough resolution (10 µs / division) and a delay of around 180 µs are selected.

Note	A Type A card responds only to every second REQA, as the ATQA switches the card into Ready state, after which it awaits an Anti-collision or Select. If, instead, a second REQA is sent, then the card detects an error and goes back to the IDLE state.

6 READER ANTENNA DESIGN

As soon as the card responds, that is as soon as we see the response on the oscilloscope, the reader must likewise detect the response and we should see the ATQA on the Trace output. If this is the case, we know that the reader's range is determined by the energy range. This is intentional, as the energy range is limited by the geometry and the power of our RC522.

It's not good if the card responds, but the receiver doesn't acknowledge it. Then, the range is being limited by the (poor) communication, which means that we're sacrificing range because our receiver is not working optimally.

With our Elektor RFID Reader, we get a range of 3.5 — 4 cm with various MIFARE cards.

> **Note** The energy range, loaded by a Reference PICC according to ISO / IEC 14443, is less than 1 cm. The supply voltage from a 'weak' USB port will collapse under that load. One must then rely on an external supply (batteries).

7
The Elektor RFID Reader

Gerhard H. Schalk

7.1 Introduction

The Elektor RFID Reader first came out in the September 2006 issue of Elektor magazine. Its hardware and software support a range of contactless 13.56 MHz cards, such as the MIFARE Ultralight, MIFARE Ultralight C, MIFARE Classic 1K and 4K, MIFARE Mini, and, to an extent, even the MIFARE Plus. The current PC software also supports the T=CL protocol. With it, it's possible to read from and write to contactless ISO/IEC 14443 Type A smartcards, such as the MIFARE DESFire EV1. In contrast to standard PC/SC readers, the Elektor RFID Reader offers the option of connecting an LCD and using the reader in a standalone mode. The Elektor RFID Reader also supports sending individual card instructions. This is the only way many card properties, such as the state diagram of an ISO/IEC 14443 Type A card, be investigated.

Figure 7.1.
The Elektor RFID Reader board with optional LCD.

7 THE ELEKTOR RFID READER

The Elektor RFID Reader can be used not only for experiments and for educational purposes with a diverse range of cards. Thanks to its powerful microcontroller with I²C, SPI, UART and USB interface, it is ideal for the development of custom applications and systems.

Contactless code-operated gate and door openers, member ID systems, storage for passwords and configuration data, payment functions for vending machines, securing electronic devices and battery pack monitoring are only some examples. The combination of secure identity, data storage and a contactless interface presents many possibilities for creative solutions.

The most important features of the Elektor RFID Reader are:

- Compatible with MIFARE and ISO/IEC 14443 Type A
- Both reading and writing are possible
- USB interface for PC connection
- Immediately usable without programming
- Free Elektor RFID Reader PC software library
- Autonomous (possibly mobile) operation using LCD
- MF RC522 reader IC
- P89LPC936 microcontroller
- I²C and SPI interfaces
- Available 8-bit I/O port
- Buffered switched output (T3)
- Customization for user-specific developments possible
- Free flash tool for firmware updates
- Assembled and tested SMD board available

Both the Reader's firmware and the PC software have been completely revised for this book. The original version of MIFARE Magic V2.x that appeared in Elektor magazine was developed in Visual Basic V6.0, which makes it much easier to develop Windows applications than, for example, MFC (Microsoft Foundation Classes) in C++.

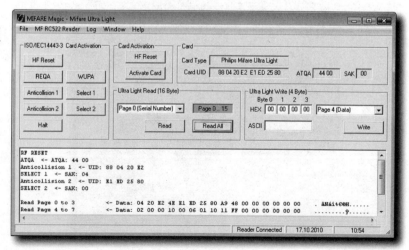

Figure 7.2.
MIFARE Magic v2.1's MIFARE Ultralight View.

INTRODUCTION 7.1

In recent years, many hardware-oriented developers have also begun to use .NET languages to create PC applications. This trend is surely encouraged by Microsoft, which offers the Express Edition free of charge.

When the .NET Framework 1.0 appeared in 2000, Microsoft made four programming languages available: Visual C# (CSharp), Visual Basic, Visual C++ and J#. Visual Basic V6.0 and Visual Basic .NET aren't really compatible, since they use fundamentally different class libraries.

The new Elektor RFID Reader library was developed entirely in C#. A major advantage of C# is that it is based on C/C++ and Java and can therefore be learned by many programmers easily. Most of the C# examples in this book should present no problem for any reader with a knowledge of C. C# allows Windows applications to be developed very easily.

The source code of the firmware and C# reader library, as well as all examples presented in this book, can be downloaded for free from the author's website *(www.smartcard-magic.net)*. All of the C# examples in this book can be compiled using the freely available Visual C# 2012 Express Edition *(http://www.microsoft.com/express/Downloads/#2012-Visual-CS, http://www.microsoft.com/visualstudio/eng/products/visual-studio-express-for-windows-desktop)*. SharpDevelop *(http://www.icsharpcode.com)* is another free IDE.

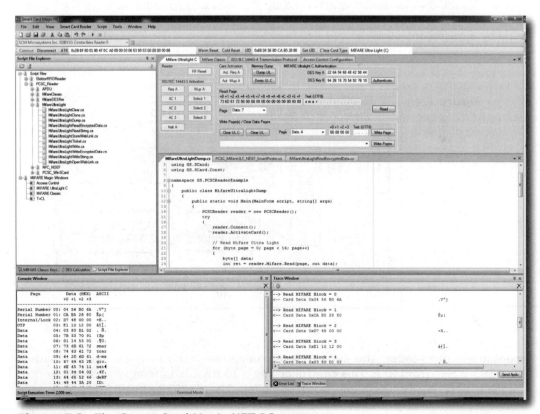

Figure 7.3. *The Smart Card Magic.NET PC program.*

7 THE ELEKTOR RFID READER

Specially for this book, a development platform was developed for MIFARE, ISO/IEC 1443, and ISO/IEC 7816 smart cards. Smart Card Magic.NET is a powerful graphic user interface supporting both the Elektor RFID Reader and commercial PC/SC reader devices. The program provides the ability to send individual instructions to different types of cards at the click of a mouse. Part of Smart Card Magic.NET is a versatile editor that allows one to create scripts, and it includes a C# compiler. Both the Trace and the Console outputs are integrated into the interface.

For learning about card types, the Elektor RFID Reader hardware, or PC/SC readers, Smart Card Magic.NET is certainly the first choice. However, if the goal is an extensive reader project or a Windows application, Visual C# 2012 Express Edition is the appropriate development environment.

7.2 Reader Hardware

Key to the development of the RFID Reader was to make the circuit as universal as possible. The reader is suitable for connection to a PC via USB as well as for autonomous operation using an LCD. Figure 7.4 shows the reader's block diagram. The actual reader functionality, that is, producing the RF field, modulation and demodulation and generating the ISO/IEC 14443 signal, is achieved by the MF RC522 reader IC from NXP Semiconductors. Simply seen, the MF RC522 is a contactless UART that is programmed directly by a microcontroller.

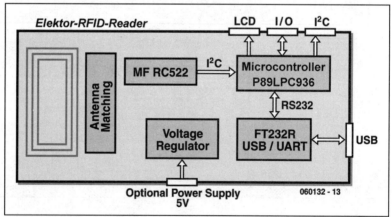

Figure 7.4. Elektor RFID Reader block diagram.

In designing the reader, the choice of microcontroller fell to the 8051-compatible P89LPC936, also from NXP. This controller has a 16 KB flash memory and can be programmed very easily using any 8051 compiler. The FT232R USB/RS-232 interface module from Future Technolgy Devices (FTDI) allows for very simple communication to a PC.

READER HARDWARE 7.2

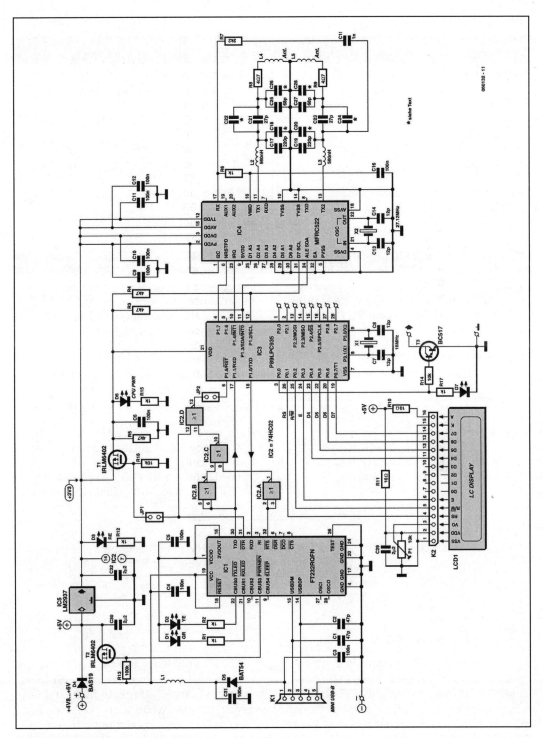

Figure 7.5. Elektor RFID Reader schematic.

7 THE ELEKTOR RFID READER

Table 7.1. Bill of Materials

Resistors (SMD 0805, 5%)	Capacitors (SMD 0805, 16 V ceramic)
R1, R2, R6, R12, R15, R17 = 1 kΩ R3, R4, R5 = 4k7 R7 = 2k7 R8, R9 = 4Ω7 R10 = 270 Ω R11 = 10 Ω R13 = 100 kΩ R14, R16 = 10 k P1 = 10 k trimmer potentiometer, SMD, 4 mm sq.	C1, C2 = 47 pF NP0 C3, C4, C5, C6, C9, C10, C11, C12, C16, C31 = 100 nF C7, C8, C13, C14 = 12 pF NP0 C15 = 1 nF NP0 C17, C19 = 220 pF NP0 C18, C20 = not fitted C21, C23 = 27 pF NP0 C22, C24 = not fitted C25, C27 = 68 pF NP0 C26, C28 = not fitted C29, C30, C32 = 2µ2 F
Semiconductors	**Miscellaneous**
D1 = SMD LED (0805) green, low-current D2 = SMD LED (0805), yellow, low-current D3, D6, D7 = SMD LED (0805), red, low-current D4 = BAS19 (200 mA, SOT23) D5 = BAT54S (30 V / 300 mA SOT23) T1, T2 = 6402 (P-channel MOSFET, 20 V / 3.7 A, SOT23) T3 = BC517 (NPN Darlington TO-92) IC1 = FT232RQFN (QFN32, FTDI) IC2 = 74HC02 (TSSOP14, Norgate) IC3 = P89LPC936FDH-S (SSOP28, NXP) IC4 = MF RC52201HN1 (HVQFN32, NXP) IC5 LM2937 = (low-drop, 3V3, SOT223)	X1 = 16 MHz crystal (18 pF parallel capacitance, 5 mm x 3.2 mm) X2 = 27.12 MHz crystal (18 pF parallel capacitance, 5 mm x 3.2 mm) K1 = mini USB-B socket, SMD, 5-pin L1 = SMD ferrite (1.5 A, 0805) L2, L3 = 560 nH inductor, SMD (0805) JP1, JP2 = 0.1" jumper (see text) LCD1 = LCD module, 2 x 16 character, backlit Enclosure, 146 x 91 x 33 (mm) with LCD windows and battery compartment for 4 AA cells PC board, 060132-91 EPS (assembled and tested, including USB cable, see Elektor Shop)

Note Because the MF RC522 is available only in a HVQFN32 package, it can only be fitted industrially. For this reason, the Elektor RFID Reader is only available fully assembled and tested.

7.2.1 Power Supply

The complete circuit diagram is shown in Figure 7.5. When connected to the PC via the mini USB connector (socket K1) the power will be supplied exclusively by the USB port. Devices that are powered by USB (bus-powered devices), are divided into high- and low-power devices. Low power devices may not exceed 100 mA, while high power devices may, after successful initialization of the USB bus (enumeration), consume up to 500 mA. The FT232R USB interface module (IC1) is already preconfigured so that the reader enumerates itself as a high-power device. Once the USB bus enumeration is completed successfully, the /PWRNEN signal at Pin 11 switches on the P-channel MOSFET, T2, so that the 5 V supply is fed to voltage regulator IC5. The LM2937 outputs 3.3 V for the P89LPC936 microcontroller (IC3) and the MF RC522 (IC4). Red LED D6 indicates the 3.3 V operating voltage. If the circuit is not powered via USB, batteries (4×AA cells) or a 5 V power supply may be used. The AC adapter must be able to deliver at least 300 mA.

7.2.2 The P89LPC936 Microcontroller

The 8051 CPU in the P89LPC936 requires only two cycles per instruction and is clocked at 16 MHz. This speed and the 16 KB flash memory is fully suitable for the realization of a wide range of reader applications. The most important features of the P89LPC936 are:

- 16 KB ISP/IAP flash, 512 bytes data EEPROM,
- 256 bytes RAM, 512 bytes AUX RAM, Dual DPTR,
- 23(26) I/O pins, all 5 V-tolerant,
- UART, I^2C, SPI, 2 timers/counters,
- analog comparators, CCU, two 8-bit 4-channel ADC/DAC, watchdog timer, RTC,
- 2.4 V to 3.6 V operating voltage.

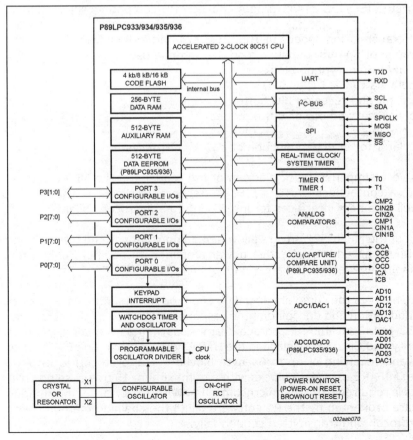

Figure 7.6.
P89LPC936 block diagram (source: NXP Semiconductors).

The P89LPC936 microcontroller comes factory-programmed with a boot loader. This is located in the upper 512 bytes of flash memory (addresses 0x3E00 to 0x3FFF). With this boot loader, all P89LPC9xx microcontrollers support in-system programming (ISP). This means that the controller can be programmed on the completed and assembled board. The boot loader can be activated immediately after power-up using three defined pulses on the reset line. So, it's always possible to flash the latest firmware update (see Section 7.3.2) into the P89LPC936.

If necessary, the reader may also be connected to an LCD (K2). A free port, P2.0, I²C and SPI interfaces on the LPC controller, as well as the switching output buffered by transistor T3, make the addition of additional hardware components very easy.

7.2.3 The MF RC522 Reader IC

The MF RC522 contactless 13.56 MHz reader IC supports both the ISO/IEC 14443 Type A protocol and the proprietary MIFARE protocol.

The MF RC522 has the following features:

- highly-integrated ISO/IEC 14443 Type A and MIFARE-compatible single-chip reader IC,
- supports RF transmission speeds of 106, 212, 424 and 848 kbit/s,
- maximum range of 50 mm, depending on the antenna used,
- integrated MIFARE encryption block,
- programmable via UART, I²C or SPI,
- 64-byte transmit and receive FIFO buffers
- programmable reset and power-down modes,
- programmable timer,
- internal oscillator for direct connection of a 27.12 MHz crystal.

Figure 7.7 shows a very simplified block diagram of the reader IC. The MF RC522 has an analog interface, a contactless UART, a FIFO buffer, program control (Status and Control SFRs) as well as a host interfaces for interfacing to a microcontroller.

The buffered output of the MF RC522 drivers allow direct connection to the transmit and receive antenna without an additional active power amplifier. In addition, only a few passive components are needed for the antenna matching.

The analog interface handles the complete encoding and modulation of the card instructions as well as demodulation and decoding of the responses returned by the card. Creation of the ISO/IEC 14443 or MIFARE protocol frames, as well as the corresponding error detection (parity and CRC) takes place in the UART's digital block. The FIFO buffer allows the sending and receiving of up to 64-byte blocks (T=CL protocol) in ISO/IEC 14443 Type A mode. MIFARE uses data blocks with a of maximum 16 bytes, so no splitting up of the blocks is required from the microcontroller. With T=CL, larger information frames (I Frames) are broken up by the PC software using the chaining method specified by the T=CL protocol, and transmitted sequentially.

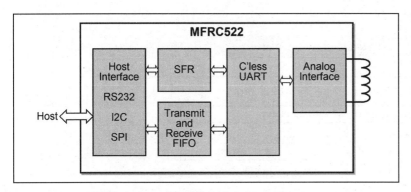

Figure 7.7.
MF RC522 reader IC simplified block diagram.

The entire procedure for contactless communication is controlled solely by the MF RC522's configuration, control and status registers. The MF RC522's Special Function Registers (SFR) are programmed via RS-232, I²C or SPI.

In Section 9.1, MF RC522 programming, by way of card activation, is presented. In this example, the MF RC522 is programmed using Smart Card Magic.NET on the PC, so it can be done by readers who have no means by which to progrm the LPC936 microcontroller.

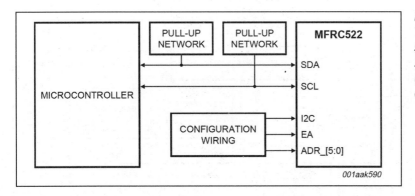

Figure 7.8.
The MF RC522 supports only the I2C Slave mode (source: NXP Semiconductors).

Since the P89LPC932 has only one serial RS-232 interface, and this is needed to communicate with the PC, the communication between the P89LPC936 and the MF RC522 takes place over I²C.

7 THE ELEKTOR RFID READER

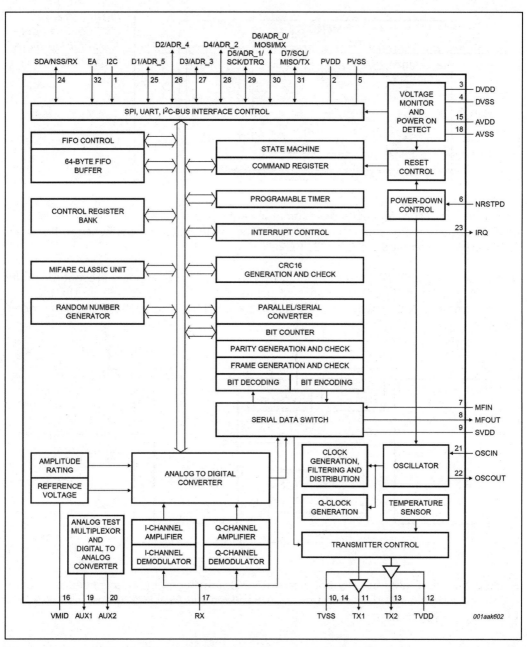

Figure 7.9. MF RC522 block diagram (source: NXP Semiconductors).

7.2.4 The FT232R USB/RS-232 Converter

Modern desktops, notebooks and netbooks usually no longer have an RS-232 interface. Therefore, the design of Elektor RFID Reader dispenses with this interface and connects to the PC via USB. The power supply issue is elegantly solved with the help of USB if the reader is connected to a PC. Therefore, a power supply can be completely dispensed with in most cases.

In practice, there are several ways of equipping a microcontroller with a USB interface. The use of a microcontroller with an integrated USB interface is very complicated and requires very good knowledge of USB. In addition to the implementation of the USB firmware, it is often necessary to develop a PC driver. This is always the case when a USB device is not assigned one of the predefined device classes. It is also necessary to support multiple operating systems, so the complexity required for a project consisting of only a few assembled units is not justified. This problem was recognized by the semiconductor industry. Since 2000, a number of USB-to-serial converter modules (bridges) have been available on the market. The most important manufacturers in this market are Silicon Labs (SiLabs – *www.silabs.com*), Future Technology Devices, Inc. (FTDI – *www.ftdi-chip.com*) and Prolific Technology, Inc. (Prolific –*www.prolific.com.tw*).

The P89LPC900 family of controllers, like most 8-bit microcontrollers, has no USB controller. FTDI's FT232R (IC1) was selected as a USB-to-serial converter for the reader circuit.

The main features of the FT232R are:

- USB 2.0 Full-speed-compatible;
- the entire USB protocol is integrated on the chip;
- RS-232, RS-422 and RS-485 with a data rate of between 300 baud and 3 Mbaud;
- different bit bang modes;
- integrated 1024-bit configuration EEPROM;
- integrated USB termination resistors;
- built-in clock generation requiring no external crystal;
- configurable Control Bus (CBUS) I/O pins;
- unique identification number (FTDIChip-ID™) ;
- free drivers for Windows, Windows CE.NET, Linux and Mac.

With the FT232R, the required data EEPROM is integrated into the chip. In addition, the FT232R uses no crystal for clock generation, as it has a highly accurate internal RC oscillator. This reduces the external component count for the USB-RS-232 converter. When a USB-serial converter was selected, special attention was paid to the programming of the P89LPC936 microcontroller. The activation of the LPC's boot mode with defined pulses at the reset pin presents no problem for the FT232R.

Of the maximum five additional CBUS I/O pins, three are used in Elektor RFID Reader. At pin 21 (CBUS1/#RXLED) of IC1 is a green LED (D1), which indicates the reception of data on the USB bus. LED D2 indicates the transmission of data to the PC via USB (Pin

7 THE ELEKTOR RFID READER

22 – CBUS0/#TXLED). CBUS3/#PWRNEN on Pin 11 controls the P-channel MOSFET, T2. This turns on the power to the rest of the circuit after successful USB enumeration. The function of the CBUS pins is configured by data in the FT232R's EEPROM.

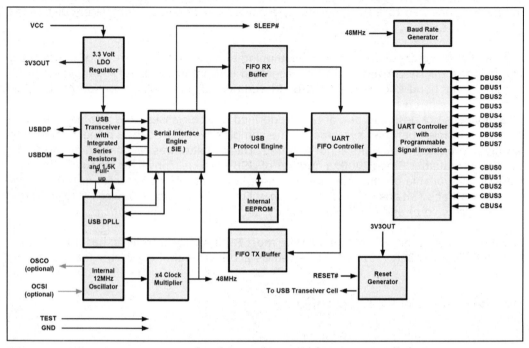

Figure 7.10. FT232R block diagram (source: Future Technology Devices International).

FTDI provides two different versions of the PC drivers. The Virtual COM Port (VCP) drivers emulate a traditional RS-232 interface (COM port). In this way, existing PC software, such as a terminal emulator, can be used.

The second driver variant, Direct Driver (D2XX), consists of a proprietary DLL (FTD2XX.DLL). This has a lot more features available than the VCP driver. Reading and writing the EEPROM configuration requires the use of the D2XX driver. Data communication is much more efficient with this driver, as it uses the Direct Block Transfer and Bulk Mode USB features.

For Windows, the installation of the VCP and D2XX drivers is done from a single package, in the Combined Driver Model (CDM). This has only been the case since 2006. Until then, it was not possible to install both drivers together on a Windows system.

For the Elektor RFID Reader, both drivers are needed. The Virtual COM Port driver is used mainly for programming the P89LPC936. The current C# reader library uses the DSXX driver DLL, due of its expanded features.

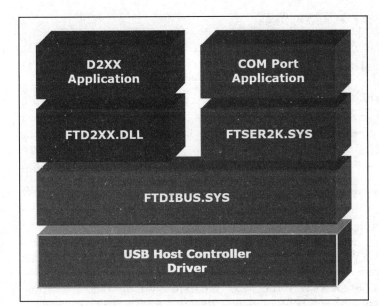

Figure 7.11. Windows CMD driver architecture (source: Future Technology Devices International).

7.2.4.1 Configuring the FT232R

When a USB device connected to the PC, the operating system carries out a USB initialization (enumeration). During this, USB configuration data is queried from the device in the form of so-called 'descriptors'. A descriptor is a data record, defined by the USB specifications, that describes the properties of the USB device.

After a power-on reset, the FT232R chip reads the device, config and string descriptors, as well as additional chip configuration data from its internal EEPROM. FTDI provides two software tools for customizing the chip configuration for your own applications. These can be downloaded from the FTDI website *(www.ftdichip.com/Support/Utilities.htm)*. A Microsoft tool, USBView, is also available there. The older EEPROM configuration software, MProg 3.5, has been replaced by FT_PROG. FT_PROG requires the installation of Microsoft. NET Framework 4.0, which can be downloaded for free from the Microsoft website. For the Elektor RFID Reader, the functionality of MProg is adequate. The FT232R on the Elektor Reader board comes correctly preconfigured.

Each USB device has a unique, global, vendor-specific ID. This consists of a 2-byte Vendor ID and a 2-byte Product ID. Custom Vendor IDs can be obtained from the USB regulatory body for around $2,000 *(www.usb.org)*. The Product ID is defined in sequential order by the manufacturer. The FTDI FT232R is supplied with the FTDI Vendor ID and Product ID. This can be used for free for your own projects. It is also possible to make use of a custom Vendor ID and Product ID, as is the case with the Elektor RFID Reader.

Configuring the FT232R is very easy to do using FT_PROG. Using the 'Scan and Parse' button, the entire EEPROM content can be read and interpreted. If data is changed, it must first be saved in a template file on the PC. Modified configuration data can only be programmed into the EEPROM from a template file.

7 THE ELEKTOR RFID READER

Figure 7.12.
Using FT_PROG, all FT232R configuration data can be read and written.

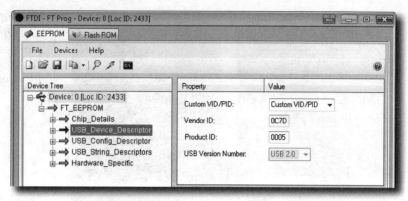

The Elektor RFID Reader comes with the Elektor Vendor ID (0x0C7D) and a Product ID (0x0005). These IDs are used by Windows to detect a new USB device and load the correct USB drivers. For this, the Windows Configuration Manager compares the Vendor ID and Product ID with a database. This consists of many `.inf` files — one for each USB device. For each USB device, a file exists with the `.inf` file extension. .inf files are simple ASCII files with a defined structure and content. These describe a USB device in detail.

Because the Elektor RFID Reader uses a modified Vendor ID and Product ID, it's not possible to use the FTDI CMD drivers from their website, but a modified version of the drivers must be created. These can be downloaded from the Elektor website. The modification consists of only two new `.inf` files. In Section 7.2.4.2, the creation of the two `.inf` files will be discussed..

> **Note** If it is not possible to install the driver correctly, it is advisable to use the original FTDI driver. For this driver, the FTDI Vendor ID (0x0403) and Product ID (0x6001) must be used. If an operating system other than Windows XP, Windows Vista or Windows 7 is used, this variant can be a simple solution.

Since the Elektor RFID Reader is supplied with power directly via USB, the USB Config Descriptor must be configured, as in Figure 7.13.

Figure 7.13.
FT_PROG: FTDI USB Configuration Descriptor dialog.

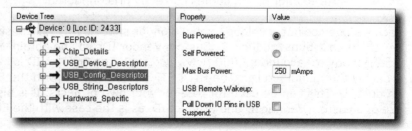

A small change to the original Elektor configuration improves the discoverability of the Elektor RFID Reader. Elektor provides the reader with the default values for Manufacturer String ("FTDI") and Product Description String ("FT232 USB UART") in place.

READER HARDWARE 7.2

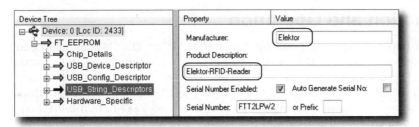

Figure 7.14.
Elektor RFID Reader: FTDI USB String Descriptor.

Many commercial devices, such as USB-to-RS-232 converters, use FTDI chips. Useful strings (for example: "Elektor" or "Elektor RFID Reader") make finding and selecting the device in the PC software easier. In the current Elektor RFID Reader software library, the Product Description String can be seen. If the Product Description String is not correctly displayed on the PC, despite correct EEPROM configuration, it is necessary to perform a a driver update in Windows Device Manager.

In a further configuration window, each CBUS pin is assigned one of the possible functions. A template file *(Elektor_RFID_Reader.xml)* is available on the Elektor website.

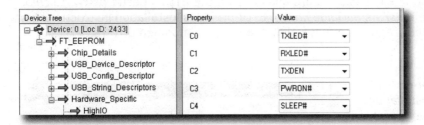

Figure 7.15.
Elektor RFID Reader: FTDI USB CBUS Configuration.

7.2.4.2 USB Driver Modification

Recently, FTDI has made a special software tool for generating manufacturer-specific `.inf` files available. The INF File Generator software is easy to use an can be downloaded for free from the FTDI website *(www.ftdichip.com/Support/Utilities.html)*. You can also find a detailed guide no the FTDI website.

The tool creates an `.inf` file for the VCP driver *(Ftdiport.inf)* and one for the D2XX driver *(Ftdibus.inf)*. Subsequently, the two original `.inf` files created by the CDM driver files are replaced. The custom driver is then ready.

7 THE ELEKTOR RFID READER

7.3 Construction and Operation

Construction of the hardware should pose no challenges. If an LCD is used, it need merely be connected to K2. Then, all that is required is to provide the reader board with power from batteries, a 5 V power supply or via USB. The two jumpers on the reader board (JP1 and JP2) are not set during normal operation. The reader is immediately ready for use and nothing stands in the way of reading of a card's serial number (UID). This happens for as long as a card appears in the reader's field. The installation of the reader board into the Elektor enclosure is a little tricky. However, the Elektor website has installation instructions available for download.

7.3.1 Installing the USB Driver

The required FTDI USB CMD driver is available on the Elektor website. It is absolutely necessary to use this driver because the FT232R has been programmed with the Elektor Vendor ID and Product ID (see Sections 7.2.4.1 and 7.2.4.2). The RFID reader is connected to a PC via USB. Windows will automatically detect a new USB device. The Elektor RFID Reader version must be selected as a USB driver. If any problems are encountered, installation instructions are available on the FTDI website *(www.ftdichip.com/Support/Documents/InstallGuides.htm)*. FTDI provides installation instructions for each operating system, which are fully valid, even for the modified drivers.

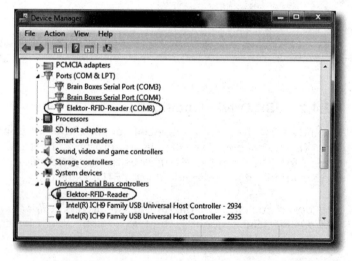

Figure 7.16.
After successful installation, the VCP driver can be found under COM & LPT, *and the D2XX driver under* USB Controllers.

7.3.2 Reader Firmware Update

All of the examples in this book use the C# reader library. This only works with reader firmware version 3.0.0. This version was developed simultaneously with the reader library, and therefore a firmware update is required.

The P89LPC936 on the reader board can be programmed directly via USB using the free PC program, Flash Magic *(www.esacademy.com)*. Flash Magic requires the VCP drivers, so these must already be installed. To program the controller, the two jumpers on the reader board are set (JP1 and JP2).

After starting Flash Magic, choose the COM port that is assigned to the USB. The correct COM port is obtained as shown in the previous section, using the Windows Device Manager. The following parameters must be correctly selected in Flash Magic: Device: *89LPC936*; Baud Rate: *19200*; Interface: *None (ISP)*. It's also important to check the "Erase blocks used by Hex file" box. The 'Browse' button brings up a file dialog for opening the .hex file.

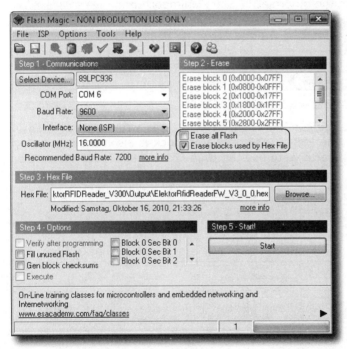

Figure 7.17. Flash Magic: required settings for firmware update.

> **Note** Flash Magic's 'Erase all Flash' option allows the entire flash memory, including boot loader, to be erased. Accidental deletion of the boot loader can be very easily overcome, as the firmware is simply compiled with the boot loader code included.

The default values in the 'Advanced Options' configuration menu are correct for the Elektor RFID Reader. It's a little confusing, but *Keil MCB 900* must be selected as the hardware.

7 THE ELEKTOR RFID READER

Figure 7.18.
Flash Magic:
Advanced
Options → Hardware Config.

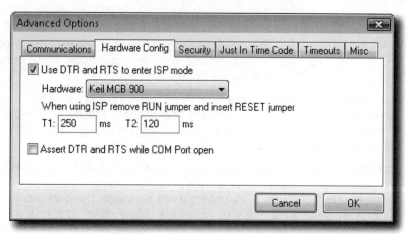

There is a critical setting in *Advanced Options* → *Security*. Make sure to check the box next to 'Protect ISP Code'. All of the other settings are unimportant.

Figure 7.19.
Flash Magic:
Advanced Options
→ Security.

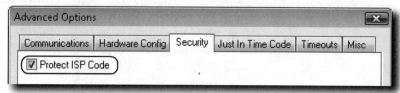

Also watch out with the *ISP/Device* configuration: selecting the wrong clock here will stop everything from working. The P89LPC936 oscillator clock is also required during programming. Therefore, it is imperative to use the *High Frequency Crystal/Resonator (4MHz-12MHz)* option.

Figure 7.20.
Flash Magic: 89LPC9xx
Configuration.

If the Elektor RFID Reader is connected via a USB cable and the two jumpers are set, the reader is in programming mode. You can the test whether the interface is working correctly by selecting *ISP* → *Blank Check* from the menu. To begin the programming process, simply click the 'Start' button. The programming process should not be interrupted under any circumstances.

After programming, you must remove the two jumpers, or else the microcontroller will remain in programming mode. For the firmware to start up correctly, it is necessary to separate the reader from the USB port briefly and then reconnect it.

7.3.3 Firmware Version Control

The current reader firmware version can easily be checked if an LCD is connected to the RFID reader. After turning on the supply voltage, the version will be displayed for about one second.

Figure 7.21.
Elektor RFID Reader firmware version.

7.4 Reader Modes

The reader firmware supports three different modes:

- *Terminal* mode,
- *PC Reader* mode,
- *Access Control* mode.

In Section 9.2.3, the Access Control mode is presented in detail. With this, an access control system is realized using the Elektor RFID Reader. The Access Control mode is an integrated part of the reader's firmware, and can be permanently activated using PC software.

7.4.1 Terminal Mode

A free MIFARE Ultralight card came with the 9/2006 issue of Elektor magazine. On such a card, a globally unique serial number, the UID, is stored. This cannot be changed after manufacture. On the basis of these numbers, Elektor initiated a lottery, and one of the attractive prizes could be won if the card's UID matched one published in the magazine. The Terminal mode was specifically included in the firmware for this competition.

Figure 7.22.
Reading a 7-byte UID.

As soon as the reader is supplied with power, it is in Terminal mode (an exception is the Access Control mode mentioned previously). Once a card enters the reader's field, the UID is read out and displayed on the LCD. This function is suitable for easily assessing a card's reading range (we know that the ranges of different cards can vary significantly). Also, the influence of metal, for example (e.g. batteries), on the reader can be investigated. With an inadequately tuned antenna, the immediate reduction in range can be seen.

7 THE ELEKTOR RFID READER

ISO/IEC 14443A cards also differ in the number of bytes in the UID. There are special cards that produce a Random ID (RID). RIDs are especially suited for electronic passports and ID cards.

Because a fixed UID has a direct relationship with a passport holder, reconstruction of movement profiles could easily be achieved, according to privacy advocates. The chip in an EU passport thus always transmits a Random ID that is valid only for the current session. Some countries, such as the United States, purposely use a fixed UID in order to create motion profiles of people. Using the Elektor RFID Reader it is very easy to check whether an electronic passport uses an RID. All that's needed is to position the passport in the reader's RF field. An RID consists of 4 bytes and starts with 0x08.

In Terminal mode, the Elektor RFID Reader continuously searches for new cards. Before each card activation, the reader firmware performs an RF reset. In the case of a card with a Random ID, this means that each new activation will produce a new UID.

Figure 7.23.
Reading an electronic passport's RID.

Upon successful card activation in Terminal mode, the ATQA, UID and SAK values are sent to the PC via USB. In the case of an Ultralight card, the entire content of the card memory is also transmitted. A simple terminal program with VT100 emulation is sufficient to receive the data on the PC. If the PC is running Windows 2000 or XP, HyperTerminal is available for the purpose. You won't find this in Windows Vista, however. The terminal program will require the following settings: Baud rate: 460800; Parity: None; 8 data bits; 1 stop bit. Since Firmware Version 2.x, the baud rate has changed from 115200 to 460800. Not all terminal programs support this transfer rate, but there are no problems when using HyperTerminal on Windows XP.

Alternatively, Smart Card Magic.NET can be used. For this, the following steps are required:

- Connect the Elektor RFID Reader to the PC.
- Start Smart Card Magic.NET.
- Click on the 'Open' button to open the USB port.
- Finally, start terminal emulation by clicking on the 'Terminal' button.

Since the terminal emulation uses the D2XX driver, no further adjustments are required.

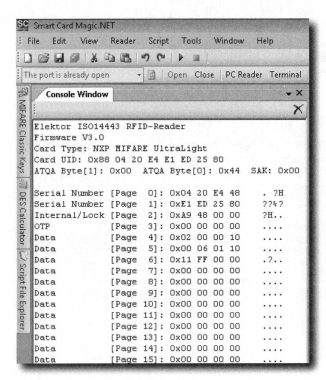

Figure 7.24.
Smart Card Magic.NET supports simple terminal emulation. The Elektor RFID Reader reads the memory content of a MIFARE Ultralight card and sends this via USB interface to the PC.

7.4.2 PC Reader Mode

The function of the PC Reader mode is much like a commercial 13.56 MHz PC reader. On such a reader, all communication with the card is controlled by a PC application. What portion the reader is responsible for depends on the manufacturer's implementation.

Most commercial readers support the Smart Card Reader PC/SC API standard (see Section 10.4). The PC/SC standard specifies a vendor- and platform-independent programming interface (API, or Application Programming Interface). The API does not support the sending of single card instructions.

The proprietary Elektor RFID Reader API does not know these limitations. The Elektor RFID Reader is thus better suited to exploring different card types. The Elektor RFID Reader API is designed so that it provides an easier entry into firmware programming. The method names in the Elektor RFID Reader library are almost identical to the names of the C functions in the reader's firmware. The PC Reader mode offers the option of developing and testing part of the reader firmware on the PC.

7.4.2.1 Activating the PC Reader Mode

As soon as the PC sends an instruction to the reader via the USB interface, the PC Reader mode is automatically activated. This also applies to Access Control mode. The firmware also supports selective activation of the Terminal and Access Control modes.

7 THE ELEKTOR RFID READER

7.5 The Firmware

The Elektor RFID Reader's firmware was developed using the Keil µVision4 development environment *(www.keil.com)*.

7.5.1 The Software Architecture

In Figure 7.25, all the C source code files and their functions are shown. Furthermore, the picture shows the Elektor RFID Reader's basic software architecture.

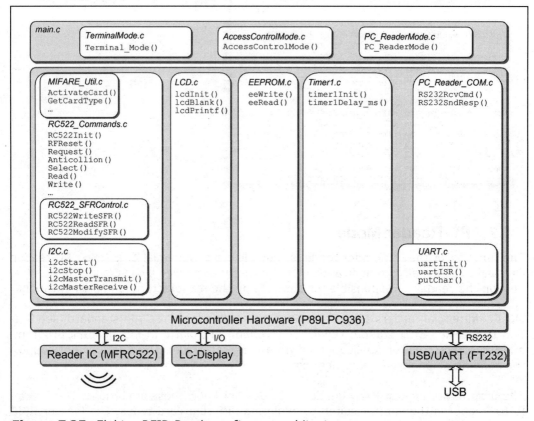

Figure 7.25. Elektor RFID Reader software architecture.

7.5.2 The Main Program

After a power-on reset, the main program on the P89LPC936 initializes the MF RC522 reader IC as well as the LCD display.

The active RFID reader mode (Terminal, PC Reader or Access Control) is stored in global variable `gbMainState`. Depending on the EEPROM configuration, the variable `gbMainState` will be assigned either the `MainState_TerminalMode` or the `MainState_`

ACCESS_CONTROL value. This ensures that, after a power-on reset, either the Terminal mode or the Access Control mode will be active.

Once the hardware is initialized, the main program will loop continuously. Depending on the current RFID reader mode, the main program loop will call one of the functions Terminal_Mode(), PC_ReaderMode() or AccessControlMode().

Listing 7.1. *Excerpt from* main.c.

```c
while(1) // MAIN LOOP
{
    switch(gbMainState)
    {
        case MainState_TerminalMode:
            Terminal_Mode();
            break;

        case MainState_PCReaderMode:
            if(gbNewCmdReceived == SET)
            {
                PC_ReaderMode();
                gbNewCmdReceived = CLEAR;
            }
            break;

        case MainState_ACCESS_CONTROL:
            AccessControlMode();
            break;

        default:
            Terminal_Mode();
            break;
    }
}
```

The reception of PC RFID Reader instructions via the RS-232 interface is interrupt-driven. For each newly-received PC RFID Reader instruction, the interrupt service routine, uartISR() is called. The UART interrupt could occur within the main program, the Terminal_Mode() function, the PC_ReaderMode() function or the AccessControlMode() function.

In the UART interrupt service routine, the global variable, gbMainState, is assigned the value PCReaderMode, and the RS232RcvCmd() function is called. Once the last instruction byte is received, the interrupt routine ends, and the global variable gbNewCmdReceived is assigned the value SET.

Thereafter, normal program execution continues. Any of the three functions, Terminal_Mode(), PC_ReaderMode() and AccessControlMode() will still be completed. Only in the next iteration of the main program loop, the function PC_ReaderMode() is called. The reader will then be in the PC Reader mode.

7.5.3 The PC_ReaderMode() Function

For performance reasons, the `RS232RcvCmd()` function stores the complete, received PC RFID Reader instruction in the global byte array, `bgaRS232CmdBuffer`. The function `PC_ReaderMode()` can then directly evaluate the instruction bytes. Which PC RFID Reader instruction is determined in the second byte. This byte is evaluated by a `switch` statement, and the corresponding RC522 reader or card function is called. The calling parameters of these functions are also contained in the PC RFID Reader instruction.

After carrying out an RC522 reader or card function, a status byte will at least be returned, and, when available, any card response. This is done by calling the `RS232Resp()` function.

Listing 7.2. *Excerpt from* PC_ReaderMode.c.

```
switch(gbaRS232CmdBuffer[1])
{
    case 0x00: // PC-Command: RF-Reset
        RFReset(gbaRS232CmdBuffer[3]);    // RF-Reset Time (1...255 ms)
        RS232SndResp(0, pRespBuffer, 0);
        break;

    case 0x01: // PC-Command: Activate Card
        status = ActivateCard(gbaRS232CmdBuffer[3], // IN: Cmd Code (REQA or WUPA)
                              &pRespBuffer[0],      // OUT: 2 Bytes ATQA
                              &pRespBuffer[2],      // OUT: 4 or 8 Byte UID
                              &pRespBuffer[12],     // OUT: 1 Byte UID Length
                              &pRespBuffer[13]);    // OUT: 1 Byte SAK
        RS232SndResp(status, pRespBuffer, 14);
        break;

    ...

    case 0x10: // PC-Command: REQA or WUPA
        status = Request(gbaRS232CmdBuffer[3], // IN: Cmd Code (REQA or WUPA)
        pRespBuffer); // OUT: 2 Bytes ATQA
        RS232SndResp(status,pRespBuffer,2);
        break;
    ...
```

7.5.3.1 The RS-232 Communication Protocol

The data exchange between PC and Elektor RFID Reader uses a very simple protocol. Since the RS-232 frames are always transmitted via a USB-to-serial converter, it is not necessary to secure the protocol frame using an XOR or CRC checksum. This happens automatically in the USB protocol.

The first byte of an RS-232 frame is always 0xA5, which marks the start of a new transmission frame (SOF). In the case of a PC RFID reader instruction, the CMD byte follows next. The CMD byte codes for the various instructions. The length byte, LEN, contains the number of optional instruction bytes or response data bytes.

THE FIRMWARE 7.5

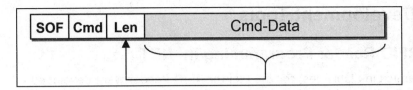

Figure 7.26.
Structure of PC RFID Reader instructions.

The structure of the PC RFID reader response differs from the PC RFID reader instruction only in the second byte. The reader or card's return code is encoded in the status byte. If needed, the PC RFID reader instruction codes can be found in the source file, `PC_ReaderMode.c`.

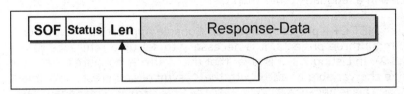

Figure 7.27.
Structure of PC RFID Reader responses.

On the PC side, USB/RS-232 communication always takes place using `GS.FTD2XXLibrary.DLL`. The FTD2XXLibrary manages the entire PC Reader protocol communication. The menu command, *Reader → FTD@xx-Trace* activates the output of debugging information in the Trace window (see Figure 7.28).

In a C# program or a Smart Card Magic.NET script, the debug output is activated with the following line of code:

```
reader.EnableFTD2XXTrace = true
```

This captures the function calls of the methods `FT_Write` and `FT_Read`. The `FT_Write` method sends a PC RFID Reader instruction to the RFID Reader. In the first step, `FT_Read` reads only the first three bytes of the PC RFID Reader response. Only if further data is indicated by the LEN byte (when LEN is greater than 0), then the data bytes are read by calling `FT_Read` a second time.

```
Trace Window
FT_Write(buf = 0xA5 00 01 05 , count = 4, bytesWritten = 4)
FT_Read(buf = 0xA5 00 00 , bytesToRead = 3, bytesReturned = 3)
    RF Reset(5 ms)
FT_Write(buf = 0xA5 03 02 00 00 , count = 5, bytesWritten = 5)
FT_Read(buf = 0xA5 00 00 , bytesToRead = 3, bytesReturned = 3)
    SetPCDBaudrate(Reader -> Card: 106 kbit/s, Reader <- Card: 106 kbit/s)
--> REQA
FT_Write(buf = 0xA5 10 01 26 , count = 4, bytesWritten = 4)
FT_Read(buf = 0xA5 00 02 , bytesToRead = 3, bytesReturned = 3)
FT_Read(buf = 0x44 00 , bytesToRead = 2, bytesReturned = 2)
<-- ATQ: 0x44 00
--> AntiCollision Cascade Level 1
FT_Write(buf = 0xA5 11 01 93 , count = 4, bytesWritten = 4)
FT_Read(buf = 0xA5 00 04 , bytesToRead = 3, bytesReturned = 3)
FT_Read(buf = 0x88 04 0C D9 , bytesToRead = 4, bytesReturned = 4)
<-- UID: 0x88 04 0C D9
```

Figure 7.28.
Using the FTD2xx-Trace output, one can debug the communication between the PC and the Elektor RFID Reader.

7.6 The PC Development Tools

7.6.1 Elektor RFID Reader Programming in .NET

All of the DLLs (Dynamic Link Libraries) for the Elektor RFID Reader were developed in C#. The Elektor RFID Reader can thus be programmed in any .NET language.

The following three examples illustrate the use of the .NET reader libraries in C/C++, Visual Basic.NET and C#. In each example, the card is activated with the `ActivateCard()` method. If a card can be successfully addressed, the card's serial number (UID) is output to the Console Window. These three examples show only the possibility that the reader can also be programmed in a language other than C#.

To create a managed, or .NET C++ project, a new CLR Console Application must be created in Visual Studio. In all three projects, it is necessary to set up a reference to the DLLs (see Section 7.6.3.2). In Listing 7.3, it is clear that the syntax is not pure C++. Thus it is necessary to define the variable `reader` with the ^ symbol and create the object with the keyword `gcnew`. This is because `ElektorISO14443Reader` is a managed class.

Listing 7.3. *Elektor RFID Reader example: managed C++.*

```cpp
#include "stdafx.h"

using namespace System;
using namespace GS::ISO14443_Reader;
using namespace GS::ElektorReader;

int main(array<System::String ^> ^args)
{
    ElektorISO14443Reader ^reader = gcnew ElektorISO14443Reader();
    reader->OpenPort();
    reader->RFReset(5);
    System::Int32 retValue = reader->ActivateCard(RequestCmd::REQA);

    if(retValue == 0)
    { // Cards successfully activated
        Console::Write("Card UID: 0x");

        for(int i = 0; i < reader->UID->Length; i++)
            Console::Write(String::Format("{0:X02} ",reader->UID[i]));
    }

    reader->ClosePort();
    return 0;
}
```

Any C# source code can easily be turned into Visual Basic .NET source code. For this, there are several tools available on the web. A code compiler is an integral part of the free development environment, SharpDevelop *(http://icsharpcode.com)*. The following example was created with this compiler. The C# source code in Listing 7.5 serves as the starting point.

Listing 7.4. *Elektor RFID Reader example: Visual Basic.NET.*

```vbnet
Imports GS.ISO14443_Reader
Imports GS.ElektorReader
Module Module1
    Sub Main()
        Dim reader As New ElektorISO14443Reader()

        reader.OpenPort()
        reader.RFReset(5)

        Dim retValue As Integer = reader.ActivateCard(RequestCmd.REQA)

        If retValue = 0 Then
            ' Cards successfully activated...
            Console.Write("Card UID: 0x")

            For i As Integer = 0 To reader.UID.Length - 1
                Console.Write([String].Format("{0:X02} ", reader.UID(i)))
            Next
        End If

        reader.ClosePort()
    End Sub
End Module
```

Listing 7.5. *Elektor RFID Reader example: C#.*

```csharp
using System;
using GS.ISO14443_Reader;
using GS.ElektorRfidReader;

namespace GS.ElektorRfidReaderExample
{
    class Program
    {
        static void Main(string[] args)
        {
            ElektorISO14443Reader reader = new ElektorISO14443Reader();

            reader.OpenPort();
            reader.RFReset(5);

            int retValue = reader.ActivateCard(RequestCmd.REQA);

            if(retValue == 0)
            { // Cards successfully activated…
                Console.Write("Card UID: 0x");

                for(int i = 0; i < reader.UID.Length; i++)
                    Console.Write(String.Format("{0:X02} ",reader.UID[i]));
            }
            reader.ClosePort();
        }
    }
}
```

7 THE ELEKTOR RFID READER

As you've probably noticed, all three programming languages use the same .NET classes. In all three examples, output to the console is done using `Console.Write()` or `Console.WriteLine()`.

For all the examples to follow, only C# will be used. Smart Card Magic.NET is a powerful integrated development environment for MIFARE and other contactless smartcards. This was developed specifically for the Elektor RFID Reader and uses very little disk space (< 3 MB). Thus, the software should also run on older PCs and is installed much faster than other C# development environments. The window management is based on Visual Studio 2008 and allows the use of netbooks with resolutions of only 1024x600 pixels.

7.6.2 Smart Card Magic.NET

This chapter provides a quick introduction to using Smart Card Magic.NET. Details of the card instructions and their correct use are covered in the next chapter.

Smart Card Magic.NET requires the installation of .NET Framework 2.0. .NET Framework is available for free from the Microsoft Website *(http://msdn.microsoft.com/en-us/vstudio/aa496123)*.

7.6.2.1 It's Usable without Programming

This interface provides the ability to test all card features without having to write a single line of code. For this purpose, Smart Card Magic.NET has the following four forms:

- MIFARE Ultralight C;
- MIFARE Classic;
- T=CL;
- Access Control Configuration.

The MIFARE Ultralight C window allows the reading and writing of MIFARE Ultralight and Ultralight C cards. The Ultralight C's triple-DES authentication is also supported. With the MIFARE Classic window, all details of a MIFARE Classic 1K or 4K can be investigated. Both windows allow the entire memory contents of a card to be read out at the click of a 'button'. This feature also allows the use of different MIFARE Classic keys (Key A/B). Simple communication with MIFARE DESFire and all the other ISO14443-4 Type A compliant smartcards is possible with the T=CL window.

The Access Control Configuration window is identical to the Windows application presented in Section 9.2.3.2, Access Control Manager.

Reading a Card's Contents
For this example, you need a MIFARE Ultralight card. The following steps are required to read out the entire memory content of the card:

- Connect the reader to the PC.
- Start Smart Card Magic.NET.
- Click the 'Open' button to open the USB port.
- Activate the card using either the 'Act Req A' or the 'Act Wup A' button.
- Read out the entire card content by clicking on the 'Dump UL' button.

Should an error occur in communication between card and reader, it's necessary to reactivate the card. After each error, the card returns to its IDLE state. In this case, the card accepts only REQA or WUPA commands. The card will not respond to any other command.

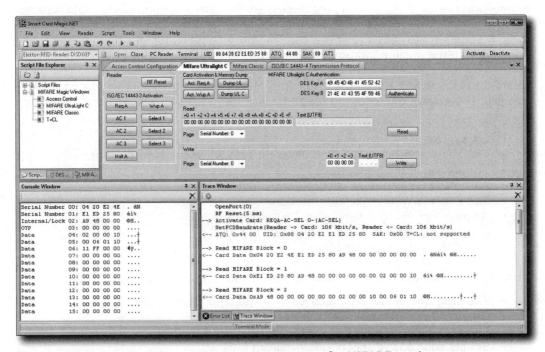

Figure 7.29. Reading the entire memory content of a MIFARE card.

The Elektor RFID Reader library logs the entire reader-card communication. This debug output it shown in the Trace Window. The 'Dump' button creates an additional Console Window.

7.6.2.2 A Scripting Tool or a C# Compiler?

If you work with MIFARE and other contactless cards frequently, it's essential to be able to cobble together a test program quickly. It's often necessary to test how a card behaves in a specific situation. One such repetitive task is reading and writing test data. Many of these small test programs can be developed as console applications, which allow for very limited ease of use, logging and error code output. Large parts of the source code must always be revised for new tasks.

Alternatively, windows applications such as MIFARE WND *(www.nxp.com)*, which send the commands to the card at the click of a mouse, may be used. These programs are very helpful for simple and quick tests, but lack the ability to create scripts. Finally, in order to personalize a MIFARE 4K card, it is necessary to write a complete program.

7 THE ELEKTOR RFID READER

The .NET Framework supports the Code Document Object Model to allow programmatic access to the C# compiler. The required classes are found in the .NET namespace, `System.CodeDOM`. This method is well documented on MSDN (Microsoft Developer Network). Smart Card Magic.NET uses this capability to compile script files.

When using the C# compiler, it is not necessary to learn a scripting language. Another advantage is that the generated script source code can be used in future C# applications. A scripts that reads the entire card memory content, for example, is especially helpful during the implementation of large projects.

Smart Card Magic.NET eases the entry into MIFARE and other contactless card technologies. It supports the following features:

- Simple user interface.
- C# syntax.
- The editor shows keywords in different colors (syntax highlighting).
- Scripts run in their own threads.
- Automatic logging of all reader-card communication.
- The Console Window is integrated into the interface.
- An interrupt mode is supported by a script method.
- Program-controlled console and trace windows.
- Low disk space requirement.

Smart Card Magic.NET has the following limitations compared to Visual Studio C#:

- The program cannot be executed in single step mode.
- Only hard-coded breakpoints are supported.
- The editor has no built-in code completion (Intellisense).
- Scripts can consist only of one program file.

In most cases, it's not a big disadvantage that scripts cannot consist of multiple files. For extensive projects, Visual Studio is the suitable choice in any event.

7.6.2.3 Our First Program: "Hello World"

For the first example, neither a card nor a reader are required. This example just shows how to create and run a script.

At first launch, the Script File Explorer window is open, as shown in Figure 7.30. The Script File Explorer window provides an easy way to manage multiple script files. A Zip file containing all of the Smart Card Magic.NET examples is available for download from this book's website. After downloading, the files must be extracted. Using the menu command *Script → Select Script Folder*, or clicking on the file folder icon opens the 'File' dialog. Select the script examples' root directory. The representation of files and folders is similar to that in Windows Explorer.

Individual subdirectories and files can be opened by double clicking. The tree structure is identical to the operating system's. Use Windows Explorer to easily move, copy and delete files.

7.6 THE PC DEVELOPMENT TOOLS

Figure 7.30.
Script File Explorer *after the first run.*

Following the tradition, the author begins by creating a "Hello World" program. You can find the sample file, `HelloWorld.cs`, in the `Basics` folder, and open it by double clicking on the file name in the Script File Explore window. Alternatively, the file can be opened using the *File → Open* menu option.

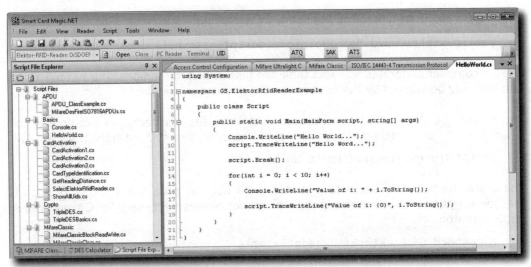

Figure 7.31. *The script file* HelloWorld.cs *in the development environment.*

Listing 7.6 is constructed in the same manner as a regular C# console application. This program has two differences to a conventional C# program, which are required for this development environment.

217

Listing 7.6. *HelloWorld.cs.*

```csharp
using System;

namespace GS.ElektorRfidReaderExample
{
    public class Script
    {
        public static void Main(MainForm script, string[] args)
        {
            Console.WriteLine("Hello World...");
            script.TraceWriteLine("Hello Word...");

            for(int i = 0; i < 10; i++)
            {
                Console.WriteLine("Value of i: " + i.ToString());
                script.TraceWriteLine("Value of i: " + i.ToString());
            }
        }
    }
}
```

The `Main()` method represents the entry point of a C# program and is automatically called when you run the script. The `Main()` method's signature is slightly different to one in a standard console application, which looks like this:

```csharp
static void Main(string[] args)
```

Smart Card Magic.NET requires the following definition:

```csharp
public static void Main(MainForm script, string[] args)
```

The C# compiler generates an executable file (assembly) in memory. The `Main()` function can only be started by the application if it is declared as `public`.

The `MainForm` class enables interaction with the user interface. An example is the `TraceWriteLine()` method. Using this method, text can be displayed in the trace window.

```csharp
script.TraceWriteLine("Hello World...");
```

These are the two differences to a standard C# console program. Another proviso is that the script has its own namespace. This is not always necessary for C# programs, but is common.

```csharp
namespace GS.ElektorRfidReaderExample
```

With `using System`, the .NET system namespace is included. This shortens the notation of the `System` namespace found in the `Console` class. Without including System, accessing the `Console` class would require fully-qualified naming. This would look as like this:

```csharp
System.Console.WriteLine("Hello World...");
```

With the `Write()` and `WriteLine()` methods, the `Console` class supports writing to the Console Window. The console output can also be cleared using the `Clear()` method. Furthermore, the `Read()` and `ReadLine()` methods allow for user input to be captured. Using `Console.Beep()`, a beep can easily be sounded.

> **Note** Because the console window is integrated into the Smart Card Magic.NET interface, none of the other methods for the .NET Console class are supported. Examples are the Title() method, which allow the window title to be changed.

7.6.2.4 Compiling and Running

To compile and execute the script file, either use the commands in the *Script* menu, or click the 'Start' icon on the menu bar.

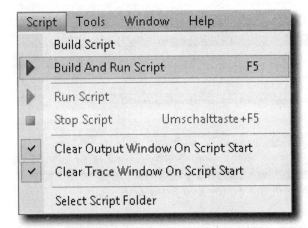

Figure 7.32.
Script File Explorer *after the first run.*

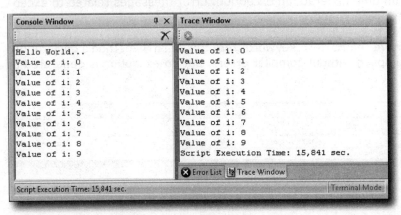

Figure 7.33.
"Hello World" example: output after program has run.

The two options, *Clear Output Window On Script Start* and *Clear Trace Window On Script Start* control whether scripts will begin on a 'clean slate'.

If there are bugs in the code, the compiler will bring up the Error List window. The Error List window will automatically appear in the foreground. The error messages will be displayed in the language/locale of the Windows operating system. In the Error List window, double-clicking on an error moves the cursor to the corresponding location in the source code. The following error message must be ignored: "GS.Elektor.Script.Main (MIFAREMagicNET.MainForm, string[]) has incorrect signature for this entry point." (or similar). If the semicolon is put in the right place on line 10, this error is resolved.

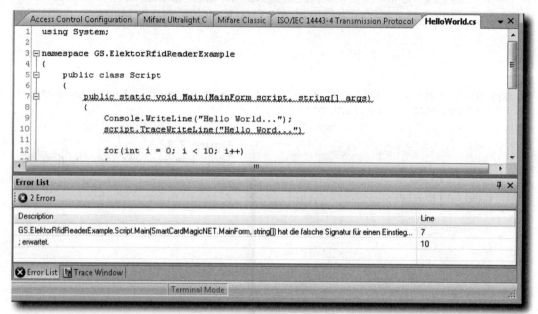

Figure 7.34. Highlighting the faulty code in the Editor and Error List.

Errors that occur at run time generate an exception. Error messages related to exceptions are output to the Trace Window.

For a simple test, remove the `public` keyword from the `Main()` method's declaration. The source code is compiled without compiler errors, but an exception is generated at runtime.

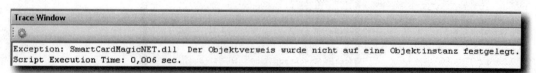

Figure 7.35. Error output in the Trace Window.

7.6.2.5 User Input from the Console Window

The `Read()` and `ReadLine()` methods allow for easy user input from the Console Window. The next example demonstrates the features of these two methods.

Listing 7.7. *ConsoleInput.cs.*

```csharp
using System;

namespace GS.ElektorRfidReaderExample
{
   public class Script
   {
      public static void Main(MainForm script, string[] args)
      {
```

`ReadLine()` expects the *Enter* key to be pressed and returns a `string`.

```csharp
   Console.Write("Your name please and press <Enter>: ");
         string name = Console.ReadLine();
         Console.WriteLine("Hello " + name);
```

The `Convert.ToInt32()` converts the input from a `string` data type to a 32-bit `int`.

```csharp
   Console.WriteLine("Now we add two numbers");
         Console.Write("Enter the first number and press <Enter>: ");
         int a = Convert.ToInt32( Console.ReadLine() );
         Console.Write("Enter the second number and press <Enter>: ");
         int b = Convert.ToInt32( Console.ReadLine() );
         Console. WriteLine("Result = {0}", a + b);
         Console.WriteLine("");
```

The `Read()` method allows a simple way for a user to make a selection.

```csharp
         Console.Write("Please enter 'a' or 'b' and press <Enter>:");
         char s = Convert.ToChar( Console.Read() );
         switch(s)
         {
            case 'a':
            case 'A':
               Console.WriteLine("Option A");
               break;

            case 'b':
            case 'B':
               Console.WriteLine("Option B");
               break;

            default:
               Console.WriteLine("Default");
               break;
         }
      }
   }
}
```

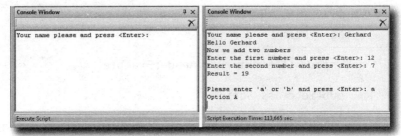

Figure 7.36.
User input with the Console Window.

> **Note** The `Console.Read()` method behaves differently to the one in the .NET Framework. This is because Smart Card Magic.NET's input and output is redirected to the Console Window. The `Console.Readline()` method, like the .NET variant, reads a character from the buffer. However, after reading, the entire buffer is emptied. For simple user input, that's an advantage as the input buffer need not be cleared manually.

7.6.2.6 Are There Really No Breakpoints?

Trying to create a breakpoint in the editor by double clicking on a line of code will be in vain. Developing a debug mode would not be an easy thing to do. In order to track down logical errors, there is a simple way to stop the program at a specific point in its execution. Instead of setting a breakpoint, a hard-coded breakpoint can be used. Calling the `Break()` method in the `MainForm` class causes the program to stop at this statement. In our actual example, this is done using `script.Break()`. Using the *F5* key, or clicking on the *Start* icon, allows the program to continue execution.

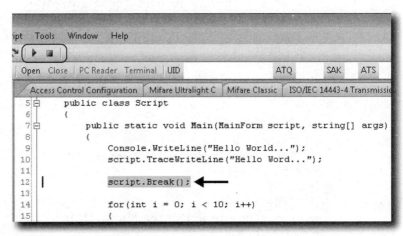

Figure 7.37.
Smart Card Magic.NET: Example of a hard-coded breakpoint.

> **Note** Visual Studio supports the same functionality using the `Debugger.Break()` method (`System.Diagnostics` namespace).

7.6.3 Visual C# 2012 Express Edition

This section shows the basic steps in using Visual C# 2012 Express Edition to create a console application. If you have never worked with Visual Studio, visit the Microsoft Web site for more information.

Visual C# 2012 Express Edition is downloadable for free from the Microsoft website at: *http://www.microsoft.com/visualstudio/eng/products/visual-studio-express-products*

7.6.3.1 Creating a Simple Console Application

In the following example, the "Hello World" C# script from Section 7.6.2 is used again for a console application. The necessary steps for porting a Smart Card Magic.NET C# script to a C# console application are universally applicable.

The *File* → *New Project* menu opens the *New Project* dialog. Choose *Console Application* from the available templates and enter the project name, "HelloWorld".

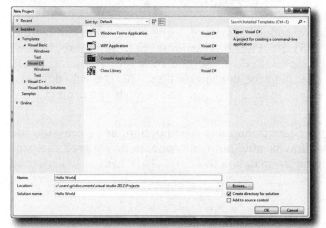

Figure 7.38.
Visual C# 2012 Express Edition: New Project dialog.

Clicking the *OK* button causes Visual Studio to automatically create a new Framework application. This already includes the correct method declaration for the `Main()` method.

> **Note** To open the sample projects from this book, you can use the menu selection *File* → *Open* Project.

Copy the source code from the `HelloWorld.cs` script's `Main()` method to the new `Main()` method. If you subsequently try to compile the project by using *BUILD* → *Build Solution* or by pressing the *F7* key, you will get three error messages. Due to the changed method declaration, the `script` variable is no longer declared. The Output Window in Visual Studio is used to log an application during execution. The Reader library logs the entire reader-card communication with the help of the .NET `Trace` class. This information is included automatically in Visual Studio's Output Window. This information is only available in the Debug mode.

7 THE ELEKTOR RFID READER

Figure 7.39.
Visual C# 2012 Express Edition:
The uncorrected source code.

In the first step, you can delete the hard-coded breakpoint, `script.Break()` completely. The `Trace` class is included with `using System.Diagnostics;`. Simply replace all `script.TraceWrite()` and `script.TraceLine()` method calls with `Trace.Write()` and `Trace.WriteLine()`, without changing the parameter lists. This is easily done using *Edit* → *Search and Replace* → *Replace in File*. Then you can compile the project without errors.

The project is run using the *Debug* → *Start Debugging* menu option, or by pressing the *F5* key. You will see the Console Window briefly, but it disappears as soon as the program ends. This problem can be solved by including the `Console.ReadLine()` method at the end of your program.

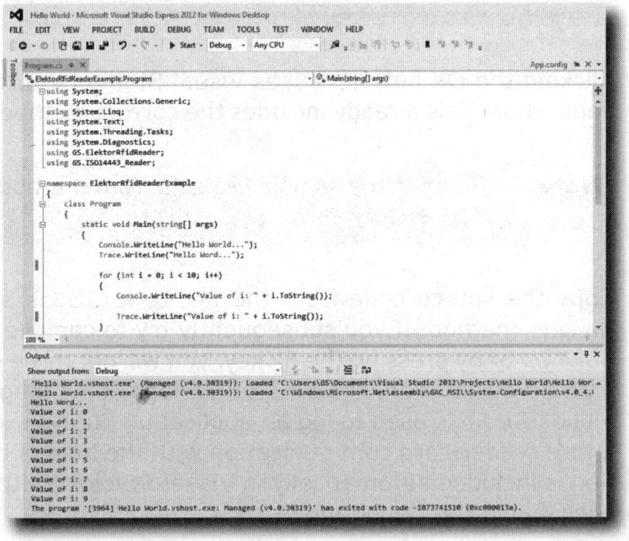

Figure 7.40.
Visual C# 2012 Express Edition:
the "Hello World" example.

The data generated by the `Trace.WriteLine()` method in the program are visible in the Output window. The Output window is opened using the menu option *View → Output Window*. Select *Debug* from the *Show Output From:* menu option to view the logged data.

If you run the project using the menu command *Debug → Run Without Debugging*, then no log will be generated.

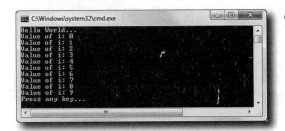

Figure 7.41.
"Hello World" application console output.

7.6.3.2 Integrating the Elektor RFID Reader Library

In this example, the source code in Listing 7.5 shows how the Elektor RFID Reader libraries can be included in a Visual C# console application. The function of the code is explained in detail in Section 8.1.4. Create a new project and add the source code in Listing 7.5 to that generated by Visual Studio. Do not change your class's namespace. This is the same as your project name.

The Elektor RFID Reader libraries consist of several DLL files. Download these from the author's website *(www.smartcard-magic.net)*. The source code for the Reader library is also available for download, but is not required for the development of a reader application.

The following statements include the Reader library's namespace:

```
using GS.ISO14443_Reader;
using GS.ElektorRfidReader;
```

If you compile the project, you get two errors, because the references to the DLLs are missing.

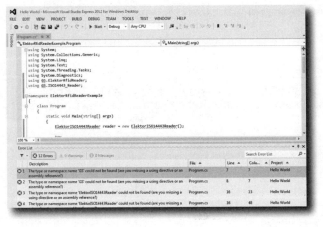

Figure 7.42.
The references to the Elektor RFID Reader libraries are missing.

7 THE ELEKTOR RFID READER

The Solution Explorer window allows you to easily add the missing library references. Right click on *References* then click on *Add References*. In this tab, you need to click *Search* and navigate to the library directory. Select all DLLs using the *Ctrl+A* key combination, then close the dialog by clicking on *OK*. Once the links are set up correctly, you can compile the project and run it without errors.

Figure 7.43.
In the Solution Explorer, *add a reference to the Elektor RFID Reader libraries.*

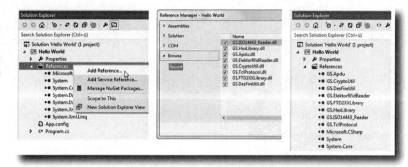

Figure 7.44.
Example: Reading a card UID with a console application.

Figure 7.45.
The library's log information is written to the Output window.

> **Note** The Smart Card Magic.NET development environment has all of the references to the Elektor RFID Reader libraries already established.

226

8
Cards and Tags in Application

Gerhard H. Schalk

In this chapter you will learn the basic techniques of programming different MIFARE and ISO/IEC 14443 Type A cards. The following topics and types of cards are covered in this chapter:

- ISO/IEC 14443-3 Type A card activation
- Card type identification
- MIFARE Ultralight card
- MIFARE Classic card
- MIFARE Ultralight C card Triple-DES authentication
- ISO/IEC 14443-4 transmission protocol (T=CL)
- MIFARE DESFire EV1 card

With the exception of the MIFARE DESFire EV1, you will become acquainted with all card commands with the help of the Elektor RFID Reader. The DESFire card's documentation is not publicly available; before downloading the documentation, a non-disclosure agreement has to be signed and sent to NXP. Therefore, it is not possible from a legal standpoint to handle all the details of DESFire card programming. Besides that, the card's extensive instruction set is beyond the scope of this book. For these reasons, we restrict ourselves to the DESFire file system, excluding encryption functions.

Readers without an Elektor RFID Reader will still benefit from the examples. The principles and programs presented are largely transferable to a regular PC/SC reader (see Section 11.4.3). Both the Elektor Library and the CsharpPCSC Library implement the IMifraeUltraLuight as well as the IMaifareClassic interfaces. The generic card utility classes presented in this chapter are reader independent.

Note	The description of the extensive Reader libraries is done in the relevant sections. The documentation of the important methods and properties can be found directly in the program examples and case studies.

Note	To improve readability, the terms 'reader' and 'card' are used instead in this chapter instead of the standard ISO/IEC 14443 terms, PCD (Proximity Coupling Device) and PICC (Proximity Integrated Circuit Card). Although this standard was originally defined for smartcards, other form factors are common.

8 CARDS AND TAGS IN APPLICATION

8.1 ISO/IEC 14443 Type A Card Activation

The card activation sequence is identical for all MIFARE family and ISO/IEC 14443 Type A-compatible products, so it can be universally implemented in software. Reader software should take the following requirements into consideration:

- there may be several cards in the reader field at one time;
- card activation should work regardless of card application;
- the number of bytes in the card's serial number (UID) is product-dependent.

After activation, it must be determined whether or not the card supports the ISO/IEC 14443 Part 4 standard T=CL transmission protocol.

In practice, an increasing number of reader systems are being designed that don't meet the first two requirements. This is due to the fact that the anti-collision is only possible with proper hardware support (reader IC). Systems with reader ICs either partially or completely omit the implementation of the anti-collision algorithm. In principle, these systems work, but they can't be extended later on. For example, sometimes only cards with 4-byte UIDs are supported, meaning that 7-byte UID cards can't be activated. Another common mistake is the strict evaluation of card responses during activation. Readers that only accept cards with specific ATQA and SAK values are not uncommon. In these cases, problems only arise when these systems need to be expanded. For installations with a large number of readers, a firmware update is only possible at high cost and then often with great logistical complexity.

8.1.1 Card Types from the Perspective of Card Activation

Smart Card Magic.NET's toolbar offers a very simple way to activate a card. After opening the USB port (*Open* button), you need only click the *Activate* button to activate the card currently in the reader's field.

Considering only card activation, we have the following distinguishing features:

- Length of the card serial number (4 bytes, 7 bytes and 10 bytes)
- UID (Unique ID) or RID (Random ID)
- Standardized (T=CL) and/or proprietary transmission protocol (MIFARE)

Figure 8.1 shows all possible card activation possibilities currently available. A larger number of bytes in the UID means that the activation time increases slightly due to an additional command sequence. Therefore, no cards with 10-byte UIDs are manufactured. The reader software (firmware) should be prepared for future card types and should therefore support all three UID lengths defined in the ISO/IEC 14443 standard.

The observant reader will have noticed that, in Figure 8.1, (Cards 4 and 5), 8 UID bytes are shown. The additional byte always has the same value (0x88) and indicates that there are twice as many UID bytes. Of these 8 bytes, only 7 are unique). So, although we speak of a 7-byte UID (double size), a total of 8 bytes is transferred between card and reader. This additional byte, known as the Cascade Tag (CT) in the ISO/IEC 14443 standard, should not be used by the reader software to detect a double- or triple-size UID, however.

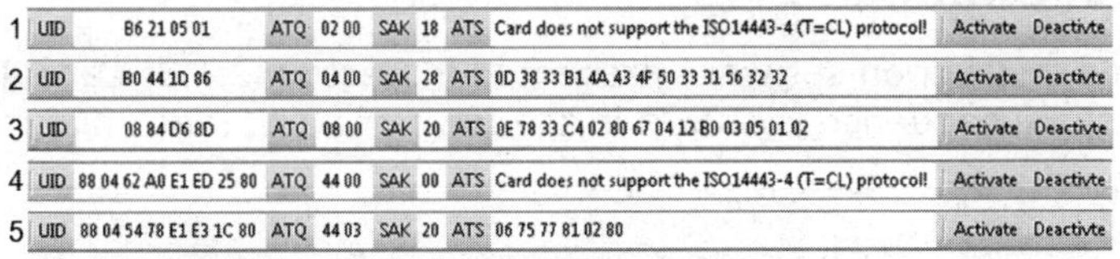

ISO/IEC 14443 TYPE A CARD ACTIVATION 8.1

1	UID	B6 21 05 01	ATQ 02 00	SAK 18	ATS	Card does not support the ISO14443-4 (T=CL) protocol!	Activate	Deactivte
2	UID	B0 44 1D 86	ATQ 04 00	SAK 28	ATS	0D 38 33 B1 4A 43 4F 50 33 31 56 32 32	Activate	Deactivte
3	UID	08 84 D6 8D	ATQ 08 00	SAK 20	ATS	0E 78 33 C4 02 80 67 04 12 B0 03 05 01 02	Activate	Deactivte
4	UID	88 04 62 A0 E1 ED 25 80	ATQ 44 00	SAK 00	ATS	Card does not support the ISO14443-4 (T=CL) protocol!	Activate	Deactivte
5	UID	88 04 54 78 E1 E3 1C 80	ATQ 44 03	SAK 20	ATS	06 75 77 81 02 80	Activate	Deactivte

Card 1 (MIFARE 4K): Single-Size UID (4-Byte); MIFARE protocol.
Card 2 (Smart Card Controller): Single-Size UID (4-Byte); T=CL & MIFARE protocol.
Card 3 (Smart Card Controller): Single-Size Random-ID (4-Byte); T=CL protocol.
Card 4 (MIFARE Ultralight): Double-Size UID (7-Byte); MIFARE protocol.
Card 4 (MIFARE DESFire EV1): Double-Size UID (7-Byte); T=CL protocol.

Figure 8.1. *Smart Card Magic. NET allows card activation via the toolbar.*

Table 8.1. *Single-size UID (4 bytes).*

Byte	UID0	UID1	UID2	UID3
Value	uu	uu	uu	uu

Table 8.2. *Single-size RID (4 bytes).*

Byte	Tag	RID0	RID1	RID2
Value	0x08	rr	rr	rr

Table 8.3. *Double-size UID (7 bytes).*

Byte	CT	UID0	UID1	UID2	UID3	UID4	UID5	UID6
Value	0x88	MID	uu	uu	uu	uu	uu	uu

Table 8.4. *Triple-size UID (10 bytes).*

Byte	CT	UID0	UID1	UID2	CT	UID3	UID4	UID5	UID6	UID7	UID8	UID9
Value	0x88	MID	uu	uu	uu	0x88	uu	uu	uu	uu	uu	uu	uu

u   –   unique number
r   –   random number
MID –   Manufacturer ID as per ISO/IEC 7816-6/AM1 (0x04 = NXP Semiconductors)
CT   –   Cascade tag as per ISO/IEC 14443-3

With a randomly-generated card serial number, or a so-called Random ID (RID), the first byte's value is always 0x08—this value indicates that it is an RID.

During an Anti-collision or Select command, 4 bytes are always transferred. For cards with a double- or triple-size UID, this command sequence is issued two or three times in a row, or cascaded. The value of the command byte is identical for each Anti-collision and Select, although it differs by cascade level (Level 1 = 0x93; Level 2 = 0x95; Level 3 = 0x97).

8 CARDS AND TAGS IN APPLICATION

8.1.2 The Activation Sequence

The card state diagram and instruction set were presented in Section 2.1.3, *Card Activation*. In Figure 8.2, the same sequence is shown in the form of a sequence diagram.

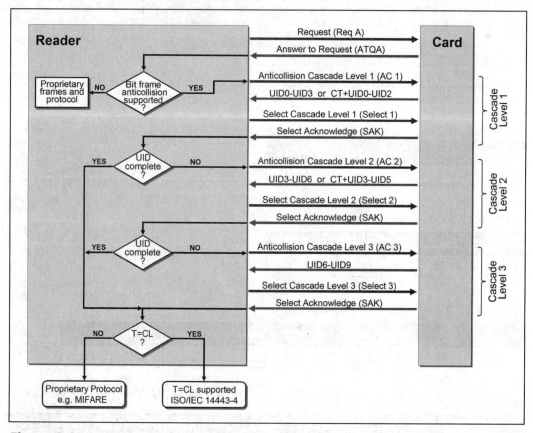

Figure 8.2. *Card activation sequence diagram.*

In the next step we look at the entire process, without addressing the details of the anti-collision algorithm, which were already presented in Section 2.1.3.4.

8.1.2.1 The Request and Wake-Up Commands

Most readers turn the RF signal off for a few milliseconds and then back on again before sending a Request (REQA) or Wake-Up (WUPA) command. This ensures that all cards in the reader field have undergone a power-on reset and are in the IDLE state. An RF reset should always be carried out if the reader has previously sent an REQB, or a request for a response from a Type B card. According to the ISO/IEC 14443 standard, a reader must wait at least 5 ms after turning on the RF signal before sending an REQA or WUPA command. Either of these are sent by the reader to determine whether there is at least one card in the field. Active cards respond with a two-byte block, the Answer To Request

ISO/IEC 14443 TYPE A CARD ACTIVATION 8.1

(ATQA). The ATQA value is not clearly defined in ISO/IEC 14443, and differs from card type to card type. A common software implementation is to poll using the sequence REQA / RF reset or WUPA / RF reset / REQB until a card responds. An RF reset prior to a REQA or WUPA is not mandatory.

Should several cards be in the reader field, they all respond synchronously with an ATQA. If the ATQA values are not identical, there is a collision. A reader always receives the superposition of the card responses. The ATQA value received by the reader will be equal to a bit-wise ORing of all of the cards' responses. To avoid erroneous interpretation, the reader software should not evaluate the ATQA bytes.

Table 8.5. Answer To Request – ATQA (LSB)

Bit	7	6	5	4	3	2	1	0
Value	UID size		r.f.u.	BitFrame anti-collision				

If only one of the bits between 1 and 4 are set to '1', the card supports the BitFrame anti-collision algorithm.
Interpretation of Bits 7 and 6: 00 = single-size UID; 01 = double-size UID, 10 = triple-size UID; 11 = r.f.u.

Table 8.6. Answer to Request – ATQA (MSB)

Bit	15	14	13	12	11	10	9	8
Value	r.f.u.				Proprietary encoding			

Bits 8 – 11 are used to encode proprietary manufacturer information.

The behavior described here can easily be verified using the Elektor RFID Reader and Smart Card Magic.NET. By simply clicking on one of the buttons, *RF Reset*, *Req A*, *Wup A*, *AC 1*, *Select 1*, *AC 2*, *Select 2* or *Halt A*, the respective command is sent to the card. The *Act. Req A* and *Act. Wup A* buttons carry out the entire activation, using either REQA or WUPA, respectively.

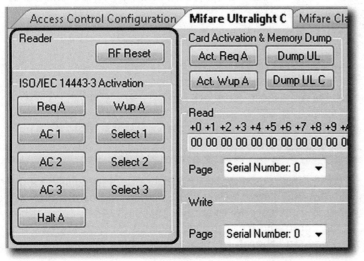

Figure 8.3.
Smart Card Magic.NET: Interface buttons for card activation.

8 CARDS AND TAGS IN APPLICATION

The reader commands (right arrow) and card responses (left arrow) can be seen in the *Trace* window. The card always sends the least-significant ATQA byte first. The byte sequence representation in the *Trace* window corresponds to the sequence in which the bytes were received (see Figure 8.4).

Figure 8.4. *If two cards respond to a REQA command, the reader receives a bit-wise OR of both cards' ATQA values.*

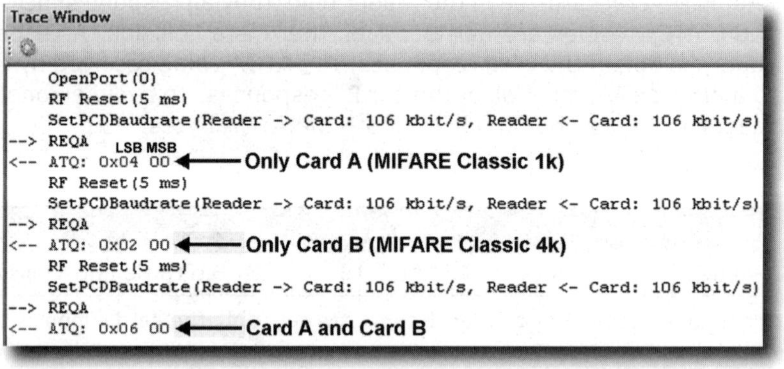

```
Trace Window
    OpenPort(0)
    RF Reset(5 ms)
    SetPCDBaudrate(Reader -> Card: 106 kbit/s, Reader <- Card: 106 kbit/s)
--> REQA         LSB MSB
<-- ATQ: 0x04 00  ◄───── Only Card A (MIFARE Classic 1k)
    RF Reset(5 ms)
    SetPCDBaudrate(Reader -> Card: 106 kbit/s, Reader <- Card: 106 kbit/s)
--> REQA
<-- ATQ: 0x02 00  ◄───── Only Card B (MIFARE Classic 4k)
    RF Reset(5 ms)
    SetPCDBaudrate(Reader -> Card: 106 kbit/s, Reader <- Card: 106 kbit/s)
--> REQA
<-- ATQ: 0x06 00  ◄───── Card A and Card B
```

Should an RF reset not precede a REQA or WUPA command, the time between two successive REQA or WUPA commands must be at least 7,000/fc, or ≈ 516 μs long. In this case, a card responds only to every second REQA or WUPA command (see Figure 8.5).

This behavior can be explained using the card state diagram. If the card is in the IDLE state, it responds to a REQA or WUPA with an ATQA and switches into the READY state. In the READY state, the card accepts only an Anti-collision command, and any REQA or WUPA command will be interpreted as an error. As in any error situation, the card switches back to the IDLE state without sending any response. This error handling is identical for all states within the card state diagram. The side effect of this is that the card has to be newly activated after every error within the reader-card communication.

Figure 8.5. *An ISO/IEC 14443 card responds only to every second REQA or WUPA command.*

```
Trace Window
    RF Reset(5 ms)
    SetPCDBaudrate(Reader -> Card: 106 kbit/s, Reader <- Card: 106 kbit/s)
--> REQA
<-- ATQ: 0x04 00
    SetPCDBaudrate(Reader -> Card: 106 kbit/s, Reader <- Card: 106 kbit/s)
--> REQA
<-- No Card Response
    SetPCDBaudrate(Reader -> Card: 106 kbit/s, Reader <- Card: 106 kbit/s)
--> REQA
<-- ATQ: 0x04 00
    SetPCDBaudrate(Reader -> Card: 106 kbit/s, Reader <- Card: 106 kbit/s)
--> REQA
<-- No Card Response
```

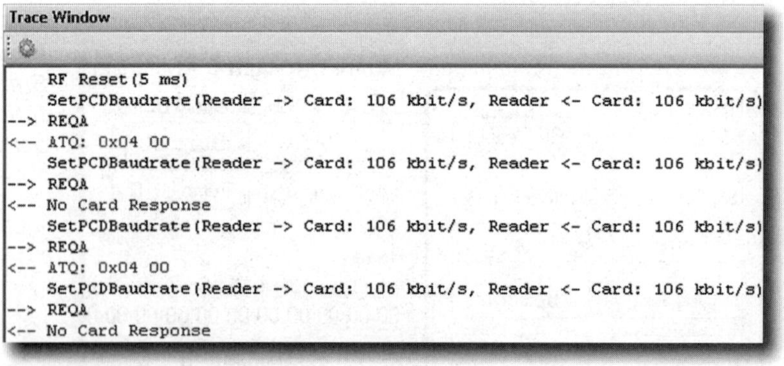

Not all ISO/IEC 14443 Type A cards and tags support the BitFrame anti-collision algorithm. Products from the Topaz company *(http://www.innovision-group.com/topaz)* are one such example. Strictly speaking, these products are not ISO/IEC 14443 compatible, as the IC supports only the REQA command from the standard. However, the NFC Forum has certified these tags anyway, and therefore most PC/SC readers support them.

ISO/IEC 14443 TYPE A CARD ACTIVATION 8.1

Basically, the tags can be read and written with the Elektor RFID Reader hardware, but, because the documentation of the proprietary commands is not publicly available, the Elektor Reader Firmware does not support them. The documentation can be obtained after signing a non-disclosure agreement with the manufacturer.

In the Topaz products' ATQA, all bits from Bit 0 to Bit 4 are set to '0' (see Figure 8.6). From this information, a reader application can see that the tag does not support the anti-collision algorithm. The card state diagram as per ISO/IEC 14443 Type A is also not supported by this tag. Therefore, the IC responds to each REQA command with an ATQA.

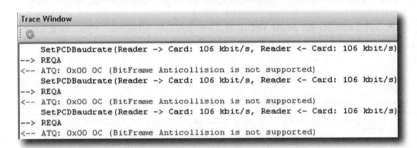

Figure 8.6.
In contrast to ISO/IEC 14443-compatible cards, the Topaz tag responds to every single REQA command.

8.1.2.2 The Anti-collision and Select Commands

When the reader receives an ATQA, it begins the anti-collision algorithm by sending the Anti-collision Cascade Level 1 (*AC 1* button). This is done completely by the firmware. Should several cards be present, the collision is resolved by the firmware. The result of this function is a 4-byte UID. The reader then sends the Select Cascade Level 1 (*Select 1* button) command and the card acknowledges with a SAK (Select AcKnowledge) byte.

As already mentioned, only 4 bytes are transferred for Anti-collision and Select commands. Whether the reader has already received the complete UID or not is encoded in Bit 2 of the SAK byte. Should the UID not be completely received, the reader sends the command sequence Anti-collision Cascade Level 2 (*AC 2* button) and Select Cascade Level 2 (*Select 2* button). The algorithm is identical to Cascade Level 1. After receiving the second SAK byte, the reader checks again whether the complete UID has been received. In this way, even triple-size UID cards can be correctly activated. Once the last SAK byte is sent, the card switches to the READY state and is fully activated.

Table 8.7. Select AcKnowledge (SAK)

Bit	7	6	5	4	3	2	1	0
Value	x	x	2.)	x	x	1.)	x	x

```
     x   =  any value (don't care)
     0   =  UID complete
[1]  1   =  UID not yet complete
[2]  Bit 5's status is only valid if Bit 2 is '0'.
     0   =  The card does not support the ISO/IEC 14443-4 (T=CL) protocol.
     1   =  The card supports the ISO/IEC 14443-4 (T=CL) protocol.
```

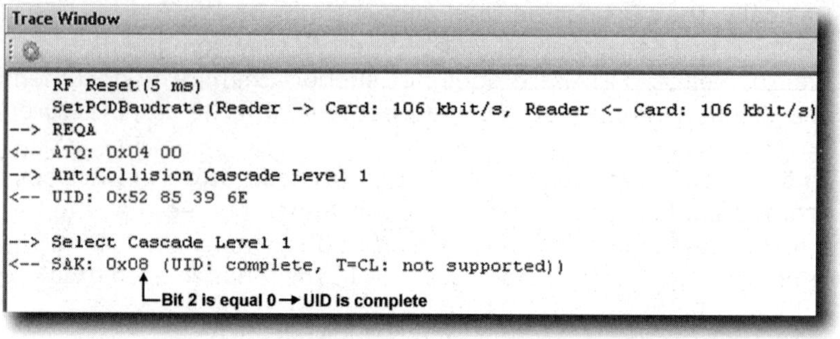

8 CARDS AND TAGS IN APPLICATION

Figure 8.7.
MIFARE 1K card activation (single-size UID).

```
Trace Window

    RF Reset(5 ms)
    SetPCDBaudrate(Reader -> Card: 106 kbit/s, Reader <- Card: 106 kbit/s)
--> REQA
<-- ATQ: 0x04 00
--> AntiCollision Cascade Level 1
<-- UID: 0x52 85 39 6E

--> Select Cascade Level 1
<-- SAK: 0x08 (UID: complete, T=CL: not supported))
              ↑
              └─ Bit 2 is equal 0 → UID is complete
```

Additionally, the last SAK byte received (see Table 8.7) encodes whether the card supports the T=CL (ISO/IEC 14443-4) protocol or not.

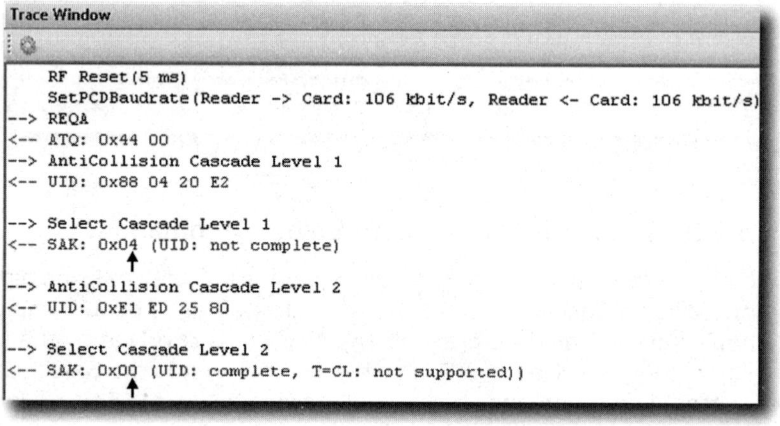

Figure 8.8.
MIFARE Ultralight card activation (double-size UID).

```
Trace Window

    RF Reset(5 ms)
    SetPCDBaudrate(Reader -> Card: 106 kbit/s, Reader <- Card: 106 kbit/s)
--> REQA
<-- ATQ: 0x44 00
--> AntiCollision Cascade Level 1
<-- UID: 0x88 04 20 E2

--> Select Cascade Level 1
<-- SAK: 0x04 (UID: not complete)
          ↑
--> AntiCollision Cascade Level 2
<-- UID: 0xE1 ED 25 80

--> Select Cascade Level 2
<-- SAK: 0x00 (UID: complete, T=CL: not supported))
          ↑
```

Whether the card supports the MIFARE protocol can be partially determined from the SAK byte. In Section 8.2, a script for identifying various card types is presented.

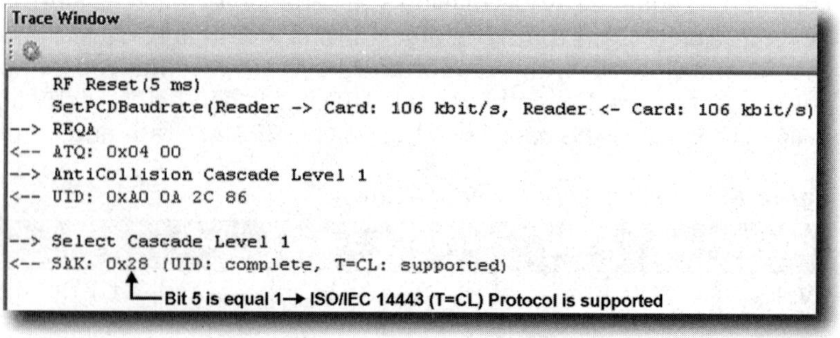

Figure 8.9.
Activating a Java card (single-size UID) that supports the T=CL protocol.

```
Trace Window

    RF Reset(5 ms)
    SetPCDBaudrate(Reader -> Card: 106 kbit/s, Reader <- Card: 106 kbit/s)
--> REQA
<-- ATQ: 0x04 00
--> AntiCollision Cascade Level 1
<-- UID: 0xA0 0A 2C 86

--> Select Cascade Level 1
<-- SAK: 0x28 (UID: complete, T=CL: supported)
              ↑
              └─ Bit 5 is equal 1 → ISO/IEC 14443 (T=CL) Protocol is supported
```

8.1.2.3 The HALT Command

In real-world applications, once the card is activated, the EEPROM memory is read and written. The reader puts the card in the HALT state using the HALT command so that the reader application doesn't have to wait until the user removes the card from the reader's field. In the HALT state, the card will only respond to a WUPA command. The reader may now look for another card in the field with a REQA command. To do this, care must be taken that the reader does not carry out an RF reset prior to sending the REQA. An RF reset causes all cards in the field to perform a power-on reset and thus automatically enter the IDLE state, where they will respond to a REQA.

Figure 8.10 shows the sequential activation of two cards in the reader's field. After an RF reset and a REQA, both cards respond with an ATQA. This triggers a collision. This is known because the ATQA values of both cards were calculated previously. Card A has an ATQA value of 0x02, while Card B's value is 0x04.

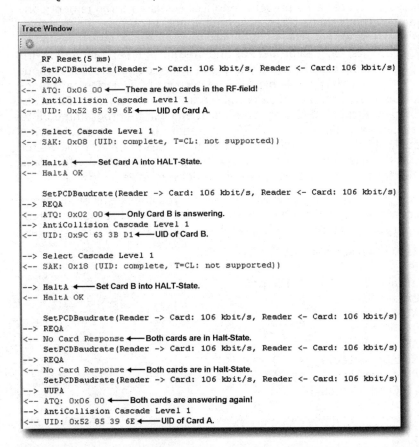

Figure 8.10. Sequential activation of two cards.

The collision is resolved using the Anti-collision Cascade Level 1 command. In our example, Card A is the 'winner' and Card B is the 'loser'. Subsequently, Card A is selected with the Select Cascade Level 1 command. Card A is now in the READY state, which is the state from which manipulation of the card memory normally takes place. In this case, however, a HALT A command is sent to the card.

Now, only Card B will respond to the next REQA command. In this example, an ATQA value of 0x04 indicates that no further collisions have taken place. After subsequent full activation, Card B is also placed in the HALT state.

8 CARDS AND TAGS IN APPLICATION

Since both cards are now in the HALT state, neither will respond to the next two REQA commands, but will respond to the WUPA command. At this point, it's again possible to first activate Card A and then Card B.

8.1.3 Elektor RFID Reader Library: Card Activation

In this section, you will learn all of the properties and methods (similar to C functions) of the reader library that are needed for card activation. Multiple interfaces declare the entire card functionality.

If you're not familiar with the concept of object orientation, you will certainly not be aware of the possibilities that interfaces present. An interface is an arrangement consisting of properties and methods that a class must implement. In our specific example, ElektorISO14443Reader implements the I_ISO14443_3_TypeA interface. In the I_ISO14443_3_TypeA interface, all properties and methods for card activation are declared, but not implemented. One may simply imagine interfaces as classes without fields (comparable with C variables) or code blocks. The purpose of declaring card functionality as an interface is that it can be implemented in another library (e.g. for other reader hardware). Complex Elektor RFID Reader applications can be programmed into an interface. This makes it very easy to use a different reader later on without having to revise the entire application source code. The use of interfaces in C# also allows for multiple inheritance. In the examples that follow, the ElektorISO14443Reader class is used directly in order for the examples to be easily understandable.

From the I_ISO14443_3_TypeA interface's declaration, you can see all of the properties and methods for card activation that the ElektorISO14443 class has implemented. The I_ISO14443_3_TypeA interface is declared in the GS.ISO14443_Reader.dll file.

```
namespace GS.ISO14443_Reader
{
    public interface I_ISO14443_3_TypeA
    {   // Properties
        byte[] ATQA { get; }
        byte[] UID { get; }
        byte? SAK { get; }
        bool ACNotSupported { get; }
        bool UidNotComplete { get; }
        bool IsTCLCard { get; }

        // Methods
        int RFReset(byte ms);
        int REQA();
        int WUPA();
        int Request(RequestCmd reqCmd);
        int AntiCollision(Cascade cascadeLevel);
        int Select(Cascade cascadeLevel);
        int HaltA();
        int ActivateCard(RequestCmd reqCmd);
    }
}
```

ISO/IEC 14443 TYPE A CARD ACTIVATION 8.1

The related enumerations are also declared in GS.ISO14443 _ Reader.dll.

```
namespace GS.ISO14443_Reader
{
   public enum RequestCmd : byte
   {
      REQA = 0x26,
      WUPA = 0x52
   }
   public enum Cascade : byte
   {
      Level_1 = 0x93,
      Level_2 = 0x95,
      Level_3 = 0x97
   }
}
```

For non-C# programmers, it's necessary to explain the properties' syntax. Properties are used to access a class's private fields. In other programming languages, it's necessary to implement special access methods, or 'getter and setter' methods. For example, the GetValue() method would serve to get the content of the private field, 'value'. It's also possible to implement getter and setter methods in C#, but properties offer a more elegant solution. These are nothing more than a syntactic shorthand for getter and setter methods. Properties also offer the same functionality. Examples include validation of data and calculation of the return value. Properties are used as public fields of a class, which greatly simplifies the syntax. Table 8.8 compares the two options.

Table 8.8. Comparison between getter and setter methods and properties.

	Getter and Setter Methods	**Property**
Read access	int myVar = obj.GetValue();	int myVar = obj.Value;
Write access	obj.SetValue(myVar);	obj.Value = myVar;

All properties of the I _ ISO14443 _ 3 _ TypeA interface are read-only, as the 'set' part is missing from the declaration. It would not make any sense to modify the ATQA, UID or SAK values in the program.

If the card supports BitFrame anti-collision, then only bits 0 to 4 in the ATQA need be set. The following source code is required for the test:

```
// Bit frame anticollision supported?
if (0 == (reader.ATQA[0] & 0x1F))
```

The ACNotSupported property simplifies this assessment and increases the code's readability.

```
// Bit frame anticollision supported?
if (reader.ACNotSupported)
```

8 CARDS AND TAGS IN APPLICATION

SAK byte, Bit 2 (see Table 8.7), indicates whether the reader has received the complete UID or not.

```
/ UID not complete?
if (0x04 == (reader.SAK & 0x04))
```

In this case, the `UidNotComplete` property could also be used.

```
/ UID not complete?
if (reader.UidNotComplete)
```

If Bit 5 of the SAK byte is a '1', this indicates that the card supports the T=CL protocol.

```
// ISO/IEC 14443-4 (T=CL) compatible card?
if (0x20 == (reader.SAK & 0x20))
```

The `IsTCLCard` property returns the Boolean value true for T=CL cards.

```
// UID not complete?
if (reader.IsTCLCard)
```

The following two tables present an overview of the `I_ISO14443_3_TypeA` interface's properties and methods.

Table 8.9. *I_ISO14443_3_TypeA interface properties.*

Property	R/W	Description
byte ATQA	R	Prerequisite: call to the `REQA()`, `WUPA()` or `ActivateCard()` method. Returns either the ATQA block as a byte array, or *null*.
byte[] UID	R	Prerequisite: call to the `AntiCollision()` method. Returns either the card serial number (UID) as a byte array, or *null*.
byte? SAK	R	Prerequisite: call to the `Select()` method. Returns either the value of the SAK byte, or *null*.
Bool ACNotSupported	R	Prerequisite: call to the `Request()` method. Returns true if the card does not support the BitFrame anti-collision algorithm per ISO/IEC 14443 Part 3.
Bool UidNotComplete	R	Prerequisite: call to the `Select()` method. Returns true if the UID has not yet been completely received.
bool IsTCLCard	R	Prerequisite: call to the `Select()` method. Returns true if the card supports the ISO/IEC 14443-4 (T=CL) protocol.

ISO/IEC 14443 TYPE A CARD ACTIVATION 8.1

Table 8.10. I_ISO14443_3_TypeA interface methods.

Method	Description
`int RFReset(byte ms)`	Turns the RF field off for a specified time (0 – 255 ms) and then back on again.
`public int ActivateCard(RequestCmd reqCmd)`	Performs the entire card activation sequence. `reqCmd` parameter: `RequestCmd.REQA` or `RequestCmd.WUPA`.
`int REQA()`	Sends a REQA command to the card.
`int Request(RequestCmd reqCmd)`	Sends either a REQA or a WUPA command. `reqCmd` parameter: `RequestCmd.REQA` or `RequestCmd.WUPA`.
`int WUPA()`	Sends a WUPA command to the card.
`int AntiCollision(Cascade cascadeLevel)`	Sends an Anti-collision Cascade Level 1, 2 or 3 command to the card. `cascadeLevel` parameter: `Cascade.Level_1`, `Cascade.Level_2` or `Cascade.Level_3`.
`int Select(Cascade cascadeLevel)`	Sends a Select Cascade Level 1, 2 or 3 command to the card. `cascadeLevel` parameter: `Cascade.Level_1`, `Cascade.Level_2` or `Cascade.Level_3`.
`int HaltA()`	Sends a HALTA command to the card.

All reader library methods that either activate the card or manipulate the card's memory have a return value. With this return value, one can evaluate the status of the reader firmware in a C# program. After executing the REQA method, the return value allows one to check if the card response was received or not. As already mentioned, these methods are all declared as interfaces. Thus, it is possible to use other reader hardware with the same software interface. Since the error codes are different from hardware to hardware, the reader methods' return values are defined using the integer data type. A user-defined data type for the error codes would be easier to use, but this increases porting complexity. In this way, only a single assumption is made regarding the error code. Every successfully executed method must return a value of 0. The use of a numerical value of the error code is not a great limitation, as it is usually only necessary to confirm whether the card has sent a valid response or not.

```
int ret = reader.REQA();
if (ret != 0)
{
    return;
}
```

In Table 8.11, the values of all of the other error codes are shown. All error codes are also summarized as constants in the enumeration type `ReaderStatus`. If software is being developed exclusively for the Elektor RFID Reader, then the `ReaderStatus` data type may also be used. This will make the source code much easier to read.

```
ret = reader.REQA();
if ((ReaderStatus)ret != ReaderStatus.OK)
{
    return;
}
```

In this case, it is necessary to explicitly convert the return value into the `ReaderStatus` data type. Finally, the return value can be checked against a constant from the `ReaderStatus` enumeration.

Table 8.11. Constants in the ReaderStatus enumeration.

Constant	Value	Description
`OK`	0x00	OK: the Reader IC card has received a valid response.
`RC52xTimeOut`	0x01	Timeout: the reader has not received a card response.
`RC52xCRC`	0x02	The reader has detected a CRC error in the card's response.
`RC52xParity`	0x03	The reader IC has detected a parity error in the card's response.
`RC52xBitcount`	0x04	The reader IC did not receive the correct number of bits.
`RC52xBitCollision`	0x06	The reader IC has detected a collision.
`MIFAREAccessDenied`	0x08	A MIFARE 1K or 4K card has denied access to a block after successful authentication. Cause of error: possibly the block is write protected (read-only).
`RC52xBufferOverflow`	0x09	The receive FIFO buffer has overflowed.
`RC522xTempError`	0x0A	The reader IC's internal temperature sensor detects excessive temperature. In this case, the antenna drivers are automatically disabled.
`RC52xProtocol`	0x0B	The reader IC has received an incorrect card response.
`RC52xFIFOWrite`	0x0D	There is a FIFO write error.
`ParameterNotSupported`	0x11	Incorrect reader IC parameter.
`CommandNotSupported`	0x12	Incorrect reader IC command.
`InvalidFormat`	0x13	The reader IC has detected a format error in the card response.
`MifareAuthFailed`	0x21	MIFARE authentication for a MIFARE 1K or 4K card was not successful. Cause of error: incorrect MIFARE key.
`MifareACKSupposed`	0x22	The reader IC received a MIFARE ACK or NACK.
`IncorrectUIDCheckByte`	0x30	The reader IC has received a UID with an incorrect XOR checksum.
`Other Error`	0x40	Unexpected error. This message should never occur.

ISO/IEC 14443 TYPE A CARD ACTIVATION 8.1

> **Note** Table 8.11 shows only error codes for the reader firmware. Should another error occur during execution, the reader library will generate an exception. For example, the `OpenPort()` method throws an exception when no reader is connected to the PC. For this reason, all instructions are enclosed in a `try-catch` block.

8.1.4 Program Examples

8.1.4.1 Card Activation

In this example, all instructions are enclosed in a `try-catch` block. Should an error occur within the protected try block (i.e. between try and catch), execution of this block is interrupted and the instructions in the `catch` block are executed. In our case, the error handling is limited to outputting the cause of the error to the console. The `finally` block is executed regardless of whether an exception has occurred or not. So, the `Close-Port()` method within the `finally` block will always be called. This ensures that an open USB port is closed properly even in the event of an error. Should the exception occur during or immediately before opening the port, a call to the `ClosePort()` method will not generate another exception.

The program sequence in Listing 8.1 corresponds to the sequence diagram in Figure 8.2. In this example, the REQA command is sent only once by the reader. It's therefore necessary that the card is already at the reader before the program is started. Otherwise, the program is terminated immediately after the REQA. The program code can initially be tested without any additional output to the console, as the entire communication between reader and card is automatically output to the *Trace* Window.

Listing 8.1. *CardActivation1.cs.*

```
using System;
using System.Diagnostics;
using GS.ElektorRfidReader;
using GS.ISO14443_Reader;

namespace GS.ElektorRfidReaderExampleExamples
{
    class CardActivation
    {
        public static void Main(MainForm script, string[] args)
        {
```

The `ElektorISO14443Reader` object must be created outside of the `try` block, so that the `ClosePort()` method within the `finally` block can be called.

```
            ElektorISO14443Reader reader = new ElektorISO14443Reader();

            try
            {
```

8 CARDS AND TAGS IN APPLICATION

```
                reader.OpenPort();
                int ret = reader.RFReset(5);
                ret = reader.REQA();
                if ((ReaderStatus)ret != ReaderStatus.OK) return;
                if (reader.ACNotSupported) return;

                ret = reader.AntiCollision(Cascade.Level_1);
                ret = reader.Select(Cascade.Level_1);

                if (reader.UidNotComplete)
                {
                    ret = reader.AntiCollision(Cascade.Level_2);
                    ret = reader.Select(Cascade.Level_2);

                    if (reader.UidNotComplete)
                    {
                        ret = reader.AntiCollision(Cascade.Level_3);
                        ret = reader.Select(Cascade.Level_3);
                    }
                }
            }
            catch (Exception ex)
            {
               Console.WriteLine(ex.Message.ToString() + Environment.NewLine);
            }
            finally
            {
                reader.ClosePort();
            }
        }
    }
}
```

When the script is started with Smart Card Magic.NET, any reader port that was manually opened previously is automatically closed. Generally, it is also possible to use more than one Elektor RFID Reader on the same PC. All FTDI devices are automatically detected using the `OpenPort()` method. If multiple FTDI devices are connected to one PC, the `OpenPort()` method automatically opens a dialog box.

Figure 8.11.
The 'Select Elektor RFID Reader' dialog.

242

The *Select Elektor RFID Reader* dialog box allows the selection of the desired device. In our case it will be an Elektor RFID Reader. In Figure 8.11, the dialog box was opened because both an Elektor RFID Reader and an RS-232-to-USB converter with an FTDI chip were connected. An overloaded `OpenPort()` method allows a specific Elektor RFID Reader to be selected usingits FTDI serial number. In this way, repeated manual selection can be avoided. For this, one need only pass the FTDI serial number as a string to the `OpenPort()` method.

```
reader.OpenPort("DiSD0EN5");
```

```
Trace Window
    OpenPort(1)
    RF Reset(5 ms)
    SetPCDBaudrate(Reader -> Card: 106 kbit/s, Reader <- Card: 106 kbit/s)
--> WUPA
<-- ATQ: 0x44 00
--> AntiCollision Cascade Level 1
<-- UID: 0x88 04 20 E4

--> Select Cascade Level 1
<-- SAK: 0x04 (UID: not complete)

--> AntiCollision Cascade Level 2
<-- UID: 0xE1 ED 25 80

--> Select Cascade Level 2
<-- SAK: 0x00 (UID: complete, T=CL: not supported))
```

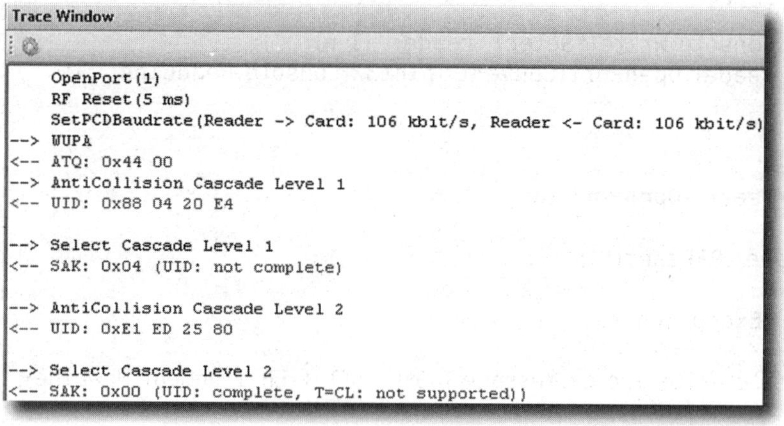

Figure 8.12.
Activating a MIFARE Ultralight card using the CardActivation1.cs script.

8.1.4.2 Reader Selection

Occasionally it is necessary to list all connected Elektor RFID Readers (or FTDI devices). The `GetReaders()` method returns an array of strings containing all active FTDI devices. For the `GetReaders()` method to work properly, all USB ports to which FTDI devices are connected must be closed.

The example in Listing 8.2 demonstrates the typical use of the `GetReaders()` method. If more than one FTDI device is detected, all devices are listed to the console. The user then has the ability to select the desired reader by inputting the device number. For this, an additional, overloaded, `OpenPort()` method is used. This allows the selection of a device using a port number (0 to *n*).

Listing 8.2. *SelectElektorRfidReader.cs.*

```csharp
using System;
using GS.ElektorRfidReader;

namespace GS.ElektorRfidReaderExample
{
    class ListAllElektorRfidReaders
    {
        public static void Main(MainForm script, string[] args)
        {
```

8 CARDS AND TAGS IN APPLICATION

```
            ElektorISO14443Reader reader = new ElektorISO14443Reader();
            try
            {
                string[] sReaderList = reader.GetReaders();
                if (null == szReaderList) return;
                if (sReaderList.Length > 1)
                {
                    for (int i = 0; i < sReaderList.Length; i++)
                    {
                     Console.WriteLine(i.ToString() + ": " + sReaderList[i]);
                    }
                    Console.Write("Select reader (1-n): ");
                    reader.OpenPort(Convert.ToInt32(Console.ReadLine()));
                }
                else
                {
                    reader.OpenPort(0);
                }
                reader.RFReset(5);
            }
            catch (Exception ex)
            {
               Console.WriteLine(ex.Message.ToString() + Environment.NewLine);
            }
            finally
            {
                reader.ClosePort();
            }
        }
    }
}
```

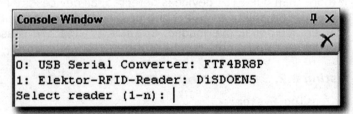

Figure 8.13.
Selecting the desired FTDI device.

8.1.4.3 Polling for Cards

In the next example, the reader searches for a card in the reader field by continuously polling using the REQA command. In contrast to the preceding example, the script waits until a card has been detected in the field. This requires only a small change to the source code from Listing 8.1.

Within a `do-while` loop, the REQA command will be repeatedly sent until the `REQA()` method returns the value `ReaderStatus.OK` (or '0'). Smart Card Magic.NET runs every script in its own thread. This has two major advantages: firstly, the user interface is not blocked by the polling loop, but is fully operational. Secondly, it's possible to terminate the script at any time by manually using the Script → Stop Script menu option. So you may, as with microcontroller programs, use an endless loop.

Once a card responds to a REQA command, it is fully activated. After that, the card parameters, that is, the ATQA, the UID and the SAK, are output to the console.

Listing 8.3. *CardActivation2.cs.*

```csharp
using System;
using GS.Util.Hex;
using GS.ElektorRfidReader;
using GS.ISO14443_Reader;

namespace GS.ElektorRfidReaderExample
{
    class CardActivation
    {
        public static void Main(MainForm script, string[] args)
        {
            ElektorISO14443Reader reader = new ElektorISO14443Reader();
            try
            {
                int ret;
                reader.OpenPort();
                Console.WriteLine("Waiting for new card...");
                do
                {
                    ret = reader.REQA();
                }
                while((ReaderStatus)ret != ReaderStatus.OK);

                Console.Clear();
                Console.WriteLine("Card detected...");

                Console.WriteLine("ATQA = 0x{0:X02} {1:X02}", reader.ATQA[0],
                                                              reader.ATQA[1]);

                if (reader.ACNotSupported) return;

                ret = reader.AntiCollision(Cascade.Level_1);
                ret = reader.Select(Cascade.Level_1);

                if (reader.UidNotComplete)
                {
                    ret = reader.AntiCollision(Cascade.Level_2);
                    ret = reader.Select(Cascade.Level_2);

                    if (reader.UidNotComplete)
```

8 CARDS AND TAGS IN APPLICATION

```
            {
                ret = reader.AntiCollision(Cascade.Level_3);
                ret = reader.Select(Cascade.Level_3);
            }
        }
        Console.WriteLine("SAK = 0x{0:X02}", reader.SAK);
        Console.WriteLine("UID = 0x" +
                    HexFormatting.ToHexString(reader.UID, true));
        if (reader.IsTCLCard)
        {
            Console.WriteLine("Card is supporting the T=CL protocol!");
        }
        else
        {
            Console.WriteLine("Card does not support the T=CL protocol!");
```

If it's not possible to activate the card correctly, the values of the ATQA, UID and SAK properties will be null. Since the ATQA, UID and SAK properties are used without the correct checking for error conditions in Listing 8.3, should card activation fail, an exception will be generated. Figure 8.14.shows the the error message that pops up (German version). Therefore, in reader applications, it is always necessary to check the ATQA, UID and SAK properties for null.

```
if (reader.SAK != null) Console.WriteLine("SAK = 0x{0:X02}", reader.SAK);
```

By calling the `REQA()`, `WUPA()` or `Halt()` methods, the ATQA, UID and SAK properties are initialized to `null`.

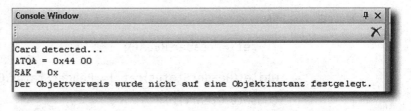

Figure 8.14. In this case, the card was passed very quickly over the reader. Therefore, the card was not fully activated or the parameters correctly read.

The UID property returns the card serial number as a byte array of the correct length. So, in the case of a single-size UID card, a 4-byte array, and for a double-size UID, an 8-byte array. If you wish to output the card's UID to the console, it's necessary to read out the individual byte values in a loop. The display output is usually in hexadecimal format.

```
Console.Write("UID = 0x");

for (int i = 0; i < reader.UID.Length; i++)
{
```

ISO/IEC 14443 TYPE A CARD ACTIVATION 8.1

```
        Console.Write("{0:X02} ", reader.UID[i]);
    }

    Console.WriteLine("");
```

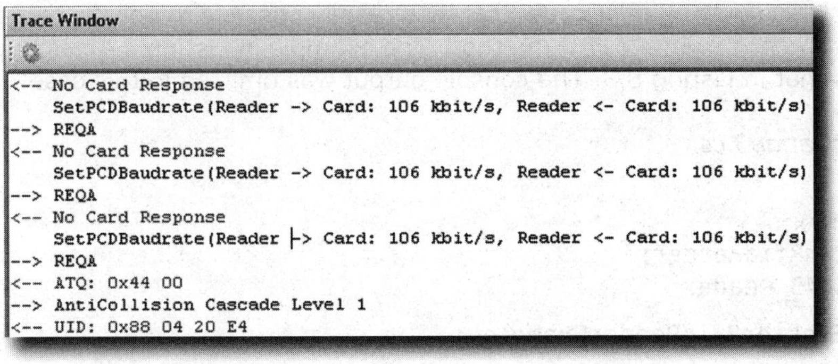

```
Trace Window
<-- No Card Response
    SetPCDBaudrate(Reader -> Card: 106 kbit/s, Reader <- Card: 106 kbit/s)
--> REQA
<-- No Card Response
    SetPCDBaudrate(Reader -> Card: 106 kbit/s, Reader <- Card: 106 kbit/s)
--> REQA
<-- No Card Response
    SetPCDBaudrate(Reader -> Card: 106 kbit/s, Reader <- Card: 106 kbit/s)
--> REQA
<-- ATQ: 0x44 00
--> AntiCollision Cascade Level 1
<-- UID: 0x88 04 20 E4
```

Figure 8.15.
Only when a card appears in the reader's field will it respond to a REQA command. Subsequently, the card is fully activated.

The output of byte arrays is one of the standard everyday programming tasks. However, the .NET Framework offers very limited support for this. For this reason, the author developed an extensive library for processing and converting hexadecimal values. This class library is encapsulated in `GS.FTD2XXLibrary.dll`. If required, the interested reader may extend the source code. With the following assignment, this file's namespace is included.

```
using GS.Util.Hex;
```

All methods of the `ByteArray`, `HexEncoding`, `HexFormatting` and `HexUtil` classes are static. In this way, all methods can be used without having to create an instance of the class. A call to a static method is done when the class name is separated from the method name by a period (.). The use of the most important methods is demonstrated in this chapter in the form of examples.

The `ToHexString()` method in the `Hexformatting` class is overloaded 11 times for different data types. This method always returns a string. The following example shows the use of this method. The optional second parameter indicates whether a blank space will be printed between individual bytes.

```
Console.WriteLine("UID = 0x" + HexFormatting.ToHexString(reader.UID, true));
```

If the UID property is a 4-byte long array containing values `0xB6`, `0x21`, `0x05` and `0x01`, the console output will look like this:

```
UID   = 0xB6 21 05 01
```

8.1.4.4 Simplified Card Activation

Regardless of the card's type, it is necessary for every reader application to activate it. For this, the reader library offers a simplified variant. The entire card activation sequence is integrated into a reader firmware function. This function can be called using the `ActivateCard()` method. This reduces the RS-232-to-USB communication between the

PC and the Reader to just one command sequence. A call to this method will be substantially quicker than calls to individual commands for card activation. The reqCmd parameter defines the command request, that is, whether a REQA or WUPA command must be sent to the card. Valid values are `RequestCmd.REQA` and `RequestCmd.WUPA`. Should the card not respond to a REQA or WUPA, the reader firmware's function terminates immediately with the `RC52xTimeOut` error code. It is thus readily possible to poll for cards in the field using only the `ActivateCard()` method. The example in Listing 8.4 has the same functionality as that in Listing 8.3. The console output was omitted in this case.

Listing 8.4. *CardActivation3.cs.*

```csharp
using System;
using GS.Util.Hex;
using GS.ElektorRfidReader;
using GS.ISO14443_Reader;

namespace GS.ElektorRfidReaderExample
{
    class CardActivation
    {
        public static void Main(MainForm script, string[] args)
        {
            ElektorISO14443Reader reader = new ElektorISO14443Reader();
            try
            {
                reader.OpenPort();
                int ret;
                Console.WriteLine("Waiting for new card...");
                do
                {
                    ret = reader.ActivateCard(RequestCmd.REQA);
                }
                while((ReaderStatus)ret != ReaderStatus.OK);
                Console.Clear();
                Console.WriteLine("Card detected...");
                reader.HaltA();
            }
            catch (Exception ex)
            {
                Console.WriteLine(ex.Message.ToString() + Environment.NewLine);
            }
            finally
            {
                reader.ClosePort();
            }
        }
    }
}
```

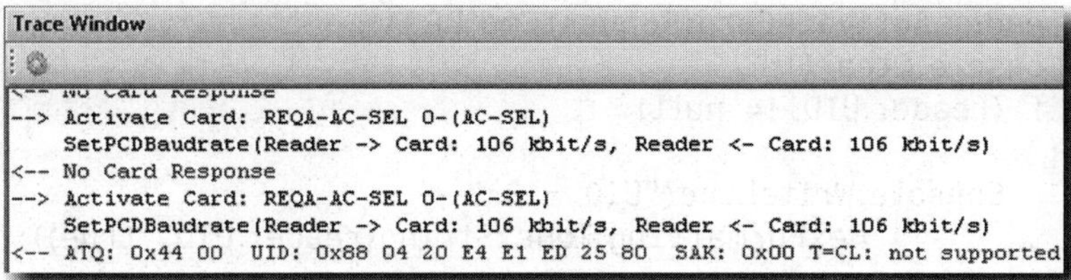

Figure 8.16.
The ActivateCard() *method outputs all card parameters to the* Trace *window.*

```
Trace Window
<-- No Card Response
--> Activate Card: REQA-AC-SEL 0-(AC-SEL)
    SetPCDBaudrate(Reader -> Card: 106 kbit/s, Reader <- Card: 106 kbit/s)
<-- No Card Response
--> Activate Card: REQA-AC-SEL 0-(AC-SEL)
    SetPCDBaudrate(Reader -> Card: 106 kbit/s, Reader <- Card: 106 kbit/s)
<-- ATQ: 0x44 00   UID: 0x88 04 20 E4 E1 ED 25 80   SAK: 0x00 T=CL: not supported
```

8.1.4.5 Testing the Reading Range

If the Elektor RFID Reader is connected to an LCD, it is very simple to ascertain a card's reading range using the Terminal mode. Naturally, the same functionality may be implemented using a script.

Listing 8.5 shows a very simple version. After opening the USB port, an endless loop attempts to activate a card. A call to the `RFReset()` method ensures that a card that is permanently in the reader field is activated anew at each iteration of the loop.

At every loop iteration, the console window is also cleared. After that, either the card UID or the text, "No card found." is output to the console. This means that the console output is not stable, due to repeated erasing and rewriting. The same problem occurs when the same code is used in reader firmware. In this case, the LCD output will flicker. In the next example, we come up with a solution to this problem. The trick is to erase and update the display only when the card status changes. With the method shown in Listing 8.5, it is also possible to have multiple cards in the field.

Listing 8.5. *GetReadingDistance.cs.*

```
using System;
using GS.Util.Hex;
using GS.ISO14443_Reader;
using GS.ElektorRfidReader;

namespace GS.ElektorRfidReaderExample
{
    public class GetReadingDistance
    {
        public static void Main(MainForm script, string[] args)
        {
            ElektorISO14443Reader reader = new ElektorISO14443Reader();

            try
            {
                reader.OpenPort();
                while (true)
                {
                    reader.RFReset(5);
```

```
                    reader.ActivateCard(RequestCmd.REQA);
                    Console.Clear();
                    if (reader.UID != null)
                    {
                        Console.WriteLine("UID = 0x" +
                                HexFormatting.ToHexString(reader.UID, true));
                    }
                    else
                    {
                        Console.WriteLine("No card found.");
                    }
                }
            }
            catch (Exception ex)
            {
                Console.WriteLine(ex.Message.ToString() + Environment.NewLine);
            }
            finally
            {
                reader.ClosePort();
            }
        }
    }
}
```

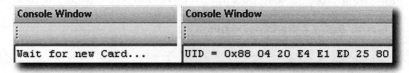

Figure 8.17. This script enables a simple way to determine the reading range for a card.

8.1.4.6 Listing All Cards in the Reader's Field

An ISO/IEC 14443 reader can usually activate several cards in the field. The number is limited only by the reader's RF output power. With the help of the following example, how many cards can simultaneously be activated in the Elektor RFID Reader's field will be checked experimentally.

Flickering of the console output can be elegantly avoided by using a dynamic data structure (collection). For example, Listing 8.6 uses the `List` class from the `System.Collections.Generic` namespace. Use of the `List` class has significant advantages over a normal array, and allows easy management of card serial numbers. For example, a UID can be added to the end of the list using the `Add()` method. All elements (in our example, card serial numbers) can be cleared from the list using the `Clear()` method. As with a conventional array, individual list elements can be referenced by index. Alternatively, use of a `foreach` loop allows iteration through the entire list.

ISO/IEC 14443 TYPE A CARD ACTIVATION 8.1

Listing 8.6. *ShowAllUids.cs.*

```
using System;
using System.Collections.Generic;
using GS.Util.Hex;
using GS.ISO14443_Reader;
using GS.ElektorRfidReader;

namespace GS.ElektorRfidReaderExample
{
    public class ShowAllUids
    {
        public static void Main(MainForm script, string[] args)
        {
            ElektorISO14443Reader reader = new ElektorISO14443Reader();

            try
            {
```

To create two generic lists for the storage of card serial numbers, the `uidList` list is used to store the currently-read UIDs. The `oldUidList` stores the 'old' serial numbers.

```
                List<byte[]> uidList = new List<byte[]>();
                List<byte[]> oldUidList = new List<byte[]>();
                int ret;

                reader.OpenPort();
```

After opening the reader port, the entire sequence is repeated in an endless loop, so, to terminate the script, it's necessary to use the *Script → Stop Script* menu option.

```
                while (true)
                {
```

At each iteration, the current `uidList` is erased and then filled with a list of all cards in the IDLE state after an RF reset.

```
                    uidList.Clear();
                    ret = reader.RFReset(5);
```

Within the `do-while` loop, the serial numbers of all the cards within the reader's field are read. At each loop iteration, a REQA command is sent to the cards. As soon as a card responds, it is activated and then placed in the HALT state. In so doing, possible collisions are resolved with the Anti-collision command. A card that is in the HALT state does not respond to the REQA in the next iteration of the loop. If no card responds to the REQA command, execution of the `do-while` loop exits. This is always the case when either all cards are in the HALT state, or when no cards are found. Using the `Add()` method, all UIDs that were read are added to the end of `uidList`.

```
            do
            {
                ret = reader.ActivateCard(RequestCmd.REQA);
                if (reader.UID != null)
                {
                    uidList.Add(reader.UID);
                    reader.HaltA();
                }
            }
            while (ret == 0);
```

The `DataAreEqual()` static method compares `uidList` with `oldUidList`. If the two lists are different, the console is erased and a `foreach` loop outputs all card UIDs to the console.

```
            if ( DataAreEqual(uidList, oldUidList) == false )
            {
                Console.Clear();
                foreach (byte[] uid in uidList)
                {
                    Console.WriteLine("UID = 0x" +
                            HexFormatting.ToHexString(uid, true));
                }
            }
```

Then, all list elements from `uidList` are copied to `OldUidList`. For this, it is necessary to first clear `oldUidList` using the `Clear()` method. The `AddRange()` method copies all of the list elements in one step.

```
            oldUidList.Clear();
            oldUidList.AddRange(uidList.GetRange(0, uidList.Count));
        }
    }
    catch (Exception ex)
    {
        Console.WriteLine(ex.Message.ToString() + Environment.NewLine);
    }
    finally
    {
        reader.ClosePort();
    }
}
```

The `DataAreEqual()` method compares the contents of two UID lists for equality. If they're equal, it returns true, else it returns false.

```
        public static bool DataAreEqual(List<byte[]> a, List<byte[]> b)
```

```
        {
            if (a == b) return true;
            if (a == null || b == null) return false;

            if (a.Count != b.Count) return false;

            int len = Math.Min(a.Count, b.Count);
            for (int i = 0; i < len; i++)
            {
                if (DataAreEqual(a[i], b[i]) == false) return false;
            }
            return true;
        }
```

Another overloading of the `DataAreEqual()` method compares the contents of two byte arrays. This method is needed to compare two card serial numbers in the `DataAreEqual()` method.

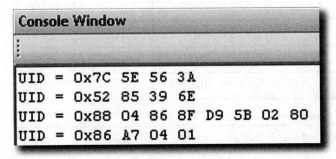

Figure 8.18.
In this case, the Elektor RFID Reader can activate four cards.

8.2 MIFARE Card-Type Detection

The ISO/IEC 14443 Standard does not define any method to uniquely identify different card types, except to distinguish whether or not the card supports the T=CL protocol. This information is stored in the card's SAK byte. Semiconductor manufacturers use proprietary bits within the ATQA and SAK bytes to encode the type of card. The ATQA value is of limited utility, as it is not always correctly received by the receiver. The previous section already showed, by means of an example, that a collision could occur during an ATQA. If two more cards respond to a REQA at the same time with two different ATQA values, the reader will always receive a superposition of all of the cards' responses. This causes the ATQA value itself to be falsified, and the reader software is no longer able to determine the card type correctly. In Table 8.12, the ATQA and SAK value for cards from the MIFARE family are listed. In the *AN10833 MIFARE Type ID Procedure* application note from manufacturer NXP, a much larger table can be found. This document can be downloaded for free from *http://www.nxp.com*.

Table 8.12.

Card Type	ATQA MSB	ATQA LSB	SAK
MIFARE Ultralight	0x00	0x44	0x00
MIFARE Ultralight C	0x00	0x44	0x00
MIFARE Mini	0x00	0x04	0x09
MIFARE 1K	0x00	0x04	0x08
MIFARE 4K	0x00	0x02	0x18
MIFARE DesFire and DesFire EV1 (2K, 4K and 8K)	0x03	0x44	0x20

From Table 8.12, it is clear that it is not possible to uniquely identify all card types. For example, the ATQA and SAK values of MIFARE Ultralight and MIFARE Ultralight C cards are the same. This problem occurs due to the fact that the number of bits available for coding the card type within the ATQA byte or SAK byte is very limited.

In real-world applications, evaluation of the ATQA should be dispensed with for the aforementioned reasons. In the NXP application note, *AN10834 PICC Selection*, a possible solution is described. This uses the SAK byte to identify different types of cards. The flowchart in Figure 8.19 represents with the diagram in the NXP application note, including the numbering of the bit positions. The least-significant bit in the SAK byte is shown in Figure 8.19 as number '0', as is the usual convention in computer science. NXP, however, uses the ISO/IEC Standard's convention, in which the bits in a byte are numbered from 1 to 8. This often leads to errors when programming.

By exclusively using the SAK byte, it is no longer possible to distinguish between a MIFARE DesFire card and a smartcard controller (e.g. SmartMX). In practice, this should not cause any great problems. With ISO/IEC 14443-4 (T=CL)-compatible cards, it is usually necessary to determine whether a card supports a specific card application or not. This distinction should be made at the application level in any event.

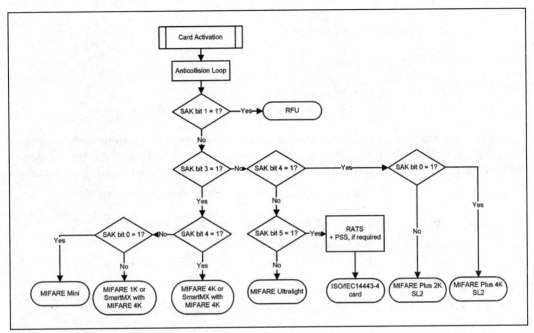

Figure 8.19. Recognition of different card types.

8.2.1 Program Example

Listing 8.7 represents the flow chart from Figure 8.19 almost identically. Only, the determination of whether the card supports the ISO/IEC 14443-4 (T=CL) protocol or not takes place at the beginning. With this small change, smartcard controllers with MIFARE emulation as well as the T=CL protocol, as well as MIFARE emulation, are represented as card types.

Listing 8.7. Excerpt from CardTypeIdentification.cs.

```
using System;
using GS.Util.Hex;
using GS.ISO14443_Reader;
using GS.ElektorRfidReader;

namespace GS.ElektorRfidReaderExample
{
    public class CardTypeIdentification
    {
        public static void Main(MainForm script, string[] args)
```

8 CARDS AND TAGS IN APPLICATION

```
    {
        ElektorISO14443Reader reader = new ElektorISO14443Reader();
        byte[] oldUid = null;

        try
        {
            reader.OpenPort();

            while (true)
            {
                reader.RFReset(5);
                reader.ActivateCard(RequestCmd.REQA);
```

Output to the console only happens when a new card is detected, or an already-detected card is removed from the field. This prevents the console from flickering. For this purpose, the current card UIDS are compared with the previous values in the `oldUid` variable.

```
                if ( DataAreEqual(reader.UID, oldUid) == false )
                {
                    Console.Clear();

                    if (reader.SAK != null)
                    {
```

SAK Byte, Bit 1, is currently reserved for future use. If this is set, the other bits might all have a different meaning. For this reason, evaluation of all the other bits is not carried out.

```
                        if (0x02 == (reader.SAK & 0x02))
                        {
                            Console.WriteLine("Card Type: Unknown");
                            continue;
                        }
```

If Bit 5 of the SAK byte is set, then the card supports the T=CL protocol. This bit is defined in ISO/IEC 14443 and thus always has the same meaning, regardless of the semiconductor manufacturer. Next, it is necessary to evaluate all bits, so that a smartcard controller's MIFARE emulation can also be correctly detected.

```
                        if (0x20 == (reader.SAK & 0x20))
                        {
                            Console.WriteLine("" +
                              "Card Type: ISO/IEC 14443-4 (T=CL) compatible card");
                        }
                        if (0x08 == (reader.SAK & 0x08))
                        {
                            if (0x10 == (reader.SAK & 0x10))
                            {
                                Console.WriteLine("Card Type: MIFARE 4K");
                            }
```

```csharp
            else
            {
                if (0x01 == (reader.SAK & 0x01))
                {
                 Console.WriteLine("Card Type: MIFARE Mini");
                }
                else
                {
                 Console.WriteLine("Card Type: MIFARE 1K");
                }
            }
        }
        else
        {
            if (0x10 == (reader.SAK & 0x10))
            {
                if (0x01 == (reader.SAK & 0x01))
                {
                    Console.WriteLine("" +
                        "Card Type: MIFARE Plus 4K");
                }
                else
                {
                    Console.WriteLine("" +
                        "Card Type: MIFARE Plus 2K");
                }
            }
            else
            {
                if (0x20 != (reader.SAK & 0x20))
                {
                    Console.WriteLine("" +
                        "Card Type: MIFARE Ultralight (C)");
                }
            }
        }
    }
}
```

At this point the value of the current card UID is stored in the oldUid variable. Should no card be found in the field, the variable is assigned the `null` value.

```csharp
            if (null == reader.UID)
            {
                oldUid = null;
            }
            else
            {
```

```
                    oldUid = new byte[reader.UID.Length];
                    for (int i = 0; i < reader.UID.Length; i++)
                    {
                        oldUid[i] = reader.UID[i];
                    }
                }
            }
        }
        catch (Exception ex)
        {
            Console.WriteLine(ex.Message.ToString() + Environment.NewLine);
        }
        finally
        {
            reader.ClosePort();
        }
    }
```

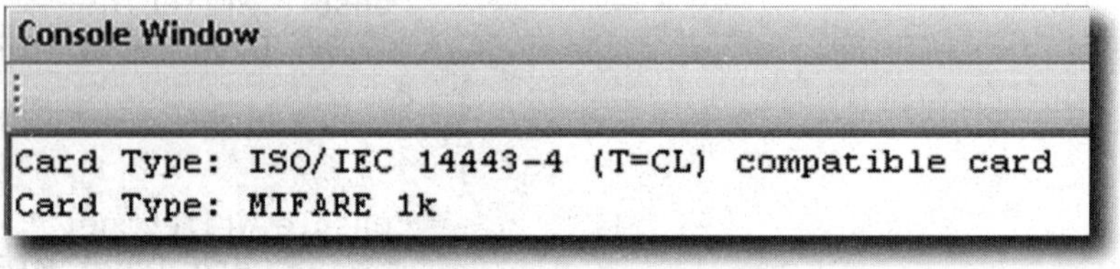

Figure 8.20. This card supports both the T=CL protocol and MIFARE 4K emulation.

```
Console Window

Card Type: ISO/IEC 14443-4 (T=CL) compatible card
Card Type: MIFARE 1k
```

8.3 The MIFARE Ultralight Card

The MIFARE Ultralight card is the smallest member of the NXP MIFARE product family.

8.3.1 Memory Organization

Figure 8.21 shows the memory organization of the 512-bit EEPROM memory. It consists of 16 pages of 4 bytes each. Every card has a 7-byte UID, programmed into Pages 0 and 1 during manufacture. The semiconductor manufacturer guarantees that this number is unique. It is not possible for the user to change this serial number.

Page 3 is a 32-bit wide one-time-programmable (OTP) storage area. Each individual bit can be irreversibly programmed from the logic '0' state to logic '1'. This means that restoring the bit to '0' is not possible. In public transportation applications the OTP area is used as a substitute for card 'points'.

Pages 4 to 15 are the 384-bit data storage. This memory can be both read and written by a reader device. With the two lock bytes, Lock0 and Lock1, it is possible to block parts of the EEPROM from write access, while reading from these locked storage areas is not restricted in any way.

8.3 THE MIFARE ULTRALIGHT CARD

Byte Number	0	1	2	3	Page
UID	UID 0	UID 1	UID 2	BCC0	0
UID	UID 3	UID 4	UID 5	UID 6	1
Internal / Lock	BCC1	Internal	Lock 0	Lock 1	2
OTP	OTP 0	OTP 1	OTP 2	OTP 3	3
User Data	Data 0	Data 1	Data 2	Data 3	4
User Data	Data 4	Data 5	Data 6	Data 7	5
User Data	Data 8	Data 9	Data 10	Data 11	6
User Data	Data 12	Data 13	Data 14	Data 15	7
User Data	Data 16	Data 17	Data 18	Data 19	8
User Data	Data 20	Data 21	Data 22	Data 23	9
User Data	Data 24	Data 25	Data 26	Data 27	10
User Data	Data 28	Data 29	Data 30	Data 31	11
User Data	Data 32	Data 33	Data 34	Data 35	12
User Data	Data 36	Data 37	Data 38	Data 39	13
User Data	Data 40	Data 41	Data 42	Data 43	14
User Data	Data 44	Data 45	Data 46	Data 47	15

Figure 8.21. Memory organization of a MIFARE Ultralight card.

8.3.2 Instruction Set

Basically, the instruction set for a MIFARE Ultralight card can be divided into two groups: commands for card activation and commands for memory manipulation.

Table 8.13. MIFARE Ultralight instruction set.

Command Type	Command Code	Card Commands	SmartCard Magic.NET Button Label
Card Activation	0x26	REQA	Req A
	0x52	WUPA	Wup A
	0x93	Anti-collision of Cascade Level 1	AC 1
	0x93	Select of Cascade Level 1	Select 1
	0x95	Anti-collision of Cascade Level 2	AC 2
	0x95	Select of Cascade Level 2	Select 2
	0x50	HALTA	Halt A
Memory Manipulation	0x30	Read	Read
	0xA2	Write	–
	0xA0	Compatibility Write	Write

All commands for card activation were already discussed at length in the previous section. The MIFARE Ultralight instruction set is fully compatible with the MIFARE Classic (1K/4K) cards, which have a larger EEPROM (either 1 KB or 4 KB), as well as additional encryption functions. For the latter, the MIFARE Classic card has additional commands for card authentication.

8 CARDS AND TAGS IN APPLICATION

After successful card activation, read and write access to the card's EEPROM is achieved using the READ and WRITE commands. With the WRITE command, an entire page, or 4 bytes, is written at once. The READ command always reads 4 pages at once, or 16 bytes. In the MIFARE 1K/4K cards, a page, which is called a 'block' in the datasheet, has a fixed size of 16 bytes. This means that reading and writing always takes place in 16-byte blocks. The MIFARE Ultralight commands, READ and COMPATIBILITY WRITE, are backwards compatible with the MIFARE Classic. Although the COMPATIBILITY WRITE command sends 16 bytes to the card, only 4 bytes are ever written to the EEPROM at a time.

With Smart Card Magic.NET, it is very easy to read from and write to the cards. For the first step, we use the *Mifare Ultralight C* window. Simply clicking on either the *Act. Req A* or the *Act. Wup A* button activates a card located within the reader's field. With the *Dump UL* button, the entire memory content of a MIFARE Ultralight card is read and output to the *Trace* window or the *Console* window (see Figure 8.23). Using the *Authenticate* button, a mutual authentication between the reader and a MIFARE Ultralight C card is performed. The MIFARE Ultralight C will first be discussed later, Section 8.5, however.

The *Read* and *Write* buttons allow the reading and writing of individual pages. It should be noted that a MIFARE Ultralight card can only address Pages 0 to 15. Selecting Pages 16 to 44 is necessary with a MIFARE Ultralight C card, as this has more memory.

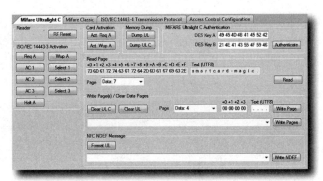

Figure 8.22. Smart Card Magic.NET: The 'MIFARE Ultralight C' window.

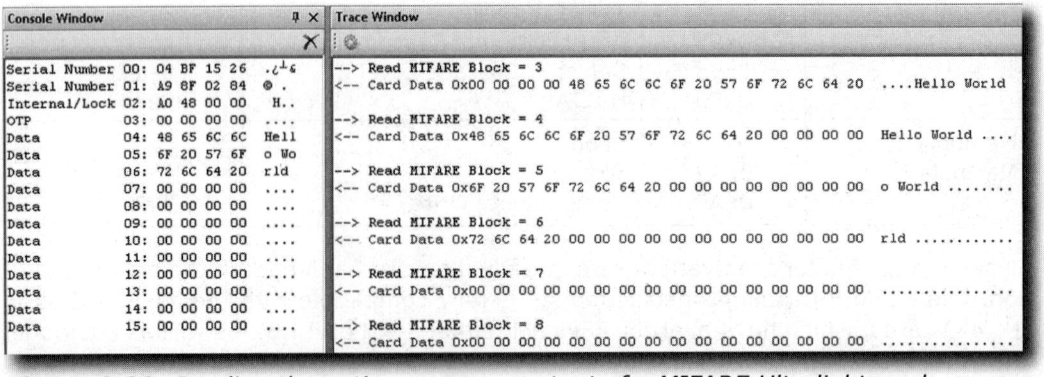

Figure 8.23. Reading the entire memory content of a MIFARE Ultralight card.

8.3.3 Function of the One-Time-Programmable (OTP) Bytes

Page 3 holds the 4-byte one-time-programmable (OTP) memory. An OTP memory can only be programmed once, and can never be erased. This behavior is similar to that of a fuse. Programming a bit is like melting a fuse wire, and thus represents an irreversible process. The OTP memory can easily facilitate the replacement of a conventional 'points' card. In this case, setting the OTP bits is like punching holes in a paper card. Figure 8.24 shows the principle use of the OTP memory as a points card.

The behavior of the OTP memory can be easily investigated with Smart Card Magic.NET. In Figure 8.25, the OTP memory's least-significant bit has been set. If one tries to write a '0' with a new write command, this command is executed without any error message. However, the value of the bit remains unchanged: '1'. In order to set another bit in the OTP memory, it is only necessary to set the selected bit from '0' to '1'. For bits that are already set, a value of either '0' or '1' (don't care) can be used.

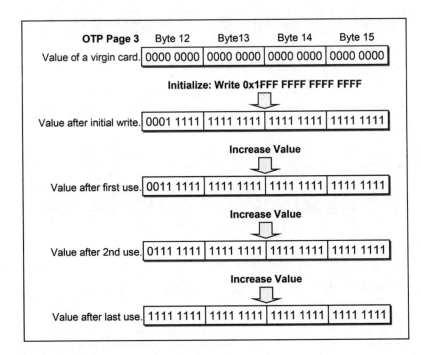

Figure 8.24. *In this example, the OTP memory is initialized with three dots and then voided.*

When the MIFARE Ultralight chip is used as a ticket for public transport (e.g. the subway), the lifecycle of the card is very short, as with conventional point cards. Therefore, the chip is integrated almost exclusively into paper cards for cost reasons.

Figure 8.25.
An OTP bit (Page 3) can only be set to '1' and never restored to to '0' again.

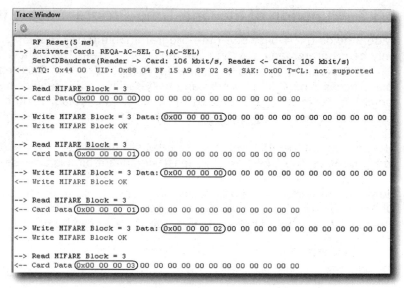

8.3.3.1 Lock Bits Functionality

The Lock bits are found in Byte 2 and 3 of Page 2. These bits are also OTP bits, and can only be set from '0' to '1'. Should an 'L' bit be set to '1', the corresponding page becomes permanently write-protected (read-only).

Figure 8.26.
MIFARE Ultralight card: lock bytes (source: NXP Semiconductors).

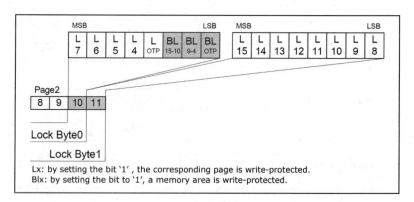

A block lock ('BL') bit activates the write protection for a memory range (for example, Pages 9 to 4). Write protection becomes active only after carrying out a REQA or WUPA command subsequent to setting a Lock or Block Lock bit.

Should one attempt to carry out a write to a write-protected area, the card responds with a MifareAuthFailed error message, and the card switches to the IDLE state. In this way, card re-activation is required prior to a subsequent memory read or write access.

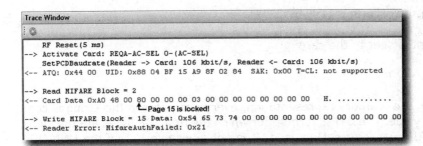

Figure 8.27.
The lock bit is set for Page 15.

8.3.4 Elektor RFID Reader Library: MIFARE Ultralight

In Table 8.4, all required methods for MIFARE Ultralight memory manipulation are listed. The ULWrite method uses only the COMPATIBILITY WRITE command. Therfore, 16 bytes are always sent to the card, even though the MIFARE Ultralight writes only 4 bytes.

Table 8.14. Ultralight Read and Write methods.

Method	Description
int Read(byte block, out byte[] data)	The Read() method reads 4 pages, or 16 bytes. The page address is passed in the block parameter.
int ULWrite(byte block, byte[] data)	The ULWrite() method writes a page, or 4 bytes. The data to be written is passed as an array of bytes.
int ULWrite (byte block, string data)	The ULWrite() method writes a page, or 4 bytes. The data to be written is passed as a string in Hex or ASCII format (see text).

So, should you want to write the value 0x11 22 33 44 to a page, this is possible in two ways. The first method uses the ULWrite() method to transfer a 4-byte array.

```
byte[] baData = {0x11, 0x22, 0x33, 0x44};
reader.Mifare.ULWrite(4, baData);
```

In the second method, the declaration and initialization of a byte array is absent. Instead, the hexadecimal values are passed directly to the ULWrite() method as a string (without any additional formatting characters). White space between the individual values increases human readability, but is not required. In the ULWrite() method, the actual conversion of the string into a byte array is done using the HexEncoding.GetBytes() method from the GS.Util.Hex namespace. You could use this method for similar tasks.

```
reader.Mifare.ULWrite(4, "11 22 33 44");
```

Should you wish to write a string to the card's EEPROM, the same method can be used. For this, it is only necessary to enclose the string in quotes. Quotes are inserted into a string using the \" escape sequence.

8 CARDS AND TAGS IN APPLICATION

```
reader.Mifare.ULWrite(6, "11\"ab\"22");
```

The following content is sent to the card:

```
0x11 61 62 22 00 00 00 00 00 00 00 00 00 00 00 00
```

As can be seen from the memory representation, every character in the string is converted into a 1-byte value. The UTF-8 format, for which the first 128 bytes are the same as those of the ASCII character set, is used. It is not advisable to use the UNICODE character set, which uses 2 bytes per character, due to the limited memory size.

8.3.5 Program Examples

8.3.5.1 Writing and Erasing Data

This example illustrates the use of `ULWrite()` method to write to different data formats to the data EEPROM of a MIFARE Ultralight.

Listing 8.8. *MifareUltraLightWrite.cs.*

```csharp
using System;
using GS.Util.Hex;
using GS.ISO14443_Reader;
using GS.ElektorRfidReader;

namespace GS.ElektorRfidReaderExample
{
    public class MifareUltraLightWrite
    {
        public static void Main(MainForm script, string[] args)
        {
            ElektorISO14443Reader reader = new ElektorISO14443Reader();
            try
            {
                reader.OpenPort();
                reader.RFReset(5);
                reader.ActivateCard(RequestCmd.WUPA);

                // Write 0x11 22 33 44 to MIFARE UL Block 4
                byte[] baData = {0x11, 0x22, 0x33, 0x44};
                reader.Mifare.ULWrite(4, baData);

                // Write 0x55 66 77 88 to MIFARE UL Block 5
                reader.Mifare.ULWrite(5, "55 66 77 88");

                // Write the string "Hello World" to UL Block 6 ... 8
                reader.Mifare.ULWrite(6, "\"Hell\"");
                reader.Mifare.ULWrite(7, "\"o Wo\"");
                reader.Mifare.ULWrite(8, "\"rld \"");
```

```
            }
            catch (Exception ex)
            {
                Console.WriteLine(ex.Message);
            }
            finally
            {
                reader.ClosePort();
            }
        }
    }
}
```

A MIFARE Ultralight card's EEPROM data can be deleted by writing 0x00. During the development phase of MIFARE Ultralight card applications, it is necessary to erase the card's entire memory many times. This feature is not directly supported by Smart Card Magic.NET, but can be achieved using a simple script. In the `MIFAREUltraLight` folder, the `MifareUltraLightClear.cs` script can be found, which writes `0x00`s to the entire data EEPROM (Pages 4 to 15), thus erasing the EEPROM.

Listing 8.9. *Excerpt from MifareUltraLightClear.cs.*

```
// Clear Mifare Ultralight Data Pages
for (byte page = 4; page < 16; page++)
{
    reader.Mifare.ULWrite(page, "00 00 00 00");
}
```

8.3.5.2 Reading the Entire Memory Contents

In Section 8.3.2, the entire memory content of an Ultralight card is read out and output to the *Console* window using the *Dump UL* button (MIFARE Ultralight C window). The source code in Listing 8.10 has the same functionality as the *Dump UL* button.

Listing 8.10. *MifareUltraLightDump.cs.*

```
using System;
using GS.Util.Hex;
using GS.ISO14443_Reader;
using GS.ElektorRfidReader;

namespace GS.ElektorRfidReaderExample
{
    public class MifareUltraLightDump
    {
        public static void Main(MainForm script, string[] args)
        {
            ElektorISO14443Reader reader = new ElektorISO14443Reader();
            try
```

8 CARDS AND TAGS IN APPLICATION

```csharp
            {
                reader.OpenPort();
                reader.RFReset(5);
                reader.ActivateCard(RequestCmd.WUPA);
```

By changing the start and end values in the `for` loop, it is easy to output only selected memory areas to the console.

```csharp
                for (byte page = 0; page < 16; page++)
                {
                    byte[] data;
                    int ret = reader.Mifare.Read(page, out data);

                    if (0 == ret)
                    {
```

Both the data content and the memory address labels are output to the console.

```csharp
                        if (page == 0 || page == 1)
                        {
                            Console.Write(
                                String.Format("Serial Number {0:D2}: ",page));
                        }
                        else if (2 == page)
                        {
                            Console.Write(
                                String.Format("Internal/Lock {0:D2}: ", page));
                        }
                        else if (3 == page)
                        {
                            Console.Write(
                                String.Format("OTP {0:D2}: ", page));
                        }
                        else
                        {
                            Console.Write(
                                String.Format("Data {0:D2}: ", page));
                        }
                        Console.WriteLine(HexFormatting.Dump(data, 4, 4));
                    }
                    else return;
                }
            }
            catch (Exception ex) { Console.WriteLine(ex.Message); }
            finally { reader.ClosePort(); }
        }
    }
}
```

8.3.5.3 Reading and Writing Strings

Reading and writing strings is a task that will always be necessary.

Listing 8.11. *Excerpt from MifareUltraLightWriteString.cs.*

```
void WriteString(byte page, string str)
{
    int strlen = str.Length;
    int offset = 0;

    while(strlen > 0)
    {
        int len = Math.Min(4, strlen);
        string s = str.Substring(offset, strlen);
        reader.Mifare.ULWrite(page, "\"" + s + "\"");
        page++;
        if (page > 15) return;

        strlen -= len;
        offset += len;
    }
}
```

The `WriteString()` method allows the writing of a string of up to 48 bytes (Pages 4 to 15) to a MIFARE Ultralight card. The following source code writes the string, "Hello MIFARE Ultralight" to the MIFARE Ultralight's EEPROM (Pages 4 to 9).

```
WriteString(4, "Hello MIFARE Ultralight");
```

Listing 8.12. *Excerpt from MifareUltraLightReadString.cs.*

```
string ReadString(byte page)
{
    byte[] baString = new byte[4 * 12];
    int offset = 0;

    while(page < 16)
    {
        byte[] baData = new byte[16];
        reader.Mifare.Read(page, out baData);
        Array.Copy(baData, 0, baString, offset, 4);
        page++;
        offset += 4;
    }
    return System.Text.Encoding.ASCII.GetString(baString);
}
```

The `ReadString()` method reads a string stored on the card.

```
string s = ReadString(4);
Console.WriteLine("String: " + s);
```

8.3.5.4 A Simple Ticket Application

In Section 8.3.3, the function of the OTP memory was introduced. In this example, a simple application to replace a traditional points card is developed.

Figure 8.28 shows a simple ticket application's basic sequence. In the first step, the OTP memory content is read. Then, a check is made to see whether the card is still valid. Should all bits be set to '1' already, then all of the points have been used and the card-holder must purchase a new card. If the card has enough points, the point value is simply incremented by 1 and stored on the card. Finally, a mechanical gate or turnstile (e.g. in a subway station) is opened.

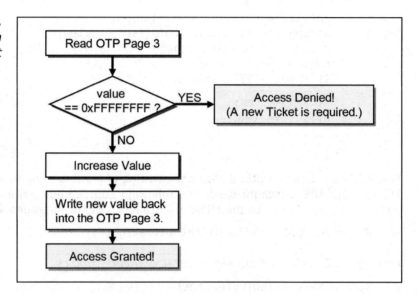

Figure 8.28. Sequence diagram for a simple ticket application.

With the script in Listing 8.13, up to 32 points can be validated. By appropriately initializing the OTP memory, any other point total under 32 can be represented.

Note In this example, data Page 4 is first used for functional testing of the script. Page 4 is eraseable, so, using this little trick, the entire software functionality can be tested without requiring a new MIFARE Ultralight card for each new trial.

Listing 8.13. *MifareUltraLightTicket.cs.*

```
using GS.ElektorRfidReader;

namespace GS.ElektorRfidReaderExample
{
    public class MifareUltraLightTicket
    {
```

THE MIFARE ULTRALIGHT CARD 8.3

By changing the value of the `OTP_PAGE` constant from 4 to 3, OTP Page 3 will be used instead of data Page 4.

```
// Use user data page 4 instead OTP page 3 for test purpose only!
const byte OTP_PAGE = 4;

public static void Main(MainForm script, string[] args)
{
    ElektorISO14443Reader reader = new ElektorISO14443Reader();
    try
    {
        int ret;
        reader.OpenPort();
```

This script polls constantly, seeking new cards in the reader's field.

```
        while(true)
        {
            do
            {
                ret = reader.ActivateCard(RequestCmd.REQA);
            }
            while (ret != 0);
```

After activating a card, the OTP page is read.

```
            byte[] baOTP;
            ret = reader.Mifare.Read(OTP_PAGE, out baOTP);
```

The OTP page is converted to an unsigned integer (`uint`) using the `OTPValueToUInt32()` method. This conversion is required for calculations with the current counter state. Then, the current value of the OTP memory is output to the console prior to any manipulation.

```
            if (ret == 0)
            {
                uint ticketValue = OTPValueToUInt32(baOTP);
                Console.Write(
                    String.Format("OTP Value: 0x{0:X08}",ticketValue));
```

Should all OTP bits already be set to '1', the card is invalid and the cardholder must obtain a new card.

```
                if (ticketValue < 0xFFFFFFFF)
                {
```

In this example, the ticket value is incremented and then written to the card. After the write operation, it is necessary to place the card in the HALT state. In this way, duplicate debiting is prevented. In real-world applications, additional measures are usually implemented. For example, with the help of the UID, a card may be blocked for a certain period after a transaction.

```
                    ticketValue = ticketValue | ( ticketValue+1 );
                       reader.Mifare.ULWrite(OTP_PAGE,GetOTPBytes(ticketValue));
                    Console.WriteLine(" Access Granted!");
                    reader.HaltA();
                }
                else
                {
```

If there are no points on the card, access is simply denied.

```
                    Console.WriteLine(" Access Denied!");
                    reader.HaltA();
                    continue;
                }
            }
        }
    }
    catch (Exception ex) { Console.WriteLine(ex.Message); }
    finally { reader.ClosePort(); }
}
```

The `OTPValueToUInt32()` method converts the OTP byte array into an unsigned integer (`uint`). In the first step, the byte order must be corrected. The actual conversion is performed by the `System.BitConverter()` method from the .NET Framework.

```
        static uint OTPValueToUInt32(byte[] baValue)
        {
            byte[] baOTP = new byte[4];

            baOTP[0] = baValue[3];
            baOTP[1] = baValue[2];
            baOTP[2] = baValue[1];
            baOTP[3] = baValue[0];

            return BitConverter.ToUInt32(baOTP, 0);
        }
```

The `GetOTPBytes()` method converts a `uint` value to a 4-byte array. The conversion is also done using the `System.BitConverter()` method. Finally, correction of the byte order is required.

```
        static byte[] GetOTPBytes(uint value)
        {
            byte[] baValue = BitConverter.GetBytes(value);

            byte[] baOTP = new byte[4];

            baOTP[0] = baValue[3];
            baOTP[1] = baValue[2];
```

```
            baOTP[2] = baValue[1];
            baOTP[3] = baValue[0];

            return baOTP;
        }
    }
}
```

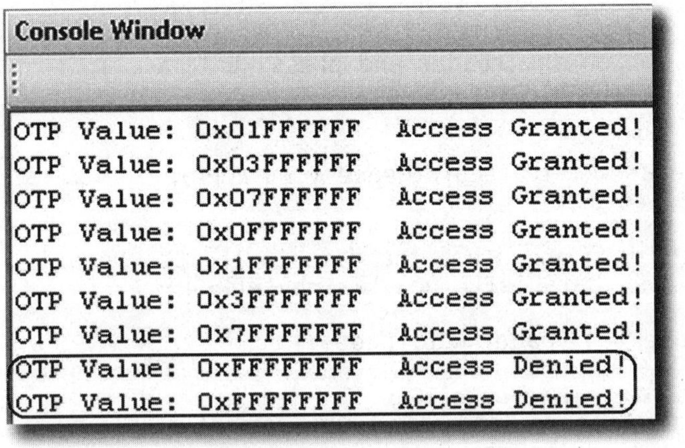

```
Console Window

OTP Value: 0x01FFFFFF    Access Granted!
OTP Value: 0x03FFFFFF    Access Granted!
OTP Value: 0x07FFFFFF    Access Granted!
OTP Value: 0x0FFFFFFF    Access Granted!
OTP Value: 0x1FFFFFFF    Access Granted!
OTP Value: 0x3FFFFFFF    Access Granted!
OTP Value: 0x7FFFFFFF    Access Granted!
OTP Value: 0xFFFFFFFF    Access Denied!
OTP Value: 0xFFFFFFFF    Access Denied!
```

Figure 8.29.
If all OTP bits are set to '1', access is denied.

8.3.5.5 Cloning the Memory Content

While the MIFARE Ultralight card has write protection for the user memory (Pages 4 to 15), this doesn't prevent the cloning of a card. Cloning means that the memory content of one card is copied 1:1 to another. Data encryption will not provide any protection against data copying. In the next section, an effective method for protecting against cloning is presented, with the help of card-specific keys.

In order to effectively demonstrate this method, it is first necessary to create a script for cloning MIFARE Ultralight cards. The script first reads the data that will be copied from the card, and saves this to a list. The data is then simply written to a blank card. Care should be taken that only one card is in the reader's field for this process. Should two cards be available, the empty card might be selected first by the anti-collision algorithm.

Listing 8.14. *MifareUltraLightClone.cs.*

```
using System;
using System.Collections.Generic;
using GS.Util.Hex;
using GS.ISO14443_Reader;
using GS.ElektorRfidReader;

namespace GS.ElektorRfidReaderExample
{
    public class MifareUltraLightClone
    {
```

```
public static void Main(MainForm script, string[] args)
{
    ElektorISO14443Reader reader = new ElektorISO14443Reader();
    try
    {
        reader.OpenPort();

        //-------------------------------------------------------
        // Read the source data
        //-------------------------------------------------------
        Console.Write("Please put only the source card " +
                      "on the reader and press <Enter>: ");
        Console.ReadLine();

        reader.RFReset(5);
        int ret = reader.ActivateCard(RequestCmd.WUPA);
        if (ret != 0) return;

        // Read the user data and store the data in the userDataList
        List<byte[]> userDataList = new List<byte[]>();

        for (byte page = 4; page < 16; page++)
        {
            byte[] baData = new byte[16];
            ret = reader.Mifare.Read(page, out baData);
            userDataList.Add(baData);
        }
        reader.HaltA();

        //-------------------------------------------------------
        // Write the data to a virgin card
        //-------------------------------------------------------
        Console.Write("Please put only the destination card " +
                      "on the reader and press <Enter>: ");
        Console.ReadLine();

        reader.RFReset(5);
        ret = reader.ActivateCard(RequestCmd.WUPA);

        if (ret != 0) return;

        for (byte page = 4; page < 16; page++)
        {
            reader.Mifare.ULWrite(page, userDataList[page - 4]);
        }
        reader.HaltA();
        Console.Write("Card copy finished.");
    }
```

Figure 8.30. Console output of the MifareUltra-LightClone.cs script.

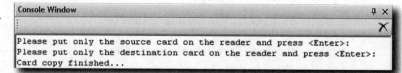

THE MIFARE ULTRALIGHT CARD   8.3

After successful execution of the script, the `MifareUltralightDump.cs` script can be used to read out and verify the memory contents.

8.3.5.6 Secure Data Storage

The use of card-specific encryption keys (key diversification – see Section 4.4.4.2) is an effective protection against the cloning of encrypted memory contents. These so-called derived keys can be generated using an encryption algorithm. A MIFARE Ultralight card's UID is a unique property of the card and cannot be changed.

Figure 8.31 shows a possible way in which a derived card-specific key can be generated from the UID. The master key is encrypted using a Single-DES algorithm. In this way, a card-specific Triple-DES key is generated. The 8-byte long UID is used for a Single-DES key. Using the card-specific key, the data (cleartext) – a string, for example – is encrypted and then stored on the card.

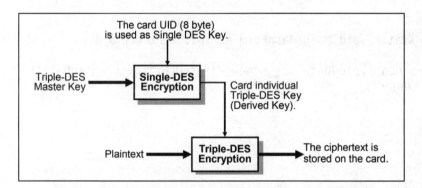

Figure 8.31. *Encryption of the card data using one of the card-specific keys.*

To demonstrate the principle with a short script, the example is kept as simple as possible. From a cryptographical perspective, using a Single-DES algorithm to generate the derived key is not suitable due to the short key length. The first two bytes of the UID present another problem. These two bytes have the same value for all MIFARE Ultralight cards. So, the 56-bit Single-DES key is reduced to 42 bits. Thus, the key is not secure enough against a brute force attack, which would try all possible key combinations.

In practice, better ways of generating card-specific keys are used. A cryptographically-secure method of generating the derived key is described in the NXP application note, *AN10922 Symmetric Key Diversifications*. Another common method is the use of a SAM (Secure Application Module) for the calculation of derived keys. The program in Listing 8.15 implements the sequence presented in Figure 8.31.

Listing 8.15. *MifareUltraLightClone.cs.*

```
using System;
using GS.Util.Hex;
using GS.CryptoUtil;
using GS.ISO14443_Reader;
using GS.ElektorRfidReader;
```

8 CARDS AND TAGS IN APPLICATION

```csharp
namespace GS.ElektorRfidReaderExample
{
    public class MifareULWriteEncryptedData
    {
```

The Triple-DES master key is stored in two byte arrays, `masterDesKeyA` and `masterDesKeyB`. The `plainString` variable contains the cleartext.

```csharp
byte[] masterDesKeyA = new byte[]{0x11,0x22,0x33,0x44,0x55,0x66,0x77,0x88};
byte[] masterDesKeyB = new byte[]{0x99,0xAA,0xBB,0xCC,0xDD,0xEE,0xFF, 0x00};
        string plainString = "I'm the secret data.";

        ElektorISO14443Reader reader;

        public MifareULWriteEncryptedData()
        {
            reader = new ElektorISO14443Reader();
        }

        public static void Main(MainForm script, string[] args)
        {
    MifareULWriteEncryptedData scr = new MifareULWriteEncryptedData();
            scr.Run();
        }

        void Run()
        {
  try
  {
    reader.OpenPort();
    reader.RFReset(5);
    reader.ActivateCard(RequestCmd.WUPA);
        Console.WriteLine("-------------------------------------------------");
        Console.WriteLine(" Step 1. Generate card individual 3-DES key... ");
        Console.WriteLine("-------------------------------------------------");
        Console.WriteLine("UID = 0x" +
                    HexFormatting.ToHexString(reader.UID, true));
```

In the first step, the two byte arrays are encrypted using a Single-DES operation and the result is saved in `cardDesKeyA` and `cardDesKeyB`. The card UID is used as a key for the Single-DES operation.

```csharp
                // Generate card individual 3-DES card key.
                SingleDESCrypto des = new SingleDESCrypto();
                des.Mode = System.Security.Cryptography.CipherMode.CBC;

                byte[] baSingleDesKey = new byte[8];

                for (int i = 0; i < 8; i++)
                {
                    baSingleDesKey[i] = reader.UID[i];
```

```
            }
            des.DesKey = baSingleDesKey;

            byte[] cardDesKeyA = des.Encrypt( masterDesKeyA );
            Console.WriteLine(HexFormatting.DumpHex("Card DesKeyA = 0x",
                            cardDesKeyA, cardDesKeyA.Length, 8));

            byte[] cardDesKeyB = des.Encrypt( masterDesKeyB );
            Console.WriteLine(HexFormatting.DumpHex("Card DesKeyB = 0x",
                            cardDesKeyB, cardDesKeyB.Length, 8));

            Console.WriteLine("");
      Console.WriteLine("---------------------------------------------");
      Console.WriteLine(" Step 2. Encrypt and write data to card...");
      Console.WriteLine("---------------------------------------------");
```

In the next step, the `plainString` string must be converted into a `byte` array (`baPlain`) for the encryption operation. Finally, `baPlain` is encrypted using a Triple-DES operation (in CBC mode) and the result is saved in the `baEncryptedData` byte array. The `ByteArrayToMifareUL()` utility method writes the encrypted data to the card. Should the number of bytes not be a multiple of 8, the remainder of the bytes will be padded with zeroes before being encrypted.

```
            TripleDESCrypto tripleDes = new TripleDESCrypto();

            tripleDes.Mode = System.Security.Cryptography.CipherMode.CBC;
            tripleDes.DesKeyA = cardDesKeyA;
            tripleDes.DesKeyB = cardDesKeyB;
            tripleDes.IV = new byte[] {0x00,0x00,0x00,0x00,0x00,0x00,0x00,0x00};

            byte[] baPlain = System.Text.Encoding.UTF8.GetBytes(plainString);
                Console.WriteLine(HexFormatting.Dump("Plain data = 0x",
                            baPlain, baPlain.Length, 8));
                byte[]baEncryptedData = tripleDes.Encrypt(baPlain);
                Console.WriteLine(HexFormatting.Dump("Enc. data = 0x",
                baEncryptedData , baEncryptedData .Length, 8));

            // Write data to card.
            ByteArrayToMifareUL(4, baEncryptedData );
        }
        catch (Exception ex)
        {
            Console.WriteLine(ex.Message);
        }
        finally
        {
            reader.ClosePort();
        }
    }
```

8 CARDS AND TAGS IN APPLICATION

Up to 48 bytes (Pages 4 to 15) can be written to a MIFARE Ultralight card using the `ByteArrayToMifareUL()` method.

```
void ByteArrayToMifareUL(byte page, byte[] value)
{
    int baLength = value.Length;
    int offset = 0;

    while(baLength > 0)
    {
        int len = Math.Min(4, baLength);

        byte[] baData = new byte[4];
        Array.Copy(value, offset, baData, 0, len);

        reader.Mifare.ULWrite(page, baData);
        page++;
        if (page > 15) return;
        baLength -= len;
        offset += len;
    }
}
```

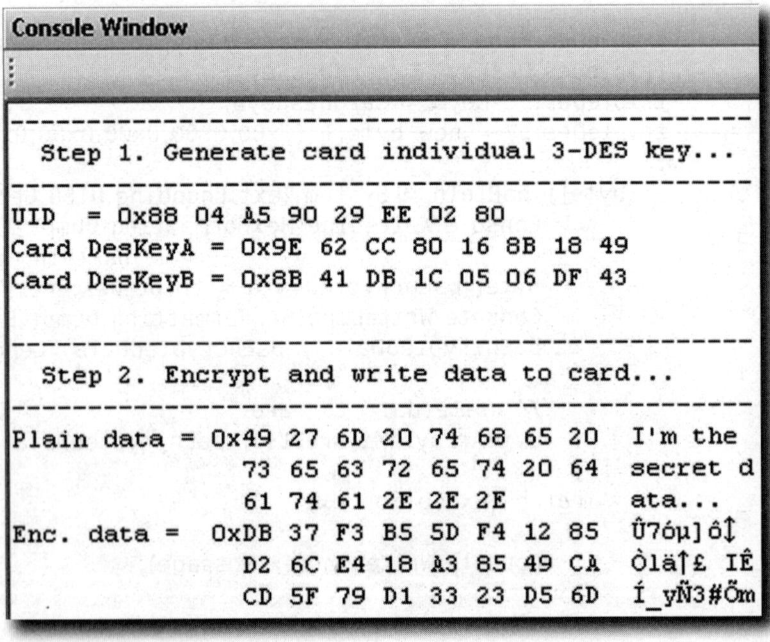

Figure 8.32. Console output of the MifareUltralightWriteEncryptedData.cs script.

```
Console Window
----------------------------------------------
 Step 1. Generate card individual 3-DES key...
----------------------------------------------
UID         = 0x88 04 A5 90 29 EE 02 80
Card DesKeyA = 0x9E 62 CC 80 16 8B 18 49
Card DesKeyB = 0x8B 41 DB 1C 05 06 DF 43

----------------------------------------------
 Step 2. Encrypt and write data to card...
----------------------------------------------
Plain data = 0x49 27 6D 20 74 68 65 20   I'm the
             73 65 63 72 65 74 20 64   secret d
             61 74 61 2E 2E 2E         ata...
Enc. data  = 0xDB 37 F3 B5 5D F4 12 85   Û7óµ]ôÎ
             D2 6C E4 18 A3 85 49 CA   Òlä↑£ IÊ
             CD 5F 79 D1 33 23 D5 6D   Í_yÑ3#Õm
```

Figure 8.33 shows the principle of decrypting the card data using the card-specific key. The first step is identical to the encryption of the card data (see Figure 8.31). The master key is also encrypted using a Single-DES algorithm here and the card-specific Triple-

DES key is derived from it. Then, the data stored on the card is read and decrypted. In the last step, the authenticity of the card is verified. For this, a simple comparison between the known cleartext and the previously encrypted card data is necessary. If the values are identical, the card has passed the authenticity test.

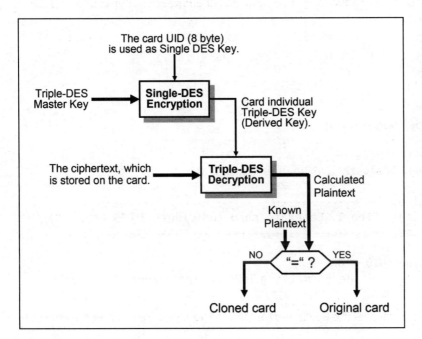

Figure 8.33.
Decrypting the card data using a card-specific key. By comparing the known cleartext with the data decrypted by the card, we can check whether it is a 'real' or a cloned card.

Listing 8.16. *MifareUltraLightReadEncryptedData.cs.*

```
using System;
using GS.Util.Hex;
using GS.CryptoUtil;
using GS.ISO14443_Reader;
using GS.ElektorRfidReader;

namespace GS.ElektorRfidReaderExample
{
    public class MifareULReadEncryptedData
    {
byte[] masterDesKeyA = new byte[] {0x11,0x22,0x33,0x44,0x55,0x66,0x77,0x88};
byte[] masterDesKeyB = new byte[] {0x99,0xAA,0xBB,0xCC,0xDD,0xEE,0xFF,0x00};

        string originalString = "I'm the secret data.";

        ElektorISO14443Reader reader;
        public MifareULReadEncryptedData()
        {
            reader = new ElektorISO14443Reader();
```

```
    }

    public static void Main(MainForm script, string[] args)
    {
   MifareULReadEncryptedData scr = new MifareULReadEncryptedData();
        scr.Run();
    }

    void Run()
    {
        try
        {
            reader.OpenPort();

            reader.RFReset(5);
            reader.ActivateCard(RequestCmd.WUPA);

Console.WriteLine("------------------------------------------------");
Console.WriteLine(" Step 1. Generate card individual 3-DES key... ");
Console.WriteLine("------------------------------------------------");

Console.WriteLine("UID = 0x" +
                  HexFormatting.ToHexString(reader.UID, true));
```

The calculation of the card-specific key is performed in the same manner as for encrypting the data.

```
                // Generate card individual 3-DES card key.
                SingleDESCrypto des = new SingleDESCrypto();
                des.Mode = System.Security.Cryptography.CipherMode.CBC;
                byte[] baSingleDesKey = new byte[8];
                for (int i = 0; i < 8; i ++)
                {
                    baSingleDesKey[i] = reader.UID[i];
                }
                des.DesKey = baSingleDesKey;

                byte[] cardDesKeyA = des.Encrypt( masterDesKeyA );
                 Console.WriteLine(HexFormatting.DumpHex("Card DesKeyA = 0x",
                                     cardDesKeyA, cardDesKeyA.Length, 8));

                byte[] cardDesKeyB = des.Encrypt( masterDesKeyB );
                 Console.WriteLine(HexFormatting.DumpHex("Card DesKeyB = 0x",
                                     cardDesKeyB, cardDesKeyB.Length, 8));
           Console.WriteLine("");
           Console.WriteLine("------------------------------------------------");
           Console.WriteLine(" Step 2. Read and decrypt card data... ");
           Console.WriteLine("------------------------------------------------");
```

In this case, the data is first read from the card and then decrypted.

```csharp
TripleDESCrypto tripleDes = new TripleDESCrypto();

tripleDes.Mode = System.Security.Cryptography.CipherMode.CBC;
tripleDes.DesKeyA = cardDesKeyA;
tripleDes.DesKeyB = cardDesKeyB;
tripleDes.IV = new byte[] {0x00,0x00,0x00,0x00,0x00,0x00,0x00,0x00};

    byte[] baEncryptedData = MifareULToByteArray(4);
    Console.WriteLine(HexFormatting.Dump("Enc. data = 0x",
            baEncryptedData, baEncryptedData.Length, 8));
    Console.WriteLine("");
    byte[] baDecData = tripleDes.Decrypt(baEncryptedData);
    Console.WriteLine(HexFormatting.Dump("Plain data = 0x",
            baDecData, baDecData.Length, 8));

string decString = System.Text.Encoding.ASCII.GetString(baDecData);

    Console.WriteLine("");
    Console.WriteLine("Decrypted String = " + decString);
```

In our example, the cleartext is known. In this way, a simple check as to whether the card is 'real' or not is possible.

```csharp
    Console.WriteLine("");
    Console.WriteLine("--------------------------------------------------");
    Console.WriteLine(" Step 3. Verify orig. data with decryp.data... ");
    Console.WriteLine("--------------------------------------------------");
            if (0 == String.Compare(originalString, 0, decString, 0,
                                        originalString.Length))
            {
                Console.WriteLine("This is an original card.");
            }
            else
            {
                Console.WriteLine("This is a cloned card !!!");
            }
        }
        catch (Exception ex) { Console.WriteLine(ex.Message); }
        finally { reader.ClosePort(); }
}
```

The `MifareULToByteArray()` method reads the entire data memory (Pages 4 to 15) from a MIFARE Ultralight card and returns the data in a 48-byte array.

```csharp
        byte[] MifareULToByteArray(byte page)
        {
            byte[] baData = new byte[4 * 12];
```

```
            int offset = 0;

            while(page < 16)
            {
                byte[] baPage = new byte[16];
                reader.Mifare.Read(page, out baPage);
                Array.Copy(baPage, 0, baData, offset, 4);
                page++;
                offset += 4;
            }
            return baData;
        }
    }
}
```

An Ultralight card can be cloned using the `MifareUltralightClone.cs` script (see Listing 8.14). Comparing the two console outputs, it can easily be seen that the cards' data are identical. However, each card has a different Triple-DES key. Therefore, the value of the decrypted ciphertext (see Figure 8.35) is incorrect, and a simple comparison with the original cleartext is enough to expose the card as a clone.

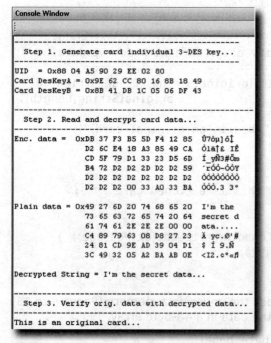

Figure 8.34. Verification of a 'real' card.   **Figure 8.35.** This card was exposed as a clone.

8.4 The MIFARE Classic Card

The MIFARE Mini and MIFARE Classic 1K and 4K cards are contactless memory cards with cryptographic functions. These members of the MIFARE family should not be used for security applications, as the proprietary encryption algorithm (Crypto 1) was cracked in 2008 (see Section 5.2.1.3). Many applications require little or very weak protection. For these, the MIFARE Classic cards are still usable. As you learned in Section 5.2.5, the memory layout of the MIFARE Plus successor is identical to the MIFARE Classic 4K card. MIFARE Plus cards are export controlled and are therefore not publicly available. The MIFARE Classic doesn't have any such limitation, and can also be ordered in very small quantities. For these reasons, MIFARE Classic cards are very suitable for a first excursion into RFID technology.

8.4.1 MIFARE Classic 1K Card Memory Organization

A MIFARE 1K card's 1 KB EEPROM memory is divided into 16 independent sectors. Each sector consists of 4 blocks. A block always has the fixed size of 16 bytes. The final block of a sector is known as a Sector Trailer. This contains the MIFARE Key A and Key B, as well as three bytes (Access Condition) for controlling access rights. A MIFARE key has a length of 48 bits, or 6 bytes. One byte is left over, which may be used for additional data storage. The remaining three blocks of 16 bytes each are used to store application data.

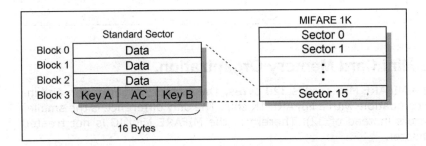

Figure 8.36. MIFARE 1K card memory organization.

During manufacture, the card serial number (UID), the UID checksum, both ATQA bytes and the SAK byte are written to the first block of the card (Block 0 of Sector 0). The remaining bytes are used to store production data, which is not specified in detail. After the manufacturing process, the data in Block 0 is read-only and cannot be changed by the user. So, a maximum of 768 bytes of user data can be written to a MIFARE 1K card.

8.4.2 MIFARE Classic 4K Card Memory Organization

Figure 8.37 shows the memory organization of a MIFARE 4K card. The memory layout of the first 2 KB is the same as that of a MIFARE 1K card. This consists of 31 sectors of 4 blocks each. A sector size of 4 blocks is known as a standard sector. A 4K card is thus fully backward-compatible with the MIFARE 1K card. The upper 2 KB is divided into 8 sectors, each of 16 blocks. These extended sectors contain 15 data blocks and a Sector Trailer.

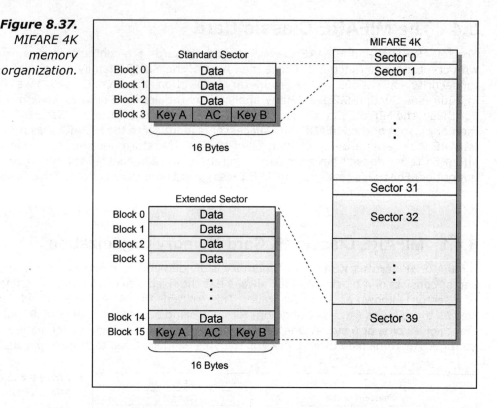

Figure 8.37. MIFARE 4K memory organization.

8.4.3 MIFARE Mini Card Memory Organization.

The memory size of a MIFARE Mini card is 320 bytes. The memory organization and instruction set is fully compatible with a MIFARE 1K chip. The only difference is the smaller EEPROM size (5 sectors instead of 32). Therefore, the MIFARE Mini IC is not treated separately in this chapter.

8.4.4 Instruction Set

The MIFARE Classic instruction set can be divided into three groups: card activation, authentication and memory manipulation. A complete list of commands is shown in Table 8.15. After card activation, authentication between reader and card takes place with the help of the authentication commands, using either Key A or Key B. The access rights for each individual block in a sector are defined by the three Access Condition bytes in the Sector Trailer. Configuration of open memory access without authentication is not supported. The default Access Condition value on a new MIFARE 1K/4K card deactivates Key B. Key B could thus theoretically be used for additional data storage. The value of both keys on a new card is 0xFF FF FF FF FF FF.

Authentication is only valid for a particular sector, but allows full access to all blocks within the sector. For example, Auth. Key A need only be sent to the card once for all blocks within the sector to be read or written. Should you need to work on a block outside this sector, a new authentication is required (see Figure 8.38).

Table 8.15. MIFARE Ultralight instruction set.

Command Type	Command Code	Card Command	SmartCard Magic.NET Button Label
Card Activation	0x26	REQA	Req A
	0x52	WUPA	Wup A
	0x93	Anti-collision of Cascade Level 1	AC 1
	0x93	Select of Cascade Level 1	Select 1
	0x50	HALTA	Halt A
Authentication	0x60	Authentication with Key A	Auth. Key A
	0x61	Authentication with Key B	Auth. Key B
Memory Manipulation	0x30	Read	Read
	0xA0	Write	Write
	0xC0	Decrement	Decrement (inclusive Transfer)
	0xC1	Increment	Increment (inclusive Transfer)
	0xC2	Restore	Restore (inclusive Transfer)
	0xB0	Transfer	–

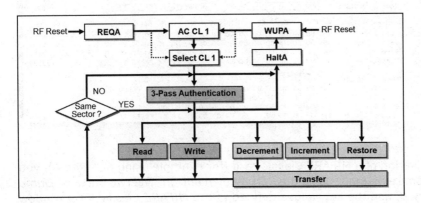

Figure 8.38. MIFARE Classic memory organization.

The Read and Write commands allow reading and writing of a block, i.e. 16 bytes. An electronic stock inventory can be realized using the Decrement, Increment, Restore and Transfer commands. These commands require a special data format (see Section 8.4.5).

With Smart Card Magic.NET, it is quite simple to get to know the properties of a MIFARE Classic card. All commands can be sent to the card by clicking on the corresponding button in the MIFARE Classic window.

8 CARDS AND TAGS IN APPLICATION

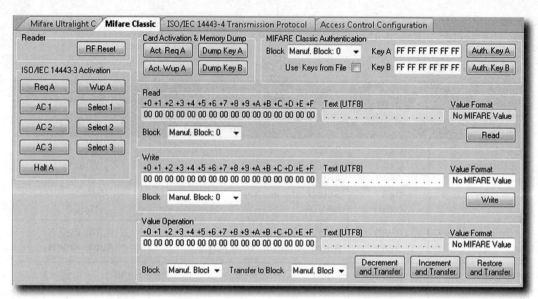

Figure 8.39. *Smart Card Magic.NET: The MIFARE Classic window.*

Figure 8.40. *Dump Key A key allows the entire memory content of a MIFARE 1K/4K card to be read.*

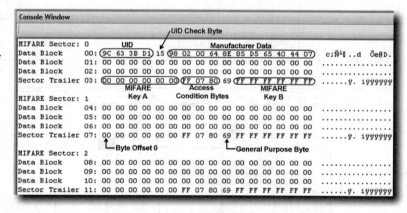

After successful activation of a MIFARE Classic card (for example using Act. Req A), you can read the card's entire memory using the *Dump Key A button*. The two buttons, *Dump Key A* and *Dump Key B*, use the key values defined in the *MIFARE Classic Keys* window for authentication. The keys can be saved to an XML file and later opened using the *Open* button if needed. When the application is closed, all changes to the *MIFARE Classic Keys* window are automatically saved to the `mifareKeys.xml` configuration file. The keys saved in the `mifareKeys.xml` file automatically populate the *MIFARE Classic Keys* window at the next program start.

The *Tools* → *Show Local User App Data Path* opens Windows Explorer in the folder that contains the Smart Card Magic.NET configuration file. If the *Use Keys from File* option is active, then the keys in the MIFARE Classic Keys window are used when the *Auth. Key A* or *Auth. Key B* buttons are clicked.

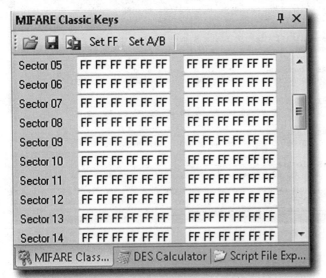

Figure 8.41.
The MIFARE Classic Key window allows the configuration of MIFARE keys A and B.

Reading and writing of data always takes place in a block. In Smart Card Magic.NET, you need only activate the card and select a block. As shown in Figure 8.42, the blocks are addressed in ascending order, beginning at 0.

For programming, these addresses are relevant, not the byte address representation in the datasheet (see also memory organization Figure 8.36 and Figure 8.37). In the data sheet, the bytes in a sector are always addressed beginning at 0. Writing invalid data to a sector trailer can result in the sector being permanently locked. Therefore, only data blocks will be used for the first experiments.

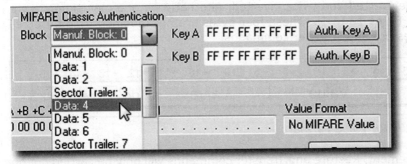

Figure 8.42.
Selecting a MIFARE block.

Figure 8.43 shows the *Trace* window for a single write and multiple read accesses. The read command for Block 8 won't work because it is outside of the authenticated sector. In this example, the reader is authenticated only for access to Sector 1, i.e. Blocks 4 to 7. As with any other incorrect command, the card switches to the IDLE state and the card must be re-activated.

Figure 8.43.
MIFARE 1K card: writing and reading data.

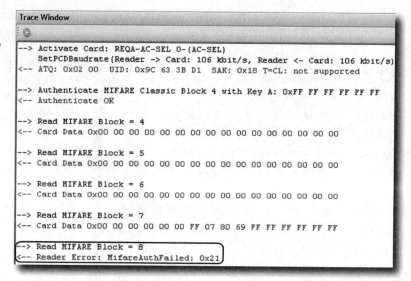

8.4.5 The MIFARE Value Format

Simple counters and electronic inventories can be realized using the Decrement, Increment, Restore and Transfer commands. For these four commands, a data block must be formatted in the MIFARE Value Format. This special data format allows the detection and correction of errors, as well as the simple implementation of a backup management system.

Figure 8.44 shows the MIFARE Value Format, in which 16 bytes is required to represent one 32-bit signed integer. Negative values are represented in 2's complement. For data redundancy, this value is stored three times in a data block (twice uninverted and once inverted. In the remaining four bytes, the sector address, which is required for the implementation of a backup management system, can be saved. The address byte is stored four times – twice uninverted and twice inverted. Alternatively, one may simply write 0x00 FF 00 FF to Bytes 12 to 15. These four bytes are not modified by the Decrement, Increment, Restore and Transfer commands.

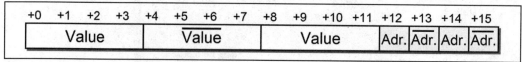

Figure 8.44. The MIFARE Value Format.

The Decrement, Increment, Restore and Transfer commands will only be carried out on a data block that was previously correctly initialized in the Value Format. Initialization of the value is quite simple using the Write command. The value is read with the Read command. Smart Card Magic.NET provides an easy way to convert an integer into the MIFARE Value Format (see Figure 8.45). The decimal value, 125, is 0x00 00 00 7D in hexadecimal (inverted: 0xFF FF FF 82). You can see in Figure 8.45 that all three integer values are represented in the so-called 'little-endian' format. In the little-endian format, the least-significant byte is stored at the lowest address.

Figure 8.45. Conversion of an integer, 125, into the MIFARE Value Format.

The `MifareClassicUtil` class in the `GS.ISO14443_Reader` namespace has three methods for working with the MIFARE Value Format. `Int32ToValueFormat()` converts the Integer data type into the MIFARE Value Format. The four address bytes are initialized with the value `0x00 FF 00 FF`.

Listing 8.17. Excerpt from MifareClassicUtil.cs.

```
    public static byte[] Int32ToValueFormat(int value, byte mifareBlock)
    {
        byte[] mifareDataBlock = new byte[16];
        byte[] intArray = BitConverter.GetBytes(value);

        mifareDataBlock[0] = intArray[0];
        mifareDataBlock[1] = intArray[1];
        mifareDataBlock[2] = intArray[2];
        mifareDataBlock[3] = intArray[3];

        mifareDataBlock[4] = (byte)~intArray[0];
            mifareDataBlock[5] = (byte)~intArray[1];
            mifareDataBlock[6] = (byte)~intArray[2];
            mifareDataBlock[7] = (byte)~intArray[3];

        mifareDataBlock[8] = intArray[0];
        mifareDataBlock[9] = intArray[1];
        mifareDataBlock[10] = intArray[2];
        mifareDataBlock[11] = intArray[3];

        mifareDataBlock[12] = mifareBlock;
        mifareDataBlock[13] = (byte)~mifareBlock;
        mifareDataBlock[14] = mifareBlock;
        mifareDataBlock[15] = (byte)~mifareBlock;

        return mifareDataBlock;
    }
```

The `IsValueFormat()` method checks to see whether a 16-byte array contains a number in the MIFARE Value Format or not. With this method, the consistency of the integer value stored in triplicate, as well as the address byte stored in quadruplicate, is validated, so you can also detect transmission errors when reading from MIFARE Value data block.

8 CARDS AND TAGS IN APPLICATION

```csharp
public static bool IsValueFormat(byte[] data)
{
    if (null == data) return false;

    if (data.Length != 16) return false;

    if ((data[0] == data[8])  &&
        (data[1] == data[9])  &&
        (data[2] == data[10]) &&
        (data[3] == data[11]) &&

        (data[0] == (byte)~data[4]) &&
        (data[1] == (byte)~data[5]) &&
        (data[2] == (byte)~data[6]) &&
        (data[3] == (byte)~data[7]) &&

        (data[12] == data[14]) &&
        (data[13] == data[15]) &&

        (data[12] == (byte)~data[13]))
    {
        return true;
    }
    else
    {
        return false
    }
}
```

The `ValueFormatToInt32()` method checks the validity of a number in the MIFARE Value Format before doing the conversion. Should an error occur, the method returns a *null* value instead of the converted integer. The return value of this method is a so-called 'nullable type', which is declared with a question mark after the data type.

A nullable type behaves like a value type, but can also take the value *null*. In this way, one can identify that a variable has not been initialized or that a function has not returned a valid value. Before using a nullable type variable, it is necessary to check if it contains a valid value or not. For this purpose, the nullable type contains the `HasValue()` method, which returns `true` if the variable contains a value. Alternatively, the variable can be compared with *null*.

```csharp
public static int? ValueFormatToInt32(byte[] data)
{
    if (IsValueFormat(data))
    {
        return BitConverter.ToInt32(data, 0);
    }
    else
    {
```

```
        return null;
    }
}
```

8.4.6 Decrement, Increment, Restore and Transfer

The Decrement and Increment commands decrease or increase the value of a data block by a specified amount and save the result to an internal card register. It is necessary to write this register value to a data block using the Transfer command. The Restore command reads the value of a MIFARE Value data block and saves it to an internal card register.

In the MIFARE Classic window, Smart Card Magic.NET offers an easy way to test the Decrement, Increment, Restore and Transfer in practice.

In the next example, a simple electronic wallet is implemented. The amount in the wallet is stored in Data Block 4. After each transaction, a copy of the value is written to Data Block 5 using the Restore and Transfer commands. Saving the value to a second data block allows the creation of a backup management system. The backup management logic must still be implemented in the reader application.

Figure 8.46. *Smart Card Magic.NET: Value Format commands.*

In the first step, the value 125 is written to the card in the Value format. Then, an amount of 15 is deducted. For this, the following steps are required:

Activate the card by clicking on *Act. Req A*.

Select Block 4 and, by clicking on *Auth. Key A*, authenticate with the standard key, 0xFF FF FF FF FF FF.

In the *Value Format* text field (*Write* group), input the number 125 and then write the value to Data Block 4 using the *Write* button.

In the *Value Operation* group, select *Block 4* from the *Block* listbox, as well as in the *Transfer to Block* listbox. An amount of 15 will be deducted when a 15 is placed in the *Value Format* text field (*Value Operation* group) and the *Decrement and Transfer* button is clicked.

To create a backup prior to a debit transaction, the Restore and Transfer commands must be used. Restore reads the value from the EEPROM, that is, from a data block, and saves the value to an internal card register. To do this, select *Block 4* in the *Block* listbox and *Block 5* in the *Transfer to Block* listbox. Then, click on the *Restore and Transfer* button.

The Read command can be used to read the values in Blocks 4 and 5. Should a value in the MIFARE Value Format be read, this is shown in the *Trace* window (see Figure 8.47).

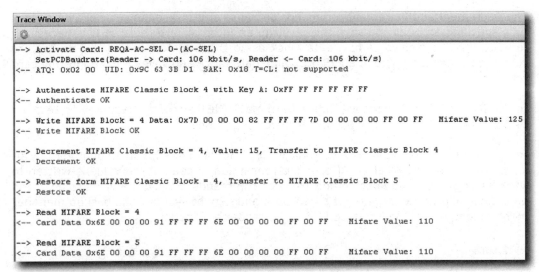

Figure 8.47. *Command sequence for an electronic wallet.*

Note	With an electronic wallet, it is necessary to manipulate monetary amounts in floating point format. For example, the price for a single ticket may be 2.65 euro. Because the MIFARE Value Format doesn't support a floating point format, it's necessary to use a fixed point number format. For example, to represent the value of 2.65, the integer 265 is saved in MIFARE Value Format and the 'imaginary' point is taken into account by the reader application.

8.4.7 Changing the Keys and Access Condition

When a MIFARE Classic card is personalized, all personal details of the cardholder, including issue date, expiry date, as well as information about the card application, are written to the card. Finally, the MIFARE keys and the access permissions can be changed by the card issuer.

Before we discuss the steps required to change the keys and access conditions individually, we consider the Sector Trailer default values for a MIFARE 4K card.

In Figure 8.48, the value of MIFARE Block 7, that is, the Sector Trailer for Sector 1, is read. For Key A, the value 0x00 00 00 00 00 00 is read, although it is saved as 0xFF

FF FF FF FF FF: the card returns a dummy value instead of the actual key value, as the Key A value can never be read from the card. This has the advantage that the key is never transferred directly between the reader and card. The default access condition value of 0xFF 07 80 deactivates Key B. In this case, Bytes 0xA – 0xF can be used for additional data storage. For this reason, when reading the Sector Trailer, the stored value of 0xFF FF FF FF FF FF is displayed. Byte 9 (General Purpose Byte) has no special meaning if a MIFARE Application Directory (MAD) is not being used. If the MAD is not used for management of different applications on the card, this byte can also be used for data storage.

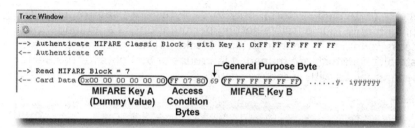

Figure 8.48. Reading a brand-new MIFARE 4K card's MIFARE Sector Trailer.

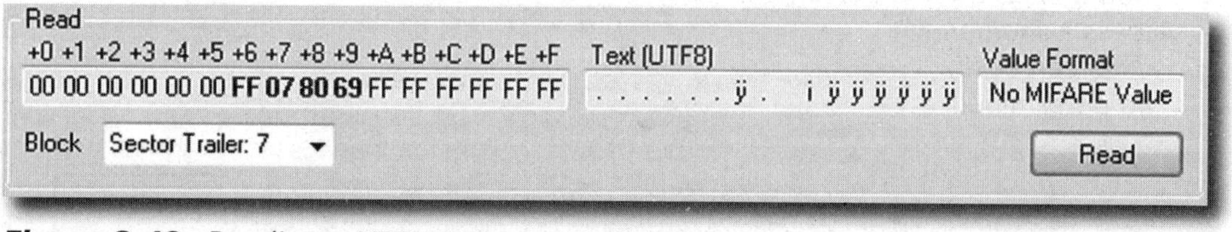

Figure 8.49. Reading a MIFARE Sector Trailer.

The MIFARE 1K/4K datasheet has extensive tables showing access condition configurations. Each group of 3 bits defines the access rights for a specific block within the sector. These three bits are stored twice – once uninverted and once inverted, within the 3 bytes. This results in 24 bits, or 3 bytes (3-bit access condition x 4 blocks x 2 = 24 bits).

Determining the correct value for an access condition from the table in the datasheet is very complicated. The MifareWnd PC software can be used to determine the correct values easily. MifareWnd is a Windows program for the NXP Pegoda Reader. You can download this software for free from the manufacturer's website. To determine the correct value for the access condition byte, no reader hardware must be connected to the PC. Using the *File* → *New Mifare* menu option, the *Mifare* window is opened. A Sector Trailer Block (e.g. 7) is then chosen and the *Edit AC* button is clicked.

In the *Access Condition* configuration dialog (see Figure 8.51), the access conditions supported by the MIFARE chip can easily be retrieved. For example, the radio button on the right of the *Block 0* label is for determining the value of the access condition for the first data block within a standard sector.

8 CARDS AND TAGS IN APPLICATION

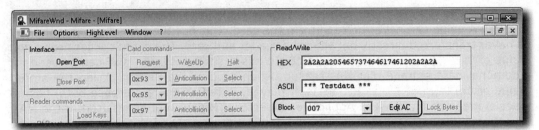

Figure 8.50. MifareWnd: the Mifare window.

Even in 16-block sectors (extended sectors), only three bytes are used to define the access condition. So, with only four possible setting permutations, it's no longer possible to assign an access condition to each individual block. In this case, there are also three settings for data blocks, which are managed in groups of 5. Rights for the Sector Trailer are defined separately, as with standard sectors.

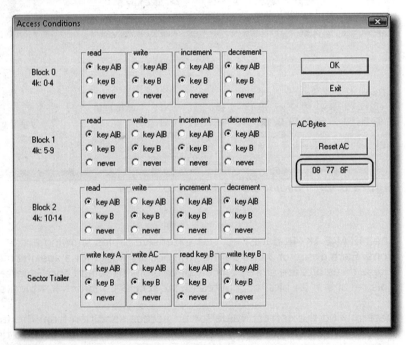

Figure 8.51. MifareWnd: Using the Access Conditions configuration dialog, the Access Condition values can easily be determined.

The two MIFARE keys, A and B, can be assigned different access rights by configuring the Access Condition. An operation (e.g. Read, Write, Decrement, and Increment) may either be permitted following authentication with either of the two keys, or only after authentication with Key B. In this way, Key B can be assigned a higher priority than that of Key A.

In the following example, both key values and the access condition of Sector 1, i.e. Data Blocks 4 and 6 and Sector Trailer 7, are modified. The value for Key A is changed to 0x11 11 11 11 11 11 and Key B to 0x22 22 22 22 22 22. Setting the Access Condition is done as per Figure 8.51 (value = 0x08 77 8F).

The following steps are required to change the MIFARE keys and the Access Condition:
- Activate the card by clicking *Act. Req A*.
- Select Block 7 and authenticate with the standard key, `0xFF FF FF FF FF FF`, by clicking on *Auth. Key A*.
- With the *Write* button, save the value `0x11 11 11 11 11 11 08 77 8F 00 22 22 22 22 22 22` to Block 7.

```
Trace Window
--> Activate Card: REQA-AC-SEL 0-(AC-SEL)
    SetPCDBaudrate(Reader -> Card: 106 kbit/s, Reader <- Card: 106 kbit/s)
<-- ATQ: 0x02 00  UID: 0x86 A7 04 01  SAK: 0x18 T=CL: not supported
--> Authenticate MIFARE Classic Block 7 with Key A: 0xFF FF FF FF FF FF
<-- Authenticate OK
--> Write MIFARE Block = 7 Data: 0x11 11 11 11 11 11 FF FF 7D 00 22 22 22 22 22 22
<-- Write MIFARE Block OK
```

Figure 8.52.
Modifying the MIFARE keys and the access conditions.

The modified keys become active after re-activating the card. Finally, authenticating Sector 1 with Key A (`0xFF FF FF FF FF FF`) causes an error (see Figure 8.53). An attempt to authenticate using the wrong key sends the card into the IDLE state, and no response is returned. Renewed card activation is then required. Authenticating using the new key values for Key A and Key B works correctly.

The modified Access Condition allows writing of data only by authenticating with Key B. Reading is possible using either Key A or Key B. This key hierarchy also affects the Decrement and Increment commands. To increase a value (Increment), authentication with Key B is required. Decrementing a value is possible using either the *Auth. Key A* or *Auth. Key B* commands.

```
Trace Window
    RF Reset(5 ms)
--> Activate Card: REQA-AC-SEL 0-(AC-SEL)
    SetPCDBaudrate(Reader -> Card: 106 kbit/s, Reader <- Card: 106 kbit/s)
<-- ATQ: 0x02 00  UID: 0xA6 EE 04 01  SAK: 0x18 T=CL: not supported
--> Authenticate MIFARE Classic Block 7 with Key A: 0xFF FF FF FF FF FF
<-- Reader Error: MifareAuthFailed: 0x21
--> Activate Card: REQA-AC-SEL 0-(AC-SEL)
    SetPCDBaudrate(Reader -> Card: 106 kbit/s, Reader <- Card: 106 kbit/s)
<-- ATQ: 0x02 00  UID: 0xA6 EE 04 01  SAK: 0x18 T=CL: not supported
--> Authenticate MIFARE Classic Block 7 with Key A: 0x11 11 11 11 11 11
<-- Authenticate OK
--> Authenticate MIFARE Classic Block 7 with Key B: 0x22 22 22 22 22 22
<-- Authenticate OK
```

Figure 8.53.
Verification of the modified key values.

In a reader application, Key B is used for personalization as well as to credit the ticket values. Therefore, the reader at a debit terminal need only store Key A. Debit terminals are usually publicly accessible, and thus much more exposed to attacks.

8 CARDS AND TAGS IN APPLICATION

Should hackers gain access to the debit terminal key, they would only be able to debit value from the card. Key security can be increased using a corresponding key management system in which a different key is used for each card (see Section 8.3.5.6).

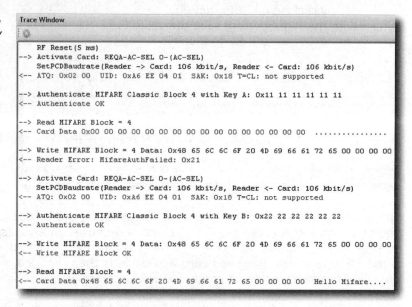

Figure 8.54. Verifying the key hierarchy.

8.4.8 Elektor RFID Reader Library: MIFARE Classic

8.4.8.1 The MifareClassicUtil Class

The most important utility classes for programming MIFARE 1K and 4K cards are found in the `MifareClassicUtil` class, in the `GS.ISO14443_Reader` namespace. All of these methods are static and are therefore called directly using just the class's name. It is not necessary to generate an instance of the class. Static methods are comparable to C functions.

Table 8.16. MifareClassicUtil class methods.

Method	Description
`static bool IsSectorTrailer( int mifareBlock)`	Returns true if the block is a Sector Trailer.
`static bool IsFirstDataBlock( int mifareBlock)`	Returns true if the block is the first block of data within the sector.
`static int GetMifareSector( int mifareBlock)`	Determines the address of the MIFARE sector.

Method (continued)	Description (continued)
`static bool IsValueFormat( byte[] data)`	Returns true of the parameter data in the byte array passed is a value in the MIFARE Value Format.
`static byte[] Int32ToValueFormat( int value)`	Converts the paramater value into the MIFARE Value Format.
`static int? ValueFormatToInt32( byte[] data)`	Converts a value in the MIFARE Value Format to an integer value. Before conversion, a format check takes place. If an error occurs, this method returns `null` instead of the converted integer value.

A single MIFARE key pair is managed within the `MifareClassicKey` class.

Table 8.17. Properties of the MifareClassicKey class.

Property	R/W	Description
`string KeyA`	R/W	Gets or sets the value of the MIFARE Classic Key A. All of the characters in this string are interpreted as hexadecimal values. White space is ignored.
`string KeyB`	R/W	Gets or sets the value of the MIFARE Classic Key B. All characters in this string are interpreted as hexadecimal values. White space is ignored.

Table 8.18. MifareClassicKey class methods.

Method	Description
`byte[] KeyAToByteArray()`	Converts the MIFARE Classic Key A into a byte array.
`byte[] KeyBToByteArray()`	Converts the MIFARE Classic Key B into a byte array.

The `MifareClassicKeys` class is used to manage all of the keys on a MIFARE 1K/4K card. Each sector is assigned a key pair. Should a new object of type `MifareClassicKeys` be created, 40 key pairs are created and initialized with the default value, `0xFF FF FF FF FF FF`. An individual key can be referenced using an index, much like the element of an array. Additionally, this class offers the ability to make the key values persistent – by saving them to an XML file.

8 CARDS AND TAGS IN APPLICATION

Table 8.19. MifareClassicKeys class method.

Method	Description
`public MifareClassicKey this[int index]`	Gets or sets the MIFARE Classic key values at the given index location.
`public void Save( string path)`	Saves all 40 key pairs to an XML file.
`public void Load( string path)`	Reads the key values from an XML file.

In the following example (`MifareClassicKeys.cs`), an object of type `MifareClassicKeys` is created. In the program, only the MIFARE keys in Sector 0 and Sector 1 are changed. Finally, all the values of all key pairs are output to the console.

Listing 8.18. MifareClassicKeys.cs.

```
using System;
using GS.Util.Hex;
using GS.ISO14443_Reader;
using GS.ElektorRfidReader;

namespace GS.ElektorRfidReaderExample
{
    public class ExampleMifareClassicKeys
    {
        public static void Main(MainForm script, string[] args)
        {
            MifareClassicKeys mifareKeys = new MifareClassicKeys();

mifareKeys[0].KeyA = "11 22 33 44 55 66"; // Sector 0: MIFARE Key A
mifareKeys[0].KeyB = "77 88 99 AA BB CC"; // Sector 0: MIFARE Key B

mifareKeys[1].KeyA = "A1 A2 A3 A4 A5 A6"; // Sector 1: MIFARE Key A
mifareKeys[1].KeyB = "B1 B2 B3 B4 B5 B6"; // Sector 1: MIFARE Key B

            for (int i = 0; i < 40; i++)
            {
                Console.WriteLine(
                    string.Format("Sector {0}: Key A = 0x{1} Key B = 0x{2}",
                        i, mifareKeys[i].KeyA, mifareKeys[i].KeyB));
            }
        }
    }
}
```

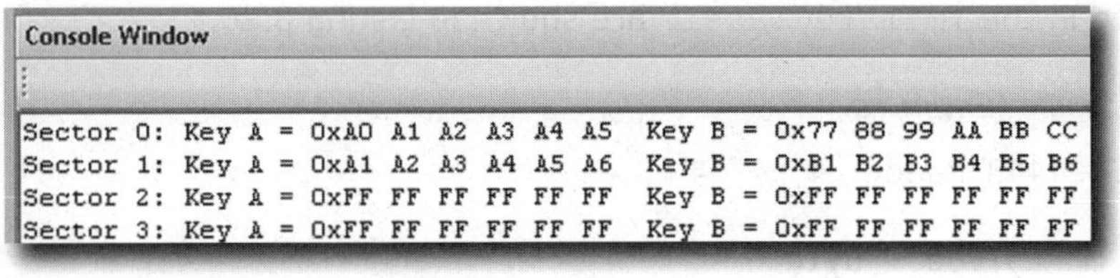

Figure 8.55.
Output of all MIFARE keys to the console.

```
Console Window
Sector 0: Key A = OxA0 A1 A2 A3 A4 A5   Key B = 0x77 88 99 AA BB CC
Sector 1: Key A = OxA1 A2 A3 A4 A5 A6   Key B = 0xB1 B2 B3 B4 B5 B6
Sector 2: Key A = 0xFF FF FF FF FF FF   Key B = 0xFF FF FF FF FF FF
Sector 3: Key A = 0xFF FF FF FF FF FF   Key B = 0xFF FF FF FF FF FF
```

8.4.8.2 The IMifareClassic Interface

All methods for MIFARE 1K/4K card memory are declared in the `IMifareClassic` interface. With the exception of `ReadBlocks()` and `WriteBlocks()`, the methods have the same purpose as the corresponding buttons in the *MIFARE Classic* window. `ReadBlocks()` and `WriteBlocks()` allow time-optimized reading and writing of up to 8 data blocks.

Listing 8.19. *IMifareClassic.cs.*

```csharp
using System;

namespace GS.ISO14443_Reader
{
    public interface IMifareClassic
    {
        int ClassicAuth(byte block, byte[] key, MFKeyType mfKeyType);
        int ClassicAuth(byte block, string key, MFKeyType mfKeyType);

        int Read(byte block, out byte[] data);
        int ReadBlocks(byte block, string key, MFKeyType mfKeyType,
                    out byte[] data, int numberOfBlocks);
        int ReadBlocks(byte block, byte[] key, MFKeyType mfKeyType,
                    out byte[] data, int numberOfBlocks);

        int Write(byte block, byte[] data);
        int Write(byte block, string data);
        int Write(byte block, int mfIntValue);
        int WriteBlocks(byte block, string key, MFKeyType mfKeyType,
                    byte[] data, int numberOfBlocks);
        int WriteBlocks(byte block, byte[] key, MFKeyType mfKeyType,
                    byte[] data, int numberOfBlocks);

    int Decrement(byte sourceBlock, int mfIntValue, byte destinationBlock);
    int Decrement(byte sourceBlock, byte[] mfValue, byte destinationBlock);
    int Increment(byte sourceBlock, int mfIntValue, byte destinationBlock);
    int Increment(byte sourceBlock, byte[] mfValue, byte destinationBlock);
        int Restore(byte sourceBlock, byte destinationBlock);
    }
}
```

8 CARDS AND TAGS IN APPLICATION

The enumerations used in the `IMifareClassic` are shown in Listing 8.20.

Listing 8.20. *MifareClassic_Enum.cs.*

```
namespace GS.ISO14443_Reader
{
    public enum MFKeyType : byte
    {
        KeyA,
        KeyB
    }

    public enum MFValueOP : byte
    {
        Decrement = 0xC0,
        Increment = 0xC1,
        Restore = 0xC2
    }
}
```

Table 8.20. *The MIFARE Classic methods for authentication.*

Method	Description

`int ClassicAuth(`
`byte block,`
`byte[] key,`
`MFKeyType mfKeyType)`
	Authenticates a MIFARE sector. A block address is passed in the block parameter, but the entire corresponding section is authenticated. The MIFARE key is passed as an array of bytes in the key parameter. The `mfKeyType` parameter determines which of the two MIFARE keys will be used. Valid values are `MFKeyType.KeyA` and `MFKeyType.KeyB`.

`int ClassicAuth(`
`byte block,`
`string key,`
`MFKeyType mfKeyType)`
	This variation of the method interprets the MIFARE key as a hexadecimal string.

Table 8.21. *The MIFARE 1K/4K reading and writing methods.*

Method	Description
`int Read(byte block,` `out byte[] data)`	Reads a data block (16 bytes).

`int ReadBlocks(`
`byte block,`
`byte[] key,`
`MFKeyType mfKeyType,`
`out byte[] data,`
`int numberOfBlocks)`
	Combines the `Read()` and `ClassicAuth()` methods, making possible the reading of up to 8 data blocks from a sector at once.

Method (continued)	Description (continued)
int ReadBlocks(byte block, string key, MFKeyType mfKeyType, out byte[] data, int numberOfBlocks)	This variant of the `ReadBlocks()` method accepts the MIFARE key as a hexadecimal string.
int Write(byte block, byte[] data);	Writes a data block (16 bytes). The data to be written is passed as an array of bytes.
int Write(byte block, string data);	In this variant, the data is sent as a string in hexadecimal or ASCII format (see text).
int WriteBlocks(byte block, byte[] key, MFKeyType mfKeyType, byte[] data, int numberOfBlocks);	Combines the two methods, `ClassicAuth()` and `Write()`, allowing the writing of up to 8 blocks to a sector at once.
int WriteBlocks(byte block, string key, MFKeyType mfKeyType, byte[] data, int numberOfBlocks);	Variant that allows the string to be passed as a hexadecimal string.

Table 8.22. The MIFARE Value Format methods.

Method	Description
int Write(byte block, int mfIntValue);	If the `Write()` method is passed an integer value, the block is initialized to the MIFARE Value Format.
int Decrement(yte sourceBlock, int mfIntValue, byte destinationBlock)	The `Decrement()` method reduces the value of the MIFARE value block (`sourceBlock` parameter) by the amount specified in the integer parameter, `mfIntValue`. The value is then stored in the data block referenced by the `destinationBlock` parameter. The `sourceBlock` and `destinationBlock` addresses must be within the same sector. This method uses the MIFARE Decrement and Transfer methods.
int Decrement(byte sourceBlock, byte[] mfValue, byte destinationBlock)	This variant of the `Decrement()` method transfers the amount as a byte array. The amount must be stored in the 16-byte array in the MIFARE Value Format.

8 CARDS AND TAGS IN APPLICATION

Method (continued)	Description (continued)

`int Increment(`
`byte sourceBlock,`
`int mfIntValue,`
`byte destinationBlock)`
	Increases the value of the MIFARE Value Block (`sourceBlock` parameter) by the amount specified in the integer parameter, `mfIntValue`). The value is then stored in the data block referenced by the `destinationBlock` parameter. The `sourceBlock` and `destinationBlock` addresses must be within the same sector. This method uses the MIFARE INCREMENT and Transfer methods.

`int Increment(`
`byte sourceBlock,`
`byte[] mfValue,`
`byte destinationBlock)`
	This variant of the `Increment()` method accepts the amount as a byte array. The amount must be stored in the 16-byte array in the MIFARE Value Format.

`int Restore(`
`byte sourceBlock,`
`byte destinationBlock)`
	The `Restore()` method copies the value of a data block to another data block, using the MIFARE Restore and Transfer commands. The source data block must be a valid number in the MIFARE Value Format. Also, the `sourceBlock` and `destinationBlock` addresses must be within the same sector.

8.4.9 Program and Case Studies

8.4.9.1 Writing and Erasing Data

Our first example for a MIFARE 1K/4K card merely demonstrates the `Write()` method, with which values in various data formats are written to the card. In this example, validation of the reader status is omitted completely.

Listing 8.21. *MifareClassicWrite.cs.*

```
using System;
using GS.Util.Hex;
using GS.ISO14443_Reader;
using GS.ElektorRfidReader;

namespace GS.ElektorRfidReaderExample
{
    public class MifareClassicWrite
    {
        public static void Main(MainForm script, string[] args)
        {
            ElektorISO14443Reader reader = new ElektorISO14443Reader();
            try
            {
                reader.OpenPort();
                reader.RFReset(5);
                reader.ActivateCard(RequestCmd.WUPA);
```

After card activation, authentication of Sector 1 is done by calling the `ClassicAuth()` method. In the first parameter, the MIFARE block address is given. The MIFARE Key A can be passed to the method either as a byte array or as a string.

```
reader.Mifare.ClassicAuth(4, "FFFFFFFFFFFF", MFKeyType.KeyA);
```

The `Write()` method has the same properties as the `ULWrite()` method. Hexadecimal values can also simply be passed as a string. The `Write()` method ensures that exactly 16 bytes are written each time. This is also the case when either fewer or more bytes are given. Undefined values are simply padded with zeroes.

```
reader.Mifare.Write(4, "11 22 33 44");
```

Writing a string takes place in the same way using `ULWrite()`, although only a string of up to 16 bytes may be stored in a block.

```
reader.Mifare.Write(5, "\"Hello MIFARE\"");
```

Unlike the `ULWrite()` method, however, the `Write()` method accepts integer values for the second parameter. In this case, the MIFARE block is initialized with a value in the MIFARE Value Format.

```
            reader.Mifare.Write(6, 125 );
        }
        catch (Exception ex)
        {
            Console.WriteLine(ex.Message);
        }
        finally
        {
            reader.ClosePort();
        }
    }
  }
}
```

8.4.9.2 Reading the Entire Memory Contents

Reading out the entire MIFARE 1K/4K card memory is substantially more complicated than with the MIFARE Ultralight card, due to the segmented memory and the required authentication.

Listing 8.22. *MifareClassicWrite.cs.*

```
using System;
using GS.Util.Hex;
using GS.ISO14443_Reader;
using GS.ElektorRfidReader;

namespace GS.ElektorRfidReaderExample
{
```

8 CARDS AND TAGS IN APPLICATION

```
           public class MifareClassicDump
           {
               public static void Main(MainForm script, string[] args)
               {
                   ElektorISO14443Reader reader = new ElektorISO14443Reader();

                   try
                   {
```

An object of the `MifareClassicKeys` class is created to manage the MIFARE keys. At object creation, all 40 MIFARE key pairs are initialized with the default value, 0xFF FF FF FF FF FF. If other keys are saved to the card, only the respective pairs of keys need to be assigned.

```
                       MifareClassicKeys mifareKeys = new MifareClassicKeys();

                       // Set MIFARE key values if required
                       // mifareKeys[0].KeyA = "FF FF FF FF FF FF";
                       // mifareKeys[1].KeyA = "FF FF FF FF FF FF";

                       Console.Clear();
                       reader.OpenPort();

                       reader.RFReset(5);
                       int ret =reader.ActivateCard(RequestCmd.WUPA);

                       if (ret != 0) return;
```

Within this for loop, all 256 blocks in a MIFARE 4K card are read.

```
                       // Read MIFARE Classic Card
                       for (int block = 0; block < 256; block++)
                       {
```

The `GetMifareSector()` method returns the current block's sector.

```
                           int sector = MifareClassicUtil.GetMifareSector(block);
```

Authentication need only be carried out for the first block within a sector.

```
                           if (MifareClassicUtil.IsFirstDataBlock(block) == true)
                           {
                               Console.WriteLine("MIFARE Sector: " + sector.ToString());
```

In this example, authentication is always done using Key A.

```
                               ret = reader.Mifare.ClassicAuth((byte)block,
                                                          mifareKeys[sector].KeyA,
```

```
                                MFKeyType.KeyA);
            if (ret != 0) return;
    }
```

Finally, the data is read and then output to the console.

```
            byte[] data;
            ret = reader.Mifare.Read((byte)block, out data);
            if (ret != 0) return;
            if (MifareClassicUtil.IsSectorTrailer(block) == true)
            {
              Console.Write(String.Format("Sector Trailer {0:D2}: ", block));
              Console.WriteLine(HexFormatting.Dump(data, 16));
              Console.WriteLine();
            }
            else
            {
              Console.Write(String.Format("Data Block {0:D2}: ", block));
              Console.WriteLine(HexFormatting.Dump(data,16));
            }
        }
    }
    catch (Exception ex)
    {
            Console.WriteLine(ex.Message);
    }
    finally
    {
            reader.ClosePort();
    }
  }
}
```

8.4.9.3 Optimizing the Read and Write Speeds

To read the entire memory of a MIFARE 4K card using the `MifareClassicDump.cs` script, you will need to be a little patient. With Smart Card Magic.NET running on the author's desktop PC, it took about 8 seconds. If you close the two output windows, the time drops to about 4.8 seconds. Both times were measured with Smart Card Magic. NET. Time measurements on a PC are not especially accurate, and also depend on other running processes. For a rough estimate, however, these measurements are usually sufficient.

Typical transaction times for a MIFARE 4K card are tabulated in Table 8.23. These values are taken from the manufacturer's datasheet. The respective transaction times reflect the card's command execution time, the reader's RF communication time and the card's RF communication time.

Table 8.23. MIFARE 1K/4K card transaction times.

Transaction	Execution Time
Card activation without collision	3.0 ms
Card activation with collision	4.0 ms
Authentication	2.0 ms
Reading a MIFARE block	2.5 ms
Writing a MIFARE block	6.0 ms
Increment and Decrement commands	2.5 ms
Transfer command	4.5 ms

Using Table 8.25, the transaction time for reading the entire memory can be calculated.

1 x	Card activation without collision	3 ms
40 x	Authentication	160 ms
256 x	Reading a block of data	640 ms
	Total:	803 ms

In this 0.8-second period, the reader firmware's execution time and the communication between PC and reader are not considered. In our example, this takes up about 80% of the total time. In order to check the result, a measurement of the communication at the RF interface is required. For such a measurement you need either a digital storage oscilloscope or an ISO/IEC 14443 protocol analyzer.

It's not always easy to measure a card command's duration with a storage scope. One problem is that it's very complicated to interpret individual byte values. Also, it the storage capacity of the oscilloscope is often not adequate for representing all of the required card commands. It is much simpler to use a special ISO/IEC 14443 protocol analyzer.

Figure 8.56.
The Proxispy protocol analyzer by Raisonance.

THE MIFARE CLASSIC CARD 8.4

The following performance measurements were carried out using the Proxispy Protocol Analyzer from Raisonance *(http://www.sc-raisonance.com)*. This analyzer enables recording and interpretation of ISO/IEC 14443, ISO/IEC 15693 and NFC protocols.

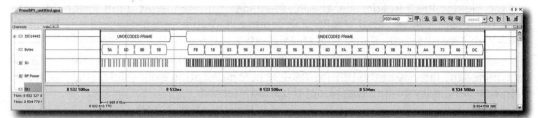

Figure 8.57. *Proxispy: Measuring the transaction time of the MIFARE Classic 'Read' command.*

In order to assess the performance of the Elektor RFID Reader, we need to compare it with measurements from another reader system. To do this, the time taken to read a standard MIFARE sector (Auth. Key A + 3 x Reads) is measured for each system.

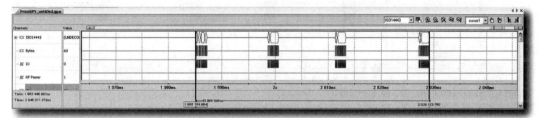

Figure 8.58. *Elektor RFID Reader: reading a standard sector using the Read() method.*

The SCM Microsystems SCL010 PC/SC Reader was used for a performance comparison. The measurements indicate that the Elektor RFID Reader takes ≈44 ms to read a MIFARE sector, while the SCL010 takes ≈25 ms.

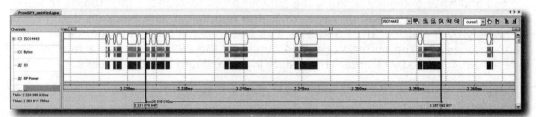

Figure 8.59. *PC/SC reader SCM SDI010: reading a standard sector.*

A comparison of the two measurements shows that the Elektor RFID Reader needs a lot more time between the individual commands than the SCL010 does. Because the card transaction time is identical for both readers, the cause of the delay lies in the firmware or in the communication with the PC. Answering this question requires a measurement of the data transfer between PC and reader. In the simplest case, one may use the RXD and TXD RS-232 signals from the microcontroller. In this way, the reader's overall execution speed can also be determined.

8 CARDS AND TAGS IN APPLICATION

The Proxispy protocol analyzer enables measurement via two digital inputs. In this way, one can measure the RS-232 communication as well as the RF interface signal simultaneously. Input 1 (see Figure 8.60) shows the RS-232 communication between the PC and the Elektor RFID Reader, i.e. the PC-to-reader commands. At Input 2, the Reader-to-PC responses can be seen. The RS-232 communication's transmission rate is 460,800 Baud. The measurement shows that this has very little influence on execution time.

The reader firmware has been optimized for speed. Also, the FT232R interface module determines the latency of the USB-to-RS-232 communication exclusively. There is, however, one very-easy-to-implement optimization method.

> **Note** If you implement PC reader commands that execute several commands in succession, overall reader performance can be improved substantially.

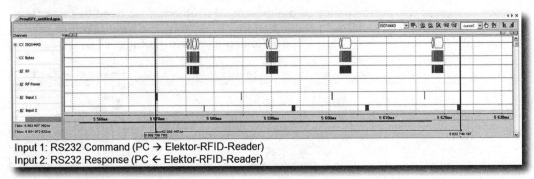

Input 1: RS232 Command (PC → Elektor-RFID-Reader)
Input 2: RS232 Response (PC ← Elektor-RFID-Reader)

Figure 8.60. Elektor RFID Reader: Additional RS-232 communication measurement.

For example, the `ReadBlocks()` and `WriteBlocks()` methods may be used to carry out MIFARE authentication and then read or write up to 8 MIFARE data blocks at once. These two PC-to-reader methods allow an entire standard sector to be read or written. To read and write extended sectors, the number of PC-to-reader commands is reduced from 17 (Auth. + 16 x Read or Write) to 2. The number of data blocks is determined exclusively by the size of the RS-232 receive and transmit buffers in the reader firmware. In our case, the buffer size is 144 bytes, as the P89LPC936 has a total of 512 bytes of RAM in its XDATA area. This results in a maximum of 8 data blocks.

A comparative measurement (Figure 8.61) shows the increase in reading speed by a factor of 2.5.

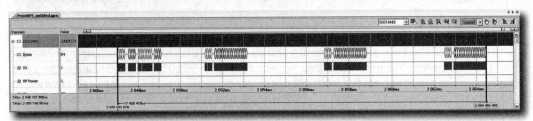

Figure 8.61. Elektor RFID Reader: Reading a standard sector using the BlockRead() method.

The Elektor RFID Reader's performance is essentially determined by the following factors:

- The LPC936 supports a maximum I²C data rate of 400 kbit/s.

Therefore, control of the MF RC522 reader IC is done only at 400 kbit/s instead of at the chip's maximum speed of 3,400 kbit/s (I²C High-speed mode).

- Use of an 8-bit microcontroller with limited resources.
- Sending the reader commands to the LCD takes about 2 ms.

8.4.9.4 Optimized Reading of the Entire Memory Contents

In the next example, the `ReadBlocks()` method is used to read the entire memory content of a MIFARE 4K card. By reading several data blocks in succession, the implementation of the console output is a little more complicated than when using the Read() method. This script takes about 3 seconds instead of 4.5 seconds.

Listing 8.23. MifareClassicFastDump.cs.

```
using System;
using GS.Util.Hex;
using GS.ISO14443_Reader;
using GS.ElektorRfidReader;

namespace GS.ElektorRfidReaderExample
{
   public class MifareClassicFastDump
   {
      ElektorISO14443Reader reader;
      MifareClassicKeys mifareKeys;

      public MifareClassicFastDump()
      {
         reader = new ElektorISO14443Reader();
         mifareKeys = new MifareClassicKeys();
      }

   public static void Main(MainForm script, string[] args)
   {
      MifareClassicFastDump readerScript = new MifareClassicFastDump();
      readerScript.Run(script);
   }

      public void Run(MainForm script)
      {
         try
         {
            reader.OpenPort();
            reader.RFReset(5);
```

```
                    reader.ActivateCard(RequestCmd.WUPA);
                    DumpMifareClassicCard(mifareKeys, MFKeyType.KeyA);
            }
            catch (Exception ex) { Console.WriteLine(ex.Message); }
            finally { reader.ClosePort(); }
    }
```

The `DumpMifareClassicCard()` method reads the entire memory contents of a MIFARE 1K/4K card and outputs it to the console. The `mfKeyType` parameter determines whether Key A or Key B is used for authentication.

```
int DumpMifareClassicCard(MifareClassicKeys mifareKeys, MFKeyType mfKeyType)
{
int n;
int block = 0;
byte[] baData = null;
byte[] baMifareDataBlock = new byte[16];
int ret = 0;
```

Within this loop, all of the data blocks are read and output.

```
            do
            {
```

In the next step, the number of data blocks to be read is calculated. Therefore, it is only necessary to check whether the current block is within a standard sector or an extended sector.

```
                if (block < 128) n = 4; else n = 8;
```

Finally, the sector's actual address is calculated. This is required to read the corresponding MIFARE keys from the `mifareKeys` object.

```
                int sector = MifareClassicUtil.GetMifareSector(block);
```

The `ReadBlocks()` method then reads either four or eight blocks.

```
                if (mfKeyType == MFKeyType.KeyA)
                {
                    ret = reader.Mifare.ReadBlocks((byte)block,
                                    mifareKeys[sector].KeyA,
                                    mfKeyType, out baData, (byte)n);
                }
                else
                {
                    ret = reader.Mifare.ReadBlocks((byte)block,
                                    mifareKeys[sector].KeyB,
                                    mfKeyType, out baData, (byte)n );
                }
                if (ret != 0) return ret;
```

Then, the data is output using the `WriteMifareBlock()` method.

```
            for (int i = 0; i < n; i++)
            {
                Array.Copy(baData, (16 * i), baMifareDataBlock, 0, 16);
                ConsoleWriteMifareBlock(baMifareDataBlock, block + i);
            }
            block += n;
        }
        while (block < 256);
        return 0;
    }
```

The `ConsoleWriteMifareBlock()` method formats then outputs the MIFARE blocks to the console.

```
    void ConsoleWriteMifareBlock(byte[] data, int block)
    {
        if (MifareClassicUtil.IsFirstDataBlock(block) == true)
        {
            Console.WriteLine("MIFARE Sector: " +
            MifareClassicUtil.GetMifareSector(block).ToString());
        }
        if (MifareClassicUtil.IsSectorTrailer(block) == true)
        {
            Console.Write(String.Format("Sector Trailer {0:D2}: ", block));
            Console.WriteLine(HexFormatting.Dump(data, 16, 16));
            Console.WriteLine();
        }
        else
        {
            if (MifareClassicUtil.IsValueFormat(data))
            {
                Console.Write(String.Format("Data Block {0:D2}: ", block));
                Console.Write(HexFormatting.DumpHex(data, 16, 16));
                Console.WriteLine(String.Format("{0}",
                MifareClassicUtil.ValueFormatToInt32(data)));
            }
            else
            {
                Console.Write(String.Format("Data Block {0:D2}: ", block));
                Console.WriteLine(HexFormatting.Dump(data, 16, 16));
            }
        }
    }
}
```

8.4.9.5 The Problem of Data Corruption

MIFARE 1K/4K and MIFARE Plus cards are often used in prepaid systems. Typical applications include electronic ticketing for public transport, road toll cards, electricity and gas cards, cards for vending machines and company canteens. In these applications, the customer initially loads a relatively small amount of money on the card and uses it, for example, to pay for the coffee from a vending machine. During a credit transaction, the system operator is paid the entire amount credited to the card. Then, the card user may spend this amount all at once or in several transactions. Finding the correct coins for the vending machine is thus no longer an obstacle. Most of these systems are very user-friendly and are rapidly gaining acceptance.

One of the main criteria of pre-payment systems is the reliability of the card. No cardholder would accept that, due to a technical defect or operator error, the money on the card is voided. The most critical time during a transaction is during the writing of data to the EEPROM. If the user removes the card from the RF field during the EEPROM write operation, the card's power supply is interrupted. This could lead to data being lost or it being incorrect. In the electronic wallet, the monetary value is also corrupted. Currently, there are no technically viable ways to maintain power to the card as long as is necessary to write to the EEPROM memory if the card is removed from the field. Special batteries for smartcards are available, but these are simply too expensive for most applications. The problem is comparable to a PC system without an uninterruptable power supply (UPS). When the PC's plug is pulled from the power socket during a hard drive write, this can destroy a file or the entire hard drive. In smart cards and RFID tags, the EEPROM assumes the role of the hard drive.

An EEPROM bit is made up of a tiny capacitor. In comparison to a normal capacitor, an EEPROM can store the data for a very long time (over ten years). The EEPROM cell's charge state determines whether it is erased or programmed. Depending on the technology, the erased state will either represent a '0' or a '1'. For our purposes, we will assume that '0' stands for the erased state. With a programming operation, individual bits can be switched from the erased state to the programmed state, or, in our example, from '0' to '1'. Should you need to return a bit from the '1' state to the '0' state, it is necessary to erase an entire page (e.g. 64 bytes) and rewrite all the '1' bits. When writing a desired value to the EEPROM, the EEPROM page is first erased, and then reprogrammed. The time required for erasing and programming is between 2 ms and 10 ms, depending on the technology. This time is consistent, whether writing a single byte or an entire page. For an EEPROM with a 64-byte page size, erasing and programming one byte takes just as long as 64 bytes, as long as all of the bytes fall within one page.

Should the card be removed from the reader during EEPROM writing, either the erase or the program cycle will be interrupted. Due to this, it is necessary to implement suitable backup management measures. Microcontroller cards with an operating system (e.g. MIFARE DESFire) program the EEPROM data atomically. This means that, either all of the bits are written, or not one single bit is changed. For this purpose, the data to be written is copied to a transaction buffer prior to writing. This is also in the EEPROM, but at another physical address. Should the actual write operation be interrupted, the operating system is able to 'roll back' the original data from the transaction buffer.

The MIFARE 1K/4K cards do not support the roll-back mechanism. The MIFARE 1K/4K cards do, however, offer an instruction set and methods involving the MIFARE Value Format to implement a backup management system in the reader hardware.

Before we discuss a possible backup management solution for the MIFARE 1K/4K card, we present an example of an EEPROM write operation being interrupted. The simple electronic wallet example (Listing 8.24) was achieved using the MIFARE Write command exclusively, and thus the MIFARE Value Format was not implemented.

Listing 8.24. *MifareClassicSimple_ePurse.cs.*

```
using System;
using GS.Util.Hex;
using GS.ISO14443_Reader;
using GS.ElektorRfidReader;

namespace GS.ElektorRfidReaderExample
{
    public class MifareClassicSimple_ePurse
    {
        public static void Main(MainForm script, string[] args)
        {
            ElektorISO14443Reader reader = new ElektorISO14443Reader();

            try
            {
```

The monetary value is stored in Byte 0 of data Block 2, and is thus capable of a maximum of 255 euro. 5 euro will be charged for each transaction.

```
                byte dataBlockCardValue = 2;
                int debitAmount = 5; // e.g. 5 Euro
                byte[] baData = new byte[16];
                int ret;

                reader.OpenPort();
                while(true)
                {
```

For an electronic wallet application, it's necessary to keep polling for a new card in the response field.

```
                    do
                    {
                        ret = reader.ActivateCard(RequestCmd.REQA);
                    }
                    while((ReaderStatus)ret != ReaderStatus.OK);
```

As soon as a card is activated, MIFARE authentication follows. If the `ClassicAuth()` method returns an error, this could be for two reasons. The card may have another MI-

8 CARDS AND TAGS IN APPLICATION

FARE key saved on it, or it could be due to a transmission error. With each error, a corresponding error message is output to the console, and the `continue` statement jumps to the next iteration of the `while(true)` loop. If the transaction is error-free, the beep does not sound.

```
ret = reader.Mifare.ClassicAuth(dataBlockCardValue,
                    "FF FF FF FF FF FF", MFKeyType.KeyA);
if ( ret != 0 )
{
    Console.WriteLine("MIFARE Authentication Error:
                    Please try" +
                    "again.");
    continue;
}
```

The amount is then read from the card and output to the console.

Should the amount not be sufficient for the transaction, a real-world application would deny access to the cardholder. In our case, a simple error message in the console window suffices.

```
if (cardValue < debitAmount)
{
    Console.WriteLine("Access Denied!");
    reader.HaltA();
    continue;
}
```

If the amount available is sufficient, 5 euro is deducted. Now, the question is, what would the effect of the card's removal from the reader's field be if this happens while the EEPROM data is being written?

```
baData[0] = (byte)(cardValue - debitAmount);
ret = reader.Mifare.Write(dataBlockCardValue, baData);
if (ret != 0)
{
    Console.WriteLine("Error while to writing the card value!"+
                    "Please try again");
    continue;
}
```

If the data is correctly written, the card will be placed in the HALT state. The `Console.Beep()` method makes a beep sound and signals that the transaction has completed successfully.

```
            reader.HaltA();
            Console.Beep();
        }
    }
    catch (Exception ex)
```

```
            {
                Console.WriteLine(ex.Message);
            }
            finally
            {
                reader.ClosePort();
            }
        }
    }
}
```

Before the behavior of the card can be tested using a script, it is necessary to load an amount (e.g. 255 euro, or 0xFF) to the card using Smart Card Magic.NET. During the debiting process, if the card is held in the reader's field until the beep is heard, the transaction was successful.

If the card is removed from the reader prematurely, various errors could arise. Errors during authentication or during reading of the card are not critical. In Figure 8.62, the card is removed from the reader during the EEPROM write operation. During the subsequent transaction, the exact same amount was read. This means that the card was removed before the actual EEPROM erase and program cycle began. Therefore, there is no data corruption.

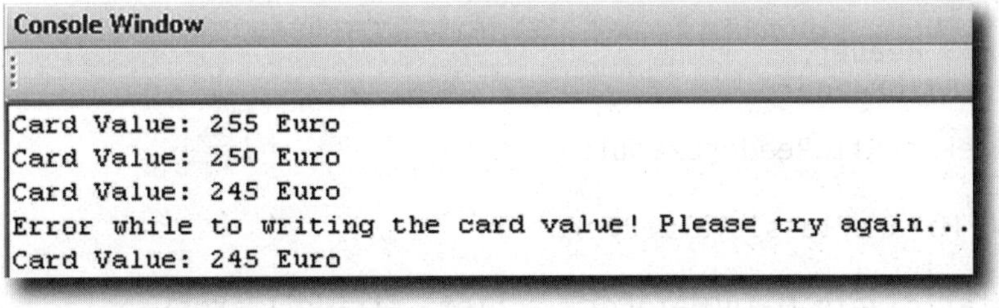

```
Console Window

Card Value: 255 Euro
Card Value: 250 Euro
Card Value: 245 Euro
Error while to writing the card value! Please try again...
Card Value: 245 Euro
```

Figure 8.62.
The card is removed from the reader field before the EEPROM erase and write cycle.

In Figure 8.63, the card was removed from the reader field either during or immediately after the EEPROM erase cycle. In this case, the 115 euro monetary value was voided, although only 5 euro was deducted. During the next transaction, the script is not able to see whether the '0' value is due to a debit transaction or data corruption.

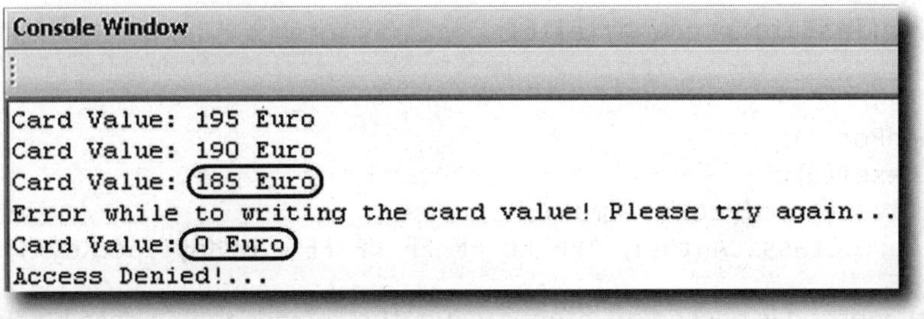

```
Console Window

Card Value: 195 Euro
Card Value: 190 Euro
Card Value: 185 Euro
Error while to writing the card value! Please try again...
Card Value: 0 Euro
Access Denied!...
```

Figure 8.63.
The card is removed from the reader field during or after the EEPROM erase cycle.

> **Note** The MIFARE Value Format enables the detection of inconsistent data (see Section 8.4.5). For this reason, the MIFARE Decrement, Increment, Restore and Transfer commands support this format exclusively.

8 CARDS AND TAGS IN APPLICATION

Use of the MIFARE Value Format cannot, however, prevent deleted or corrupt data if the card's power supply is interrupted during an EEPROM write operation. In contrast to the previous example, the MIFARE Value Format does allow for a consistency check of the EEPROM data. In the MIFARE Value Format, a value of 0 is represented as follows:

    0x00 00 00 00 FF FF FF FF 00 00 00 00 00 FF 00 FF

The reader software can then tell that the data does not represent a valid value.

    0x00 00 00 00 00 00 00 00 00 00 00 00 00 00 00 00

In other words, while the MIFARE Value Format does not solve the problem of inconsistent data, it does make possible its detection.

8.4.9.6 The MIFARE Value Format Methods

Before we implement the Decrement, Increment, Restore and Transfer MIFARE commands in the next example, we look at an example using the corresponding methods (see Table 8.23) from the reader library.

Listing 8.25. *MifareClassicValueOperation.cs.*

```
using System;
using GS.Util.Hex;
using GS.ISO14443_Reader;
using GS.ElektorRfidReader;

namespace GS.ElektorRfidReaderExample
{
    public class MifareClassicValueOperation
    {
        public static void Main(MainForm script, string[] args)
        {
            ElektorISO14443Reader reader = new ElektorISO14443Reader();
            try
            {
                int ret;
                byte[]baData = new byte[16];

    Console.Clear();
    reader.OpenPort();
    reader.RFReset(5);
    reader.ActivateCard(RequestCmd.WUPA);
    reader.Mifare.ClassicAuth(4, "FF FF FF FF FF FF", MFKeyType.KeyA);
```

Each time the `Write()` method is passed an integer value, the corresponding data block is initialized in the MIFARE Value Format. Then, the data block is read.

```
                reader.Mifare.Write(4, 125);
                reader.Mifare.Read(4, out baData);
```

The `Decrement()` method is used to reduce the amount in Data Block 4 by 15 and write the result back to the same block. The Restore() method is used to copy the value from Block 4 to Block 5, using the MIFARE 1K/4K Restore and Transfer commands.

```
reader.Mifare.Decrement(4, 15, 4);
reader.Mifare.Restore(4, 5);
```

The data block is again read, for control purposes.

```
reader.Mifare.Read(4, out baData);
reader.Mifare.Read(5, out baData);
```

The `Increment()` method is used to increase the value in the data block by 28 and write the result back to the same block.

```
reader.Mifare.Increment(4, 28, 4);
reader.Mifare.Restore(4, 5);
```

Again, for control, the two data blocks are re-read.

```
            reader.Mifare.Read(4, out baData);
            reader.Mifare.Read(5, out baData);
        }
        catch (Exception ex) { Console.WriteLine(ex.Message); }
        finally { reader.ClosePort(); }
      }
   }
}
```

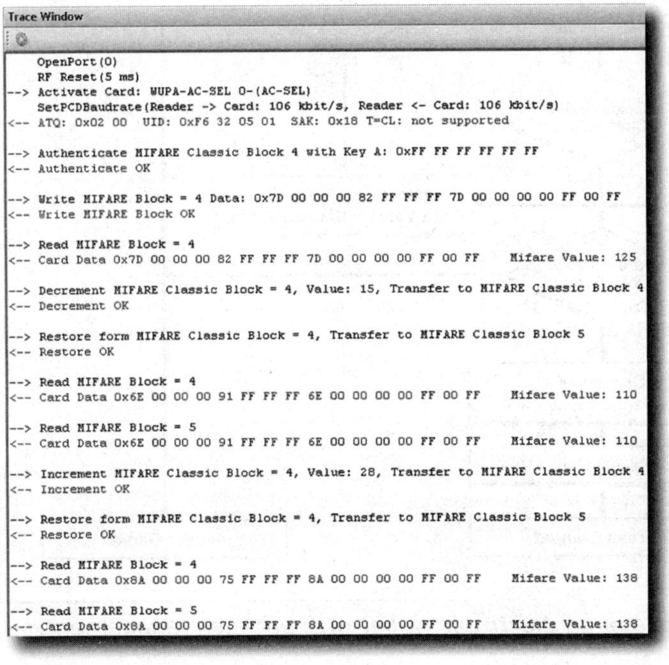

Figure 8.64.
The script's Trace output.

8 CARDS AND TAGS IN APPLICATION

8.4.9.7 Electronic Purse with Backup Management

Now that we're familiar with the MIFARE Value Format, we can create an electronic purse with backup management based on these methods.

In addition to a card balance, this example stores a backup value on the card. Before the debit process, the data integrity of both values is checked and, optionally, rolled back. In this way, it is guaranteed that, should the card be prematurely removed from the reader field during a transaction, at least either the Card Balance or the Backup Value will be valid.

Figure 8.65 shows the basic transaction flow. A description of the details is shown in Listing 8.26.

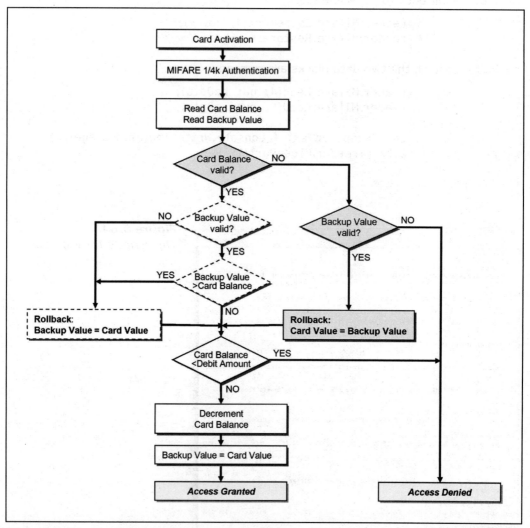

Figure 8.65. The basic sequence of an electronic wallet with backup management.

In order to debit a floating point amount (e.g. 2.65 euro), conversion to fixed-point notation is necessary (see Section 8.4.6). Monetary amounts are usually represented using two decimal places. Therefore, every amount must be multiplied by 100 before saving to the card. In turn, when the amount is output to the console, division by 100 is required.

Listing 8.26. *MifareClassic_ePurseDebit.cs.*

```csharp
using System;
using GS.Util.Hex;
using GS.ISO14443_Reader;
using GS.ElektorRfidReader;

namespace GS.ElektorRfidReaderExample
{
    public class MifareClassic_ePurseDebit
    {
        public static void Main(MainForm script, string[] args)
        {
            ElektorISO14443Reader reader = new ElektorISO14443Reader();

            try
            {
```

The Card Value is saved in Block 4, and the Backup Value in Block 5. In this example, every transaction deducts 2.65 euro.

```csharp
                byte balanceDataBlock = 4;
                byte backupDataBlock = 5;
                int debitAmount = 265; // 2.65 Euro
                byte[] baBalance = new byte[16];
                byte[] baBackup = new byte[16];
                int ret;

                reader.OpenPort();
```

Polling for a card in the reader's response field.

```csharp
                while(true)
                {
                    do
                    {
                        ret = reader.ActivateCard(RequestCmd.REQA);
                    }
                    while((ReaderStatus)ret != ReaderStatus.OK);
```

If a card was successfully activated, MIFARE authentication follows. As with the previous example, any error condition causes the continue statement to branch to the next iteration of the `while(true)` loop. In the event of an error, the user is denied access. In our example, only an error is printed, and no beep is sounded.

```
        ret = reader.Mifare.ClassicAuth(balanceDataBlock, "FFFFFFFFFFFF",
                                        MFKeyType.KeyA);
        if ( ret != 0 )
        {
           Console.WriteLine("MIFARE Authentication Error: " +
                             "Please try again.");
           continue;
        }
```

In the next step, Block 4 and Block 5 (Card Value and Backup Value), are read. The subsequent error checking ensures that no transmission error occurred during reading of the data blocks.

```
        ret = reader.Mifare.Read(balanceDataBlock, out baBalance);
        if ( ret != 0 )
        {
           Console.WriteLine("Error while reading the card balance: " +
                             "Please try again.");
           continue;
        }
        ret = reader.Mifare.Read(backupDataBlock, out baBackup);
        if ( ret != 0 )
        {
           Console.WriteLine("Error while reading the card balance: " +
                             "Please try again.");
           continue;
        }
```

For further processing, it is necessary to convert both the Card Value and the Backup Value to integer values. If one of these values was corrupted during a transaction, the `ValueFormatToInt32()` method will return null. The return value of this method is a nullable type, so it is declared with a question mark after the data type.

```
        int? cardBalance = MifareClassicUtil.ValueFormatToInt32(baBalance);
        int? backupValue = MifareClassicUtil.ValueFormatToInt32(baBackup);
```

If the `cardBalance` variable is also *null*, the amount in the previous transaction was not saved correctly to the card. The backup management then attempts a roll-back.

```
           if (cardBalance == null)
           {
              Console.WriteLine("Balance Data Block is corrupted! " +
                                "Try to use Backup Data Block");
```

The backup value is also stored in the MIFARE Value Format. The data integrity of the Backup Value can thus also be checked. The roll-back is achieved using the `Restore()` method. In our case, the value in Block 5 (Backup Value) is copied to Block 4 (Card Value).

```
              if ( backupValue != null )
              {
```

```
                    ret = reader.Mifare.Restore(backupDataBlock,
                                                balanceDataBlock);
                    if ( ret != 0 )
                    {
                        Console.WriteLine("Error while to restoring the " +
                                "Balance Data Block! Please try again.");
                        continue;
                    }
                    cardBalance = backupValue;
                }
                else
                {
                    Console.WriteLine("The Backup Data Block is also " +
                                    "corrupted! Please try again.");
                    reader.HaltA();
                    continue;
                }
            }
```

If the card is removed just before the `backup value` is updated, the backupValue variable will contain a readable, valid value, but the amount will be too high by the value of one transaction. Therefore, it's necessary to carry out a roll-back of the backup value, if the value is either corrupt or the amount is too large.

This step must take place in any event prior to the debit transaction. Otherwise, there is the danger that the larger backup value is used to roll back the actual amount.

```
            else
            {
                if (backupValue == null || (backupValue > cardBalance))
                {
                    ret = reader.Mifare.Restore(balanceDataBlock,
                                                backupDataBlock);
                    if (ret != 0)
                    {
                        Console.WriteLine("Error while to restoring the " +
                            "Backup Data Block! Please try again.");
                        continue;
                    }
                }
            }
```

If the amount is not sufficient for the transaction, an error message is printed and access is denied.

```
                if (cardBalance < debitAmount)
                {
                    Console.WriteLine(string.Format("Access Denied! " +
                        "card balance ({0:##0.00} Euro)",(cardBalance / 100)));
```

```
            reader.HaltA();
            continue;
        }
```

At this point, the amount is deducted. Premature removal of the card is detected by the reader's firmware. In this case, the `Decrement()` method returns an error code. The `continue` statement after the error output ensures that no change to the backup value occurs.

```
        ret = reader.Mifare.Decrement(balanceDataBlock,
                                      debitAmount,
                                      balanceDataBlock);
        if (ret != 0)
        {
            Console.WriteLine("Error while to decrement the " +
                              "card balance! Please try again.");
            continue;
        }
```

The amount is deducted from the card only if the entire process has completed without error. In this example, the update amount is already output to the console, although the transaction is still not yet complete.

```
    Console.WriteLine(string.Format("Access Granted! Actual card " +
                         "balance: {0:##0.00} Euro",
                         (cardBalance - debitAmount) / 100.0));
```

The final step is to update the backup value.

```
        ret = reader.Mifare.Restore(balanceDataBlock, backupDataBlock);
        if (ret != 0)
        {
            Console.WriteLine("Error while updating the backup value!")
        }
        reader.HaltA();
```

Only if the entire transaction was error-free, a beep will sound.

```
            Console.Beep();
        }
    }
    catch (Exception ex)
    {
        Console.WriteLine(ex.Message);
    }
    finally
    {
        reader.ClosePort();
    }
}
```

    }
}

It may only be possible to remove the card precisely during execution of the Decrement command by making several attempts. In Figure 8.66, the amount can be restored during the next transaction.

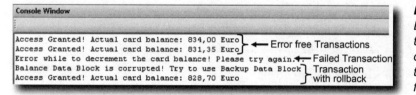

Figure 8.66.
During the third transaction, the card was prematurely removed from the reader.

Figure 8.67 shows the details of a rollback transaction. The card is removed from the reader immediately after the EEPROM erase cycle. Therefore, the value of all 16 bytes in data block 4 (value) is 0. Therefore, the reader application may perform a rollback. Only if both the card value and the backup value are valid, will the debit transaction take place (Decrement command). The backup value is then updated using the Restore and Transfer commands.

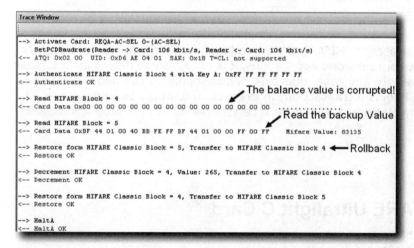

Figure 8.67.
Command sequence of a transaction with a rollback of the amount.

In Figure 8.68, the card was removed shortly before updating the backup value.

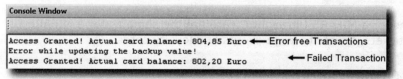

Figure 8.68.
The card is removed shortly before updating the backup value.

Figure 8.69 shows the next transaction, including the rollback of the backup value. In this case, the backup value was not changed during the previous transaction. Therefore,

8 CARDS AND TAGS IN APPLICATION

the backup value is 2.65 euro larger than the actual value in the wallet. It is necessary to correct the backup value prior to the debit transaction so that, should the power be interrupted again during the deduction of the monetary value, this will not cause inconsistency between amounts.

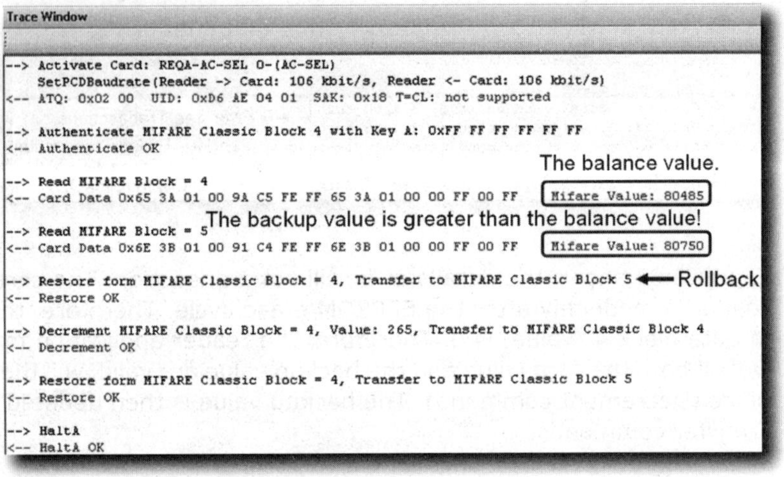

Figure 8.69. Command sequence of a transaction with a rollback of the backup value.

```
Trace Window
--> Activate Card: REQA-AC-SEL 0-(AC-SEL)
    SetPCDBaudrate(Reader -> Card: 106 kbit/s, Reader <- Card: 106 kbit/s)
<-- ATQ: 0x02 00  UID: 0xD6 AE 04 01  SAK: 0x18 T=CL: not supported
--> Authenticate MIFARE Classic Block 4 with Key A: 0xFF FF FF FF FF FF
<-- Authenticate OK
                                                          The balance value.
--> Read MIFARE Block = 4
<-- Card Data 0x65 3A 01 00 9A C5 FE FF 65 3A 01 00 00 FF 00 FF   Mifare Value: 80485
--> Read MIFARE Block = 5  The backup value is greater than the balance value!
<-- Card Data 0x6E 3B 01 00 91 C4 FE FF 6E 3B 01 00 00 FF 00 FF   Mifare Value: 80750
--> Restore form MIFARE Classic Block = 4, Transfer to MIFARE Classic Block 5  <-- Rollback
<-- Restore OK
--> Decrement MIFARE Classic Block = 4, Value: 265, Transfer to MIFARE Classic Block 4
<-- Decrement OK
--> Restore form MIFARE Classic Block = 4, Transfer to MIFARE Classic Block 5
<-- Restore OK
--> HaltA
<-- HaltA OK
```

This example demonstrates the MIFARE 1K/4K reader application's backup management logic in a simple way. For real-world applications, it is necessary to determine if further measures are required. In any case, duplicate debits should be avoided. In our case, each new activation begins a transaction. In practice, this would be undesirable, and this is preventable by blocking the card for a certain period of time after a successful transaction.

8.5 The MIFARE Ultralight C Card

The MIFARE Ultralight C offers active protection against card cloning as an additional advantage over the MIFARE Ultralight. For this purpose, the MIFARE Ultralight C chip uses mutual authentication based on a Triple-DES algorithm. The authentication procedure is similar to a microcontroller card with an operating system.

The main applications for a MIFARE Ultralight C card are the same as for a MIFARE Ultralight card: travel tickets, tickets to large events and loyalty cards.

8.5.1 Memory Organization

Figure 8.70 shows the memory layout of the 1,536-byte EEPROM memory. The first 512 bits are fully compatible with the MIFARE Ultralight (see Section 8.3.1).

8.5 THE MIFARE ULTRALIGHT C CARD

Figure 8.70. MIFARE Ultralight C card memory organization.

Byte Number	0	1	2	3	Page
UID	UID 0	UID 1	UID 2	BCC0	0
UID	UID 3	UID 4	UID 5	UID 6	1
Internal / Lock	BCC1	Internal	Lock 0	Lock 1	2
OTP	OTP 0	OTP 1	OTP 2	OTP 3	3
User Data	Data 0	Data 1	Data 2	Data 3	4
User Data	Data 4	Data 5	Data 6	Data 7	5
⋮					
User Data	Data 136	Data 137	Data 138	Data 139	38
User Data	Data 140	Data 141	Data 142	Data 143	39
Lock	Lock 2	Lock 3	rfu	rfu	40
Counter	CNT	CNT	rfu	rfu	41
Auth. Config.	Auth 0	rfu	rfu	rfu	42
Auth. Config.	Auth 1	rfu	rfu	rfu	43
Triple DES Key A	KeyA 0	KeyA 1	KeyA 2	KeyA 3	44
Triple DES Key A	KeyA 4	KeyA 5	KeyA 6	KeyA 7	45
Triple DES Key B	KeyB 0	KeyB 1	KeyB 2	KeyB 3	46
Triple DES Key B	KeyB 4	KeyB 5	KeyB 6	KeyB 7	47

Pages 4 to 39 hold the 144-byte data storage. This provides three times the memory capacity of the MIFARE Ultralight. The two bytes Lock2 and Lock3 offer the ability to lock parts of the extended EEPROM data memory to write access.

For ticket applications, Page 41 has an additional 16-bit counter.

The two Triple-DES keys, Key A and Key B, are stored in Pages 44 to 47. For security reasons, the key values can be written, but never read. Auth 0 in Page 42 defines from which page onward authentication is required. Valid Auth 0 page addresses are between 3 and 48 (0x03 to 0x30). A value of 48 (0x30) allows for disabling of the authentication.

The Auth 1 byte in Page 43 defines whether authentication is required only for writing or for reading as well.

Table 8.24. AUTH1

Bit	7	6	5	4	3	2	1	0
Value	1.)	1.)	1.)	1.)	1.)	1.)	1.)	2.)

[1] X = Any value (don't care).
[2] 0 = Authentication is required for writing and reading data.
    1 = Authentication is required for writing only.

8.5.2 Instruction Set

The MIFARE Ultralight C card uses additional authentication commands over and above the MIFARE Ultralight instructions. The authentication method is discussed in detail in the following sections.

Table 8.25. MIFARE Ultralight C instruction set.

Command Type	Command Code	Card Commands	SmartCard Magic.NET Button Label
Card Activation	0x26	REQA	Req A
	0x52	WUPA	Wup A
	0x93	Anti-collision of Cascade Level 1	AC 1
	0x93	Select of Cascade Level 1	Select 1
	0x95	Anti-collision of Cascade Level 2	AC 2
	0x95	Select of Cascade Level 2	Select 2
	0x50	HALTA	Halt A
Authentication	0x1A	Authenticate Step 1	Authenticate
	0xAF	Authenticate Step 2	
Memory Manipulation	0x30	Read	Read
	0xA2	Write	–
	0xA0	Compatibility Write	Write

8.5.3 Triple-DES Authentication

The basics of mutual authentication were already addressed in Section 4.4.1.

A method of authentication serves to validate the identity and authenticity of a communication partner. In our case, the card must authenticate the reader and the reader must authenticate the card.

The principle of operation is basically that one communication partner generates a random number, presenting a random question ('challenge'), and the other partner must encrypt it and send it back ('response'). The receiver simply decrypts this and compares it with the original random number. If it's identical, both participants must know the shared secret. In our case, it is a Triple-DES key.

This type of authentication is known as the challenge-response method. In mutual authentication, both are authenticated using an encryption protocol.

The MIFARE Ultralight C's authentication method is a modified variant of the ISO/IEC 7816-8 mutual authentication command. The ISO/IEC 7816-8 specifies commands for smartcard operating systems.

The detailed two-step process for the MIFARE Ultralight is shown in Figure 8.71. In the first step, the reader sends the Authenticate Step 1 command to the card. The card then produces an 8-byte random number (Rnd B) and encrypts this. In this case, the card returns an encrypted card challenge. This consists of a status code (0xAF) and the 8-byte encrypted random number (enc(Rnd B)).

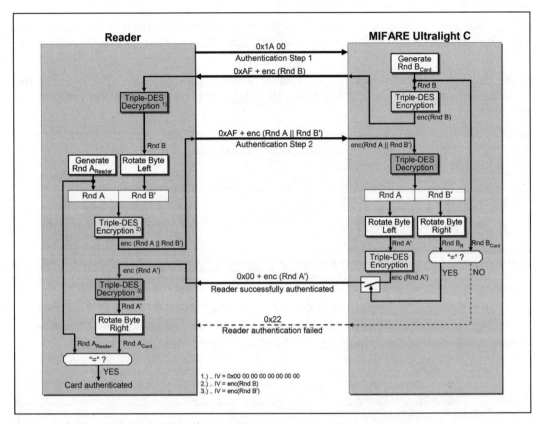

Figure 8.71. *MIFARE Ultralight C card mutual authentication process.*

The reader decrypts the encrypted random number received from the card (enc(Rnd B)) and rotates it by one byte. The left-rotated decrypted random number from the card is called Rnd B'. Subsequently, the reader also generates an 8-byte random number (Rnd A). The two random numbers, Rnd A and Rnd B', are concatenated into a single 16-byte block and encrypted (enc(Rnd A||Rnd B')). The Authentication Step 2 command consists of the command code (0xAF) and enc(Rnd A||Rnd B'). The Authenticate Step 2 command thus includes both the reader challenge and the reader response.

After receiving the Authenticate Step 2 command, the card decrypts the random number and gets the original Rnd A||Rnd B'. The card rotates Rnd B' by one byte to the right, restoring the reader-calculated random number, Rnd BReader. If this is identical to the card-generated random number, Rnd BCard, the reader is then successfully authenticated. In this case, the card then produces a card response. This is done by rotating the reader's random number (Rnd A) one byte to the left and then encrypting it (enc(Rnd A')). The card response consists of error code 0x00 (OK) and enc(Rnd A').

The reader checks the authenticity of the card by decrypting the card response (enc(Rnd A')) and rotating it one byte to the right, getting Rnd ACard. In the final step, this is compared with the random number produced by the reader, Rnd AReader. If the two are the same, the card is authenticated.

Should authentication of the reader fail, the card responds with error code 0x22 and returns to the IDLE state. So, after failed authentication, a new card activation is required.

Triple-DES encryption is used in CBC (Cipher Block Chaining) mode. The value of the initialization vector (IV) is 0 at the first calculation, and then it takes the value of the last pass.

Smart Card Magic.NET allows easy authentication of a MIFARE Ultralight card. For this, it is only necessary to enter the correct Triple-DES key into the MIFARE Ultralight C window and then click on *Authenticate*.

The Triple-DES key for a MIFARE Ultralight C engineering sample card is:

- DES Key A = 0x49 45 4D 4B 41 45 52 42;
- DES Key B = 0x21 4E 41 43 55 4F 59 46.

Figure 8.72. *Smart Card Magic.NET: MIFARE Ultralight C authentication.*

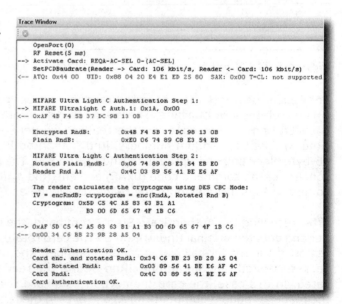

Figure 8.73. *MIFARE Ultralight C authentication.*

A complete program for MIFARE Ultralight C authentication is shown in Listing 8.72.

8.5.4 Elektor RFID Reader Library: MIFARE Ultralight C

Prior to authentication, the MIFARE Ultralight C instruction set is identical to that of the MIFARE Ultralight. The methods for reading and writing data have already been discussed in Sections 8.3.4 and 8.3.5.

An additional method, `ULCAuth()`, (see Table 8.26) authenticates both the card and the reader.

Table 8.26. The MIFARE Ultralight C authentication methods.

Method	Description

`int ULCAuth(`
`byte[] desKeyA,`
`byte[] desKeyB)`
	Performs mutual authentication between the Elektor Reader an a MIFARE Ultralight C card. If `ULCAuth()` returns 0x00, both sides are successfully authenticated. Triple-DES keys A and B are passed as an array of bytes (parameters deskeyA and deskeyB).

`int ULCAuth(`
`string desKeyA,`
`string desKeyB)`
	This variant of the `ULCAuth()` method accepts the Triple-DES keys as hexadecimal strings.

8.5.5 Programming Examples

8.5.5.1 The MIFARE Ultralight C Authentication Sequence

Before we use the `ULCAuth()` method to authenticate a MIFARE Ultralight C card, we will implement this method (entire authentication sequence) using a script.

Listing 8.27. MifareClassicValueOperation.cs.

```
using System;
using System.Collections.Generic;
using System.Security.Cryptography;
using GS.ISO14443_Reader;
using GS.ElektorRfidReader;
using GS.CryptoUtil;
using GS.Util.Hex;
using GS.Util.ByteArray;

namespace GS.ElektorRfidReaderExample
{
    public class MifareUltraLightCAuth
    {
        public static void Main(MainForm script, string[] args)
        {
            ElektorISO14443Reader reader = new ElektorISO14443Reader();
            reader.EnableFTD2XXTrace = false;
            reader.Raw.EnableRAWExchangeTrace = true;
```

The Triple-DES in CBC mode algorithm is performed by the `TripleDESCrypt` class from the `GS.CryptoUtil` namespace. In this example, the Triple-DES key from a MIFARE Ultralight C engineering sample is used.

```
TripleDESCrypto tdes = new TripleDESCrypto();
tdes.Mode = CipherMode.CBC;

tdes.DesKeyA = new byte[] {0x49,0x45,0x4D,0x4B,0x41,0x45,0x52,0x42};
tdes.DesKeyB = new byte[] {0x21,0x4E,0x41,0x43,0x55,0x4F,0x59,0x46};

try
{
    Console.Clear();
```

We opening the reader port and performe the subsequent card activation:

```
reader.OpenPort();
reader.RFReset(5);
reader.ActivateCard(RequestCmd.WUPA);
```

In the first step, the Authenticate Step 1 command is sent to the card. The `ExchangeFrame()` method from the Raw class sends both command bytes, 0xA1 00, and receives the card's response. This consists of the status code 0xAF and an 8-byte random number.

```
// --------------------------------------------------
// Mifare Ultra Light C Authentication Step 1
// --------------------------------------------------
Console.WriteLine("Mifare Ultra Light C Authentication Step 1:");

// send authentication request
Console.WriteLine("--> 0x1A 00");
byte[] sndBuf = {0x1A, 0x00};
byte[] recBuf = new byte[16];
int recLen = recBuf.Length;
int retValue = reader.Raw.ExchangeFrame(sndBuf, sndBuf.Length,
                                        out recBuf, out recLen);
if (retValue != 0 || recBuf.Length != 9 || recBuf[0] != 0xAF)
    {
    Console.WriteLine(" Get Encrypted RndB Failed.");
    return;
}
Console.WriteLine(HexFormatting.DumpHex("<-- 0x", recBuf,
                        recBuf.Length, 9));
```

The encrypted random number (enc(Rnd B)) is extracted from the card's response and saved to the `baEncRndB` variable.

```
byte[] baEncRndB = new byte[8];
byte[] baRndB;
Array.Copy(recBuf,1,baEncRndB,0,8); // Extract Encrypted Rnd B
```

```
Console.WriteLine(HexFormatting.DumpHex(" Encrypted RndB:" +
                  "         0x", baEncRndB, baEncRndB.Length, 8));
```

The card's random number (Rnd B) is obtained by decrypting `abEncRndB`. The initialization vector's value is 0x00 00 00 00 00 00 00 00 for the first iteration of the Triple-DES calculation.

```
// Rnd B = Dec(EncRndB)
tdes.IV = new byte[]{0x00, 0x00 ,0x00, 0x00, 0x00, 0x00, 0x00, 0x00};
baRndB = tdes.Decrypt(baEncRndB);
Console.WriteLine(HexFormatting.DumpHex(" Plain RndB:" +
                  "         0x", baRndB, baRndB.Length, 8));
```

Rotating Rnd B left by one byte is done using the `RotateLeft()` method from the `ByteArray` class (`GS.Util` namespace).

```
// ---------------------------------------------------
// Mifare Ultra Light C Authentication Step 2
// ---------------------------------------------------
Console.WriteLine("\nMifare Ultra Light C Authentication Step 2:");
byte[] baRndB_Rotated = ByteArray.RotateLeft(baRndB);
Console.WriteLine(HexFormatting.DumpHex("   Rotated Plain RndB:" +
                  "0x", baRndB_Rotated,
                  baRndB_Rotated.Length, 8));
```

In the next step, the reader produces 8-byte random number Rnd A. The `Rng` class (`GS.CryptoUtil`) implements a cryptographic random number generator based on the .NET class library.

```
Console.WriteLine();
byte[] baReaderRndA = Rng.GetBytes(8);
Console.WriteLine(HexFormatting.DumpHex(" Reader Rnd A:" +
                  "         0x", baReaderRndA, baReaderRndA.Length, 8));
```

Before encrypting the two random numbers, it is necessary to create a block containing (Rnd A || Rnd B').

```
List<byte> baList = new List<byte>();
baList.AddRange(baReaderRndA);       // RndA
baList.AddRange(baRndB_Rotated);     // RndB'
```

The 16-byte block generated by a list, (Rnd A || Rnd B'), is encrypted. For every further Triple-DES calculation, the last 8 bytes of the response received from the card are used to populate the initialization vector (IV), in this case enc(Rnd B).

```
tdes.IV = baEncRndB;
byte[] baCryptogram = tdes.Encrypt( baList.ToArray() );
Console.WriteLine();
Console.WriteLine("    The reader calculates the cryptogram using " +
                  "DES CBC Mode:\n IV = encRndB; cryptogram = " +
```

```
                        "enc(RndA, Rotated Rnd B)");
Console.WriteLine(HexFormatting.DumpHex(" Cryptogram: 0x",
                baCryptogram, baCryptogram.Length, 8));
```

The Authenticate Step 2 command consists of the command byte, 0xAF and (Rnd A || Rnd B').

```
        baList.Clear();
        baList.Add(0xAF);
        baList.AddRange(baCryptogram);

        Console.WriteLine();
        Console.WriteLine(HexFormatting.DumpHex("--> 0x", baList.ToArray(),
                                baList.Count, 17));
        recLen = recBuf.Length;
        retValue = reader.Raw.ExchangeFrame(baList.ToArray(), baList.Count,
                                out recBuf, out recLen);
```

The reader is successfully authenticated when `ExchangeFrame()`'s return value is 0 and the number of received bytes equals 9 (card response: 0x00 + enc(Rnd A')).

```
        if (retValue == 0 && recBuf.Length == 9)
        {
            Console.WriteLine(HexFormatting.DumpHex("<-- 0x", recBuf,
                                recBuf.Length, 9));
            Console.WriteLine("    Reader Authentication OK.");
        }
        else
        {
            Console.WriteLine(" Reader Authentication Failed.");
            return;
        }
```

In the final step, the card's authenticity is checked. The 8-byte enc(Rnd A') is extracted from the card response.

```
        byte[] baCardEncRotatedRndA = new byte[8];

        // Extract Encrypted rotated Rnd A
        Array.Copy(recBuf, 1, baCardEncRotatedRndA, 0, 8);
        Console.WriteLine(HexFormatting.DumpHex("    Card enc. and rotated" +
        " RndA: 0x", baCardEncRotatedRndA, baCardEncRotatedRndA.Length, 8));
```

The initialization vector, IV, for the calculation of Rnd A' consists of the last 8 bytes from the Authenticate Step 2 command, or enc(Rnd B').

```
        byte[] baIV = new byte[8];
        Array.Copy(baCryptogram, 8, baIV, 0, 8);
        tdes.IV = baIV; // IV = enc(Rnd B')
        byte[] baCardRotatedRndA = tdes.Decrypt(baCardEncRotatedRndA);
```

```
            Console.WriteLine(HexFormatting.DumpHex(" Card Rotated RndA:" +
            "           0x", baCardRotatedRndA, baCardRotatedRndA.Length, 8));
```

By rotating Rnd A' one byte to the right, we get Rnd ACard.

```
            byte[] baCardRndA = ByteArray.RotateRight( baCardRotatedRndA );
            Console.WriteLine(HexFormatting.DumpHex(" Card RndA:" +
            "           0x", baCardRndA, baCardRndA.Length, 8));
```

Should the two random numbers, Rnd ACard and Rnd AReader, be identical, the card is also authenticated.

```
            if (ByteArray.DataAreEqual(baReaderRndA, baCardRndA))
            {
                Console.WriteLine(" Card Authentication OK.");
            }
            else
            {
                Console.WriteLine(" Card Authentication Failed.");
            }
        }
        catch (Exception ex)
        {
            Console.WriteLine(ex.Message);
        }
        finally
        {
            reader.ClosePort();
        }
    }
  }
}
```

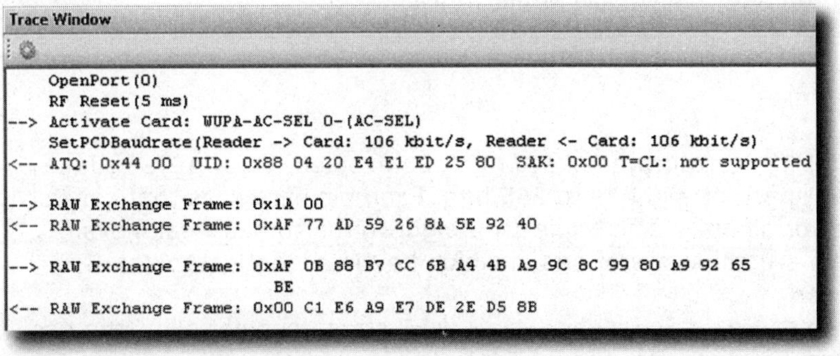

Figure 8.74.
Successful MIFARE Ultralight C authentication.

Figure 8.74 shows that only two command sequences are required for authentication of the card by the reader and authentication of the reader by the card. The time taken to calculate the respective card response in both Authenticate Step 1 and Authenticate Step 2 commands is about 850 μs.

8 CARDS AND TAGS IN APPLICATION

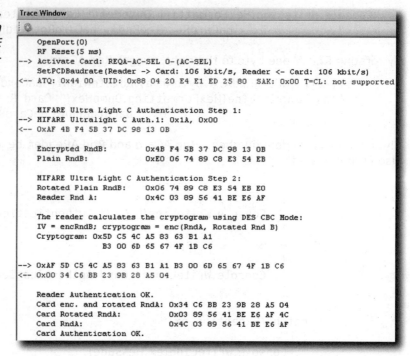

Figure 8.75.
Console output of a successful MIFARE Ultralight C authentication.

> **Note** The Microsoft .NET class library provides several classes for symmetric and asymmetric encryption, in the `System.Security.Cryptography` namespace. In this example, the calculation of the Triple-DES encryption was done using the with the `TripleDESCrypto` class. The class encapsulates only the .NET `TripleDESCryptoServiceProvider` class and presents an API optimized for smartcard applications.

> **Note** When using DES, certain weak keys must be avoided. Weak keys reduce the DES key space of 256 bits to 255 bits. Examples are keys that consist of all zeroes or all ones. The Triple-DES algorithm also has weak keys. Among them are all permutations in which Key A is identical to Key B.
>
> The `TripleDESCryptoServiceProvider` .NET class does not allow the use of weak keys. For example, should Triple-DES encryption be attempted using Key A = Key B = 0x00 00 00 00 00 00 00 00, a `Cryptographic-Exception` is triggered. In practice, this leads to problems only when they key of a factory-condition card has a weak key at random.

8.5.5.2 MIFARE Ultralight C Card Personalization

During card personalization, either all personal data of the cardholder or application-specific data are written to the card. Finally, the access conditions and the card-specific keys are written.

The following example will focus on writing new keys and changing the access conditions. The script writes the following Triple-DES keys to the card:

- DES Key A = 0x00 11 22 33 44 55 66 77
- DES Key B = 0x88 99 AA BB CC DD EE FF

The keys will be stored in the card's memory in the format shown in Figure 8.76.

Auth. Config.	0x03	rfu	rfu	rfu	42
Auth. Config.	0x00	rfu	rfu	rfu	43
Triple DES Key A	0x77	0x66	0x55	0x44	44
Triple DES Key A	0x33	0x22	0x11	0x00	45
Triple DES Key B	0xFF	0xEE	0xDD	0xCC	46
Triple DES Key B	0xBB	0xAA	0x99	0x88	47

Figure 8.76.
Memory dump: configuration of the access conditions and the Triple-DES key values.

Because '3' is written to the Auth 0 byte and '0' to the Auth 1 byte, authentication will be required from Page 3 onward.

Brand-new MIFARE Ultralight C cards require authentication to write data, as the value of Auth 0 is 48. This value is outside the page address space and allows access to the memory without authentication, according to the specifications.

- A MIFARE Ultralight C engineering sample card requires authentication for writing from page address 40 onward, using the following keys:
- DES Key A = 0x49 45 4D 4B 41 45 52 42
- DES Key B = 0x21 4E 41 43 55 4F 59 46

The script in Listing 8.28 allows the personalization of both brand-new cards and sample cards. For a brand-new card, the source code for Step 1 need only be commented out.

Listing 8.28. *MifareUltraLightC_Perso.cs.*

```
using System;
using GS.ISO14443_Reader;
using GS.ElektorRfidReader;
using GS.Util.Hex;

namespace GS.ElektorRfidReaderExample
{
    public class MifareUltraLightC_Perso
    {
        ElektorISO14443Reader reader;
```

```csharp
            public MifareUltraLightC_Perso()
            {
                reader = new ElektorISO14443Reader();
            }

            public static void Main(MainForm script, string[] args)
            {
                MifareUltraLightC_Perso scr = new MifareUltraLightC_Perso();
                scr.Run(script);
            }

            void Run(MainForm script)
            {
                try
                {
                    int retValue;
```

Generating the Triple-DES key for a MIFARE Ultralight C engineering sample:

```csharp
                string sampleCardKeyA = "49 45 4D 4B 41 45 52 42";
                string sampleCardKeyB = "21 4E 41 43 55 4F 59 46";
```

Generating the new Triple-DES card key:

```csharp
    byte[] newKeyA = new byte[]{0x00,0x11,0x22,0x33,0x44,0x55,0x66,0x77};
    byte[] newKeyB = new byte[]{0x88,0x99,0xAA,0xBB,0xCC,0xDD,0xEE,0xFF};

    Console.Clear();
```

Opening the reader port, and card activation:

```csharp
                reader.OpenPort();
                reader.RFReset(5);
                reader.ActivateCard(RequestCmd.WUPA);
```

Authentication is only required if a MIFARE Ultralight C sample card is used. With a brand-new card, only this part of the source code need be commented out.

```csharp
        // ----------------------------------------------------
        // Step 1: Auth. if card is not virgin card
        // ----------------------------------------------------
        retValue = reader.Mifare.ULCAuth(sampleCardKeyA, sampleCardKeyB);

        if (retValue != 0)
        {
            Console.Write("Authentication failed!");
            return;
        }
```

The new key values are written to the card using the `UpdateTripleDesKeys()` method.

```csharp
// -------------------------------------------------
// Step 2: Update Triple-DES Keys
// -------------------------------------------------
retValue = UpdateTripleDesKeys(newKeyA, newKeyB);
if (retValue != 0)
{
    Console.Write("Update Triple-DES Keys failed!");
    return;
}
```

After updating the Triple-DES key, a check is done to see if authentication can be carried out correctly. Because the card allows authentication only after card activation, the card is re-activated after an RF reset. Authentication with the new Triple-DES key then follows. This test ensures that an error during the writing of the new key is detected.

```csharp
// -------------------------------------------------
// Step 3: Step 3: Verify Triple-DES Key Update
// -------------------------------------------------
reader.RFReset(5);
reader.ActivateCard(RequestCmd.WUPA);
retValue = reader.Mifare.ULCAuth(newKeyA, newKeyB);
if (retValue != 0)
{
    Console.Write("Authentication failed!");
    return;
}
```

Only when we're sure that authentication is possible using the new keys, the two new access condition bytes are written.

```csharp
// -------------------------------------------------
// Step 4: Update Access Configuration
// -------------------------------------------------
retValue = UpdateAC(3, 0); // Page=3, RD & WR restricted
if (retValue != 0)
{
    Console.Write("Update Access Configuration failed!");
    return;
}
}
catch (Exception ex)
{
    Console.WriteLine(ex.Message);
}
finally
{
```

```
            reader.ClosePort();
        }
    }
```

The `UpdateTripleDesKeys()` method writes the new key bytes to the card in the correct byte order.

```
        int UpdateTripleDesKeys(byte[] desKeyA, byte[] desKeyB)
        {
            int retValue;
            byte[] baData = new byte[4];

            // Update Triple-DES Key A: Write Key Bytes 0 to 3 to Page 44
            baData[3] = desKeyA[4];
            baData[2] = desKeyA[5];
            baData[1] = desKeyA[6];
            baData[0] = desKeyA[7];
            retValue = reader.Mifare.ULWrite(44 , baData);
            if (retValue != 0) return retValue;

            // Update Triple-DES Key A: Write Key Bytes 4 to 7 to Page 45
            baData[3] = desKeyA[0];
            baData[2] = desKeyA[1];
            baData[1] = desKeyA[2];
            baData[0] = desKeyA[3];
            retValue = reader.Mifare.ULWrite(45, baData);
            if (retValue != 0) return retValue;

            // Update Triple-DES Key B: Write Key Bytes 0 to 3 to Page 46
            baData[3] = desKeyB[4];
            baData[2] = desKeyB[5];
            baData[1] = desKeyB[6];
            baData[0] = desKeyB[7];
            retValue = reader.Mifare.ULWrite(46 , baData);
            if (retValue != 0) return retValue;

            // Update Triple-DES Key B: Write Key Bytes 4 to 7 to Page 47
            baData[3] = desKeyB[0];
            baData[2] = desKeyB[1];
            baData[1] = desKeyB[2];
            baData[0] = desKeyB[3];
            retValue = reader.Mifare.ULWrite(47, baData);
            return retValue;
        }
```

The first parameter, `pageStart`, and the `UpdateAC()` method, define from which page authentication is required. The second parameter, `rdWr`, defines the access type. The values for `rdWr` can be found in Table 8.24.

```
            int UpdateAC(byte pageStart, byte rdWr)
            {
                int retValue;
                byte[] baData = new byte[4];

                // Update Auth 0 (Page 42)
                baData[0] = pageStart;
                baData[1] = 0;
                baData[2] = 0;
                baData[3] = 0;
                retValue = reader.Mifare.ULWrite(42, baData);
                if (retValue != 0) return retValue;

                // Update Auth 1 (Page 43)
                baData[0] = rdWr;
                baData[1] = 0;
                baData[2] = 0;
                baData[3] = 0;
                retValue = reader.Mifare.ULWrite(43, baData);
                return retValue;
            }
        }
    }
```

After running the script in Listing 8.28, authentication is required for reading and writing data from Page 3 onward.

8.6 The T=CL Transmission Protocol

The communication between the reader and the card takes place on a half-duplex master-slave basis, in which the card is the slave and the reader is the master.

ISO/IEC Part 3 defines two transmission protocols (T=0 and T=1) for contact-type microcontroller smartcards. The byte-oriented T=0 protocol is the oldest and most-used. T=0 is mainly used in GSM/UMTS mobile phone SIM cards. The more modern, block-oriented, T=1 protocol supports a clear separation of the application layer and the data link layer as per the OSI (Open System Interconnection) reference model for data communication. The block-oriented transfer protocol specified in ISO/IEC 14443 Part 4 is a modified version of the T=1 protocol adapted to the requirements of contactless communication. The T=CL protocol, like the T=1 protocol, features a clear separation between layers. Based on the T=0 and T=1 protocols, the term T=CL (contactless) is common.

The general structure of the commands to the smart card and the responses thereto is defined in Part 3 of ISO/IEC 7816. These commands are known as Application Protocol Data Units (APDUs). Transportation of the application data takes place on the transfer

8 CARDS AND TAGS IN APPLICATION

layer, i.e. the T=0, T=1 and T=CL transmission protocols. The transmission layer's data records are referred to as transport protocol data units (TPDUs).

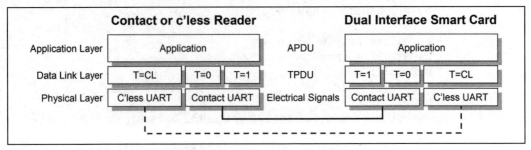

Figure 8.77. Standard communication model of a dual-interface card.

Most card operating systems (COS) support only specific applications, for example, GSM, EMV (Europay, MasterCard, Visa) or ICAO (International Civil Aviation Organization). The commands of the different application-specific standards are always constructed according to ISO/IEC 7616 Part 4. These differ slightly in their coding. In addition, most operating systems support proprietary commands. The MIFARE DESFire EV1 card supports both manufacturer-specific (DESFire Native) and ISO/IEC 7816-3 APDU command structures. Both command structures support the same functions.

The following sections build on the theoretical introduction to the T=CL protocol from Section 2.1.4.

8.6.1 T=CL Protocol Activation and Deactivation

Figure 8.78 shows a T=CL card's entire communication sequence.

The card activation command sequence (ISO/IEC 14443 Part 3) takes place in the same way as described in Section 8.1.2. The information as to whether or not the card supports the T=CL protocol is encoded in the most-recently received SAK byte. The T=CL protocol defines various reader and card transmission parameters. The reader and card parameters are exchanged during T=CL protocol activation. For this purpose, the reader sends the Request for Answer to Select (RATS) command to the card. The reader's maximum receive buffer size (FSD – Frame Size, Device) and the card's logical identification number (CID) are encoded in the RATS. The card returns the following parameters in its Answer To Select (ATS):

- The maximum card receive buffer size (FSC – Frame Size, Card)
- The maximum transmission rate (DS – Divisor Send; DR – Divisor Receive)
- The guard time after sending the ATS (SFGT – Startup Frame Guard Time)
- The maximum card response time (FWT – Frame Waiting Time)
- Support for the logical identification number (CID – Card IDentifier)
- Support for the logical node address (NAD – Node ADdress)
- The historical bytes, within which the card operating system is usually encoded

8.6 THE T=CL TRANSMISSION PROTOCOL

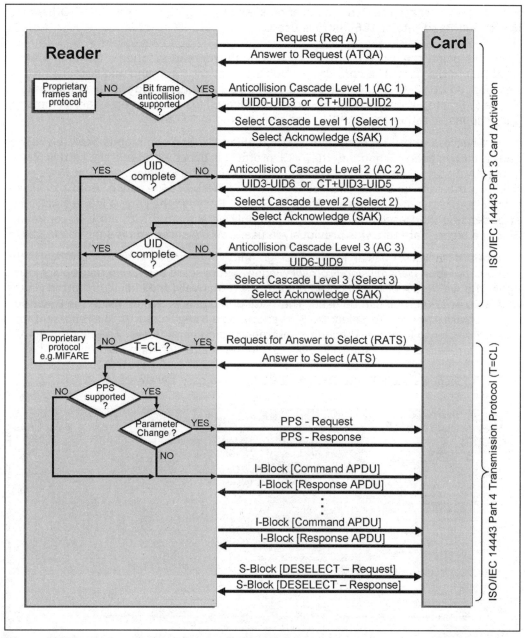

Figure 8.78. Sequence diagram for T=CL card activation and communication.

The entire communication, including reception of the ATS, takes place at 106 kbit/s. If both reader and card support a faster transmission rate, the reader can switch the Baud rate using a Protocol and Parameter Selection Request (PPS-Request). The card acknowledges this request with a PPS Response (Protocol and Parameter Response). Then,

8 CARDS AND TAGS IN APPLICATION

both parties increase their communication speed to 212, 424 or 848 kbit/s. Otherwise, the default Baud rate of 106 kbit/s is used.

From this point, the T=CL protocol is active, and the reader sends a Command APDU (C-APDU) to the card. The card carries out the relevant command and responds with a Response APDU (R-APDU). C-APDUs and R-APDUs are transported within an Information Frame (I-Frame). With MIFARE DESFire and MIFARE Plus cards, proprietary commands and responses, rather than APDUs, are transmitted in I-Frames.

After a successful transaction, that is, when all application commands between reader and card have been exchanged, the T=CL protocol is deactivated and the card is placed in the HALT state. This is done when the reader sends the Deselect Request T=CL command to the card, and the card responds with a Deselect Response. Both blocks are transmitted within a Supervisory Block (S-Block). Because the card is in the HALT state after sending the Deselect Response, it can only be activated with a WUPA command. Even if a higher data rate was previously in use, a WUPA command is sent at 106 kbit/s.

Smart Card Magic.NET has an easy-to-use graphical interface for the T=CL protocol in the *ISO/IEC 14443-4 Transmission Protocol* window. Simple card activation takes place using either the *Act. Req A* or the *Act. Req B* button. Alternatively, the individual buttons in the *ISO/IEC 14443-3* group can be used. The T=CL protocol is activated using the RATS button. The card's Answer To Select (ATS) is shown as a tree structure. This representation makes it easier to interpret the individual transmission parameters. The reader and ATS card parameters are then shown in plain text in the *Reader Info* and *Card Info* text boxes.

Figure 8.79 shows the MIFARE DESFire EV1 ATS and card parameters.

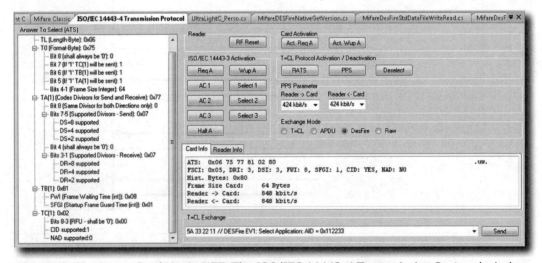

Figure 8.79. *Smart Card Magic.NET: The ISO/IEC 14443-4 Transmission Protocol window.*

Provided that the card supports higher data rates, the speed can be changed immediately after receiving the ATS, by using the *PPS* button. The maximum data rate on the Elektor RFID Reader is 424 kbit/s.

8.6.1.1 Multi-Card Activation

Multi-card operation is currently of limited importance, and thus most commercial PC/SC readers don't support it. Therefore, this function is not supported in the *ISO/IEC 14443-4 Transmission Protocol* window implementation. Section 8.6.4.2 has a simple program example that clarifies the basic idea of multi-card activation.

8.6.2 Data Exchange

All of the T=CL protocol functionality is implemented in the Elektor RFID Reader library (GS.TclProtocol.dll). So, all of the blocks (I-Frame, R-Frame and S-Frame) are generated by the PC software. Monitoring of the Frame Waiting Time is done by the reader firmware using the timers integrated in the MF RC522 reader IC. If there is a timeout, error handling is done according to the T=CL protocol.

Smart Card Magic.NET's *ISO/IEC 14443-4 Transmission Protocol* window also supports the sending and receiving of card commands. In the *T=CL Transmission Protocol* group, commands can be entered in hexadecimal format in the *Send* ComboBox. For readability, white space may be entered between individual instruction bytes. Optionally, a comment may be added after the last command byte. This is preceded by a double slash (//). If a command has a string in it, the characters may be included directly by enclosing them in quotation marks. The command is sent to the card by clicking the *Send* button or by pressing Enter.

The *Send* ComboBox remembers all new commands that are entered, so it's not necessary to retype them all In the command history, a previously used command can be selected from the drop-down list, or by using the ↓ and ↓ keys. When you exit Smart Card Magic.NET, the entire history is stored in the application's local User directory, in the comboBoxSend.his file.

Figure 8.80. Send ComboBox command history.

The *Tools* → *Show Local User App Data Path* menu option opens Windows Explorer in the file's folder. The comboBoxSend.his file is a normal text file, which can be edited with any text editor. This makes it possible to add or edit commands using a text editor.

Right-clicking on the *Send* ComboBox brings up a context menu (see Figure 8.81). This offers additional functions for deleting individual commands or the entire history. Also, the current command history can be saved under a different file name. The commands loaded from a file are either added to or replace the current history.

8 CARDS AND TAGS IN APPLICATION

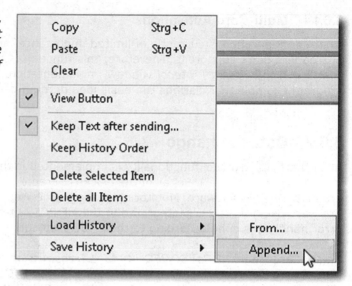

Figure 8.81.
The Send ComboBox context menu supports the management and archival of card commands.

8.6.2.1 Smart Card Magic.NET – Exchange Mode

The T=CL & DesFire, APDU and Raw Exchange options in the *Exchange Mode* groupBox specify the interpretation of the command bytes and the output to the *Trace* window.

T=CL Option
With this option, an I-Frame is sent to the card. The data in the *Send* ComboBox is interpreted as an information field (INF) within the I-Frame. Therefore, this option supports the sending of manufacturer-specific commands (e.g. DESFire Native), as well as the sending of APDU commands. In the *Trace* window, all details of the T=CL protocol, with the exception of the Epilog field (the two CRC bytes), are shown. In the Elektor RFID Reader, the CRC is automatically calculated by the MF RC522 reader IC and automatically appended to the end of the frame.

Figure 8.82.
Activating a MIFARE DESFire card, sending the Get Version command, and deactivating the card.

```
Trace Window

    RF Reset(5 ms)
--> Activate Card: REQA-AC-SEL 0-(AC-SEL)
    SetPCDBaudrate(Reader -> Card: 106 kbit/s, Reader <- Card: 106 kbit/s)
<-- ATQ: 0x44 03   UID: 0x88 04 96 66 A1 76 1B 80   SAK: 0x20 T=CL: supported

    Raw: SetFWT(FWI 10, FWTM 1)
--> TCL RATS: 0xE0 50
<-- TCL ATS: 0x06 75 77 81 02 80

    Raw: SetFWT(FWI 8, FWTM 1)
--> TCL: PPS Request: 0xD0 11 0A
<-- TCL: PPS Response: 0xD0

    SetPCDBaudrate(Reader -> Card: 424 kbit/s, Reader <- Card: 424 kbit/s)
--> TCL: I-Block: 0x02 60                             <-- Information Field
<-- TCL: I-Block: 0x02 AF 04 01 01 01 00 16 05
                                                      <-- Protocol Control Byte
--> TCL: S-Deselect Request: 0xC2
<-- TCL: S-Deselect Response: 0xC2
```

APDU Option

This option also sends an I-Frame to the card. The *Send* comboBox's data is also interpreted as an information field (INF) within the I Frame. However, in contrast to the T=CL option, this option is only used to send APDU commands. In the *Trace* window, only the C-APDU and R-APDU data is output. Full T=CL protocol details (Chaining, Waiting Time Extension, Error Handling, etc.) are not output. The status word, SW1SW2, is also output on a separate line.

DesFire Option

This option supports the sending of DESFire commands exclusively. Data transfer takes place within an I-Frame. The output of the trace information is in the proprietary DESFire command format. The T=CL protocol details are also hidden here. The DESFire card response always contains a status byte. This status byte is shown on its own line.

Raw Option

The Raw Exchange option allows the manual creation of TCL Blocks (I-Frame, R-Frame and S-Frame). Unlike the other three options, this one requires that the PCB byte and the optional CID, NAD and user data be entered into the *Send* ComboBox. Only the two CRC bytes are added automatically by the reader firmware. Among other things, this option enables the creation of frames containing errors, which allows for negative card protocol testing.

8.6.2.2 Block Chaining

An I-Frame's maximum block size is determined by the two parameters, FSD and FSC. Chaining allows the transmission of longer card commands or card responses, even when these exceed the size of the card or reader receive buffer. For this purpose, the data is simply divided into blocks whose maximum size matches the receive buffer size. The I-Frame's PCB (Protocol Control Block) indicates whether chaining is active or not.

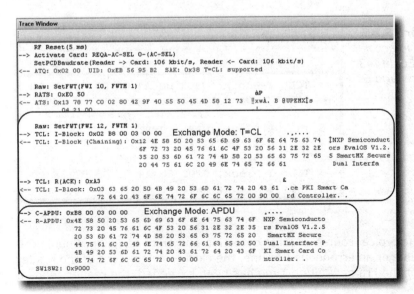

Figure 8.83.
In this example, the card response overwrites the reader's receive buffer. The data is thus divided into blocks (chaining).

8.6.2.3 Waiting Time Extension

The card has to respond within the time specified in the ATS parameter (FWT – Frame Waiting Time). Should a timeout occur, the reader begins with error handling in accordance with ISO/IEC 14443-4 rules. If the card needs more time than the ATS-prescribed FWT, the card may request a time extension using an S(WTX) request.

Figure 8.84. In this example, the card needs more time than specified in the ATS, and request a time extension by calling S(WTX) twice.

```
Trace Window

--> TCL: I-Block: 0x02 B8 FB 00 00 05 11 22 33 44 55 05
<-- TCL: S-Block (WTX Request): 0xF2 01

--> TCL: S-Block (WTX Response): 0xF2 01
<-- TCL: S-Block (WTX Request): 0xF2 01

--> TCL: S-Block (WTX Response): 0xF2 01
<-- TCL: I-Block: 0x02 11 22 33 44 55 90 00
```

8.6.2.4 Error Detection and Correction

Interference during the transmission of large amounts of data happens repeatedly in practice. The ISO/IEC 14443 Part 4 standard defines a fairly extensive set of rules for error detection and recovery, so that an error does not require the re-activation of the card. Error detection at the byte level is done with a parity bit (odd parity). The 16-bit CRC checksum is calculated across the entire transmission frame of an I-Frame, R-Frame or S-Frame. This also detects errors when several bits within a frame fail. The reader also monitors the card response times (SFGT and FWT).

Figure 8.85. Error correction in accordance with ISO/IEC 14443 Part 4 – Protocol Scenario 6.

```
Trace Window

--> TCL: I-Block: 0x02 00 A4 00 00 02 3F 00
<-- TCL: No Card Responce

--> TCL: R-Block (NAK): 0xB2
<-- TCL: R-Block (ACK): 0xA3

--> TCL: I-Block: 0x02 00 A4 00 00 02 3F 00
<-- TCL: I-Block: 0x02 90 00
```

Figure 8.85 shows the typical course of error detection and error handling. Should the card receive no I-Frame or an erroneous I-Frame, the card remains silent. If the reader software detects that the Frame Waiting Time has been exceeded, the reader sends a

NAK (Negative AcKnowledge) to the card. If the card is still in the reader field, it responds with an ACK (ACKnowledge), after which the reader repeats the I-Frame. ACK and NAK are sent within an R-Frame.

8.6.3 Elektor RFID Reader Library: T=CL

All T=CL protocol classes are encapsulated in the `GS.TclProtocol.dll` library file. In order for this library to be reusable in other reader projects, no concrete methods from `GS.ElektorReader.dll` are used. The methods necessary for communication with the card are declared in the `I_RAWCardExchange` interface. A reader that uses the `TCL_Protocol` class must at least implement these methods.

Table 8.27. Methods in the I_RAWCardExchange interface.

Method	Description
`int SetPCDBaudrate ( DivisorInteger dri, DivisorInteger dsi)`	This method sets the reader Baud rate. The `DRI` and `DSI` parameters are of the `DivisorInteger` type. `DRI` transfer direction: reader → card. `DSI` transfer direction: card → reader. `DivisorInteger` is an enumeration with the following values: `Baud_106K`, `Baud_212K`, `Baud_424K` and `Baud_848K`.
`int SetFWT ( byte fwti, byte fwtm);`	Sets the Frame Waiting Time in the reader firmware. `fwti`: Frame Waiting Time Integer (range: 0 – 14). `fwtm`: Frame Waiting Time Multiplier (range: 1 – 59). The time in seconds is calculated with the following formula: = FWT = `fwtm` × (256 × 16 / fC) × 2 × `FWTI`.
`int ExchangeFrame ( byte[] cmdData, int cmdLength, byte[] respData, out int respLength)`	Exchanges I-Frames, R-Frames and S-Frames. CRC checksum calculation takes place in the reader firmware. If there is a timeout (FWT) or a communication error, the method must return an error code other than 0x00. A return value of 0x00 means OK.

When creating an object of the `TCL_Protocol` class, it is necessary to pass a `Reader` object to the constructor, which all methods of the `I_RawCardExchange` interface implement. The `ElektorISO14443Reader` class implements the `I_RAWCardExchange` interface methods in the embedded `RawCardExchange` class. This class is accessed through the public class variable, `Raw`.

```
ElektorISO14443Reader reader = new ElektorISO14443Reader();
TCL_Protocol tcl = new TCL_Protocol(reader.Raw);

tcl.PcdParameter.CID = 0;
tcl.PcdParameter.FSDI = 5;
tcl.PcdParameter.DRI_Max = DivisorInteger.Baud_424K;
```

8 CARDS AND TAGS IN APPLICATION

```
    tcl.PcdParameter.DSI_Max = DivisorInteger.Baud_424K;
    tcl.PcdParameter.DriEqualDsi = DriEqualDsi.YES;

    reader.OpenPort();
    reader.ActivateCard(RequestCmd.WUPA);
    tcl.RATS();
```

In the `TCL_Protocol` class, the two helper classes, `PCDParam` and `PICCParam`, manage the properties of the reader and the card. The default values of the `PCDParam` class are selected so that they match the properties of the Elektor RFID Reader hardware. An application may access these properties via the `PcdParameter` and `PiccParameter` class variables.

An object of the `TCL_Protocol` class is instantiated in the `ElektorISO14443Reader` class, so the user of the Elektor RFID Reader library (`GS.ElektorReader.dll`) does not have to deal with the details of `TCL_Protocol` object creation. Access is via the `Tcl` instance variable. In the `ElektorISO14443Reader` class, only one instance of the `TCL_Protocol` object is created. To manage a T=CL card, a `TCL_Protocol` object is required.

```
    ElektorISO14443Reader reader = new ElektorISO14443Reader();

    reader.OpenPort();
    reader.ActivateCard(RequestCmd.WUPA);

    reader.Tcl.RATS();
```

For this reason, the `ElektorISO14443Reader` class does not support multi-card activation. To manage several physical card instances, it's necessary to create an array of `TCL_Protocol` objects (see Listing 8.30).

Table 8.28. *Properties of the PCDParam class.*

Property	R/W	Description
CID	R/W	The logical card number (CID), which is required for multi-card activation. Range: 0 – 14; Default: 0.
DivisorInteger DRI_Max	R/W	Maximum transmission rate in the reader → card direction. DivisorInteger is an enumeration with the following values: Baud_106K, Baud_212K, Baud_424K and Baud_848K. Default value: Baud_424K.
DivisorInteger DSI_Max	R/W	Maximum transmission rate: card → reader.
DriEqualDsi	R/W	Indicates that the reader does not support different transmission rates in both directions. The two values of the DriEqualDsi enumeration are YES and NO. Default value: NO.

Property (continued)	R/W	Description (continued)
FrameSize FSD	R/W	The reader's maximum receive buffer size. The FrameSize enumeration has the following values: FS_16, FS_24, FS_32, FS_40, FS_48, FS_64, FS_96, FS_128 and FS_256. Default value: FS_64, or 64 bytes. If the FSD property is changed, the FSDI property's value is updated.
byte FSDI	R/W	ISO/IEC 14443 Part 4 defines a conversion table between the FSDI value (RATS command) and the FSD value. The PCDParam class supports the ability to read and write the maximum receive buffer size via FSDI. During write access, an update of the FSD property also takes place. Range: 0 – 8; default: 5, or 64 bytes.

Table 8.29. Properties of the PICCParam class.

Property	R/W	Description
byte[] ATS	R/W	ATS bytes received from the card.
byte[] AtsHistBytes	R/W	The historical bytes of the ATS.
Supported CidSupported	R/W	Indicates whether the card supports the logical identification number (CID), and thus multi-card activation. CidSupported is an enumeration with the values YES and NO.
Supported NadSupported	R/W	Indicates whether the card supports the logical node address (NAD).
DivisorInteger DRI_Max	R/W	Maximum transfer rate: reader → card. DivisorInteger is an enumeration with the following values: Baud_106K, Baud_212K, Baud_424K and Baud_848K.
DivisorInteger DSI_Max	R/W	Maximum transfer rate: card → reader.
DriEqualDsi DREqualDS	R/W	Defines whether the card supports different transmission rates in each direction. The two values of the enumeration DriEqualDsi are YES and NO.
FrameSize FSC	R/W	The card's maximum receive buffer size. The FrameSize enumeration has the following values: FS_16, FS_24, FS_32, FS_40, FS_48, FS_64, FS_96, FS_128 and FS_256.
byte FSCI	R/W	Frame Size Card Integer.
byte FWI	R/W	Frame Waiting Time Integer.
byte SFGI	R/W	Startup Frame Guard Time Integer.

8 CARDS AND TAGS IN APPLICATION

Table 8.30. *The TCL_Protocol class properties.*

Property	R/W	Description
`int RecBufferSize`	R/W	Sets the size of the card receive buffer at the application level (APDU buffer size). The default value is 1,024.
`Bool EnableApduTrace`	R/W	Enables or disables the SCard trace output.
`Bool EnableTclTrace`	R/W	Enables or disables the T=CL trace output.
`PcdParam PcdParameter`	R/W	Reads or writes the reader parameter of type `PcdParameter`.
`PiccParam PiccParameter`	R/W	Prerequisite: calling the `RATS()` method. The card returns the communication parameters in the ATS. The individual values can be read from the `PiccParameter` property.

A broad overview of the TCL_Protocol class methods and properties is in the tables below.

Table 8.31. *TCL_Protocol: methods for T=CL protocol activation and deactivation.*

Method	Description
`TclRetCodes RATS()`	Prerequisite: successful card activation. Sends a RATS command to the card and receives the ATS. The `PiccParameter.ATS` property holds the ATS value.
`TclRetCodes PPS()`	Prerequisite: successful call to `RATS()`. Sends a PPS request command to the card and receives the PPS response. If no PPS parameter is passed to the `PPS()` method, the `PcdParameter` and `PiccParameter` properties will determine the maximum transfer rate in both directions.
`TclRetCodes PPS ( DivisorInteger dri, DivisorInteger dsi)`	This variant of the `PPS()` method offers targeted selection of the communication speeds. If one of the two parameters is not supported by the reader or the card, the highest possible value is used. `DivisorInteger` is an enumeration of the following values: `Baud_106K`, `Baud_212K`, `Baud_424K` and `Baud_848K`.
`TclRetCodes Deselect()`	Prerequisite: successful call to the `RATS()`. Sends a Deselect command to the card and receives the Deselect response.

The overloaded Exchange() methods in Table 8.36 are for sending and receiving I-Frames. Using these methods, one may send both manufacturer-specific (DESFire Native) and ISO/IEC 7816 standard APDU commands to the card.

Table 8.32. *TCL_Protocol: extract of the general Exchange() methods.*

Method	Description

`int Exchange(`
`byte[] sndBuffer,`
`out byte[] recBuffer);`
	The data in the `sndBuffer` byte array are sent to the card in an I-Frame. The method creates a byte array of the data received from the card and returns this in the out `recBuffer` parameter. The size of the byte array is identical to the number of bytes received from the card, including the status information. Before this method creates an array for the card response, all received bytes are copied to an internal, temporary receive buffer. The size of this buffer is defined by the `RecBufferSize` property.

`int Exchange(`
`byte[] sndBuffer,`
`byte[] recBuffer,`
`int recOffset,`
`ref int recLength);`
	The `recBuffer` parameter passes an instance of a byte array. The `recOffset` parameter defines the start index within this receive buffer. The free receive buffer size is passed as a reference parameter, `ref recLength`, in the method call. In the same `ref recLength` parameter, the method returns the number of bytes received from the card.

`int Exchange(`
`string sndBuffer,`
`out byte[] recBuffer);`
	These two overloaded `Exchange()` methods take the card command as a hexadecimal string.

`int Exchange(`
`string sndBuffer,`
`byte[] recBuffer,`
`int offset,`
`ref int recLength);`	

`int Exchange(`
`byte sndNAD,`
`byte[] sndBuffer,`
`ref byte recNAD,`
`out byte[] recBuffer);`
	Additionally, the node address (NAD) of the receiver is exchanged. This is to establish a logical connection between a reader and a card. The node address is defined in ISO/IEC 7816 Part 3 for T=1. The node address is almost never used in contact-type or contactless cards. This would require the reader to have several logical card slots. Nevertheless, in practice, this feature is supported by most protocol libraries.

`int Exchange(`
`byte sndNAD,`
`byte[] sndBuffer,`
`ref byte recNAD,`
`byte[] recBuffer,`
`int offset,`
`ref int recLength);`	

In Sections 8.8.3 and 8.8.4, special APDU and MIFARE DesFireExhange() methods are presented.

8 CARDS AND TAGS IN APPLICATION

> **Note** The `TCL_Protocol` class was developed exclusively for the Elektor RFID Reader. Its functionality has only been tested with different card types. Therefore, there is the possibility that the implementation deviates slightly from the ISO/IEC 14443 standard. The `GS.TclProtocol.dll` source code can be downloaded from the Elektor website.

8.6.4 Example Programs

8.6.4.1 T=CL Protocol Activation and Deactivation

This example shows the activation and deactivation of a single T=CL card (single card activation). After successful T=CL protocol activation, the Select Master File APDU command is sent to the card. This command selects the card's root directory, i.e. the master file. For this example, it doesn't matter whether the card supports this command or not. The card responds with its status code regardless. The smartcard file system and associated commands are presented in Sections 8.7.3 and 8.8.3.

Listing 8.29. TCL_ProtocolActivation.cs.

```csharp
using System;
using GS.ISO14443_Reader;
using GS.ElektorRfidReader;

namespace GS.ElektorRfidReaderExample
{
    class TclProtocolActivation
    {
        public static void Main(MainForm script, string[] args)
        {
            ElektorISO14443Reader reader = new ElektorISO14443Reader();
```

The `EnableTclTrace` property controls debug output in the Trace window. If the value is true, all details of the T=CL protocol are output.

```csharp
            reader.Tcl.EnableTclTrace = true;

            try
            {
```

The card activation is identical to the activation of a MIFARE Ultralight or MIFARE 1K/4K card.

```csharp
                int ret;

                reader.OpenPort();
                ret = reader.RFReset(5);
```

```csharp
            ret = reader.ActivateCard(RequestCmd.WUPA);
            if (ret != 0) return;
```

All cards that support the T=CL protocol are placed in the HALT state, as prescribed in the ISO/IEC 14443 standard.

```csharp
            if (reader.IsTCLCard == false)
            {
                reader.HaltA();
                return;
            }
```

After successful T=CL protocol activation, the `ToString()` method from the PiccParam class can be used to output the card parameter to the console.

```csharp
            ret = (int)reader.Tcl.RATS();
            if (ret != 0) return;

            Console.WriteLine(reader.Tcl.PiccParameter.ToString());
```

If the card supports baud rates higher than 106 kbit/s, the transmission rate is increased, using a Protocol and Parameter Selection, to the maximum possible value.

```csharp
      ret = (int)reader.Tcl.PPS();
                if (ret != 0) return;
```

As an example, the Select Master File APDU command is sent to the card.

```csharp
            byte[] sndBuffer = {0x00, 0xA4, 0x00, 0x00, 0x02, 0x3F, 0x00};
            byte[] recBuffer;
            ret = reader.Tcl.Exchange(sndBuffer, out recBuffer);
```

In the last step, all cards are deactivated. These will then be in the HALT state.

```csharp
             ret = (int)reader.Tcl.Deselect();
        }
        catch (Exception ex)
        {
          Console.WriteLine("Exception from " + ex.Source.ToString() +".dll: "
                      + ex.Message.ToString() + Environment.NewLine);
        }
        finally { reader.ClosePort(); }
        }
      }
    }
```

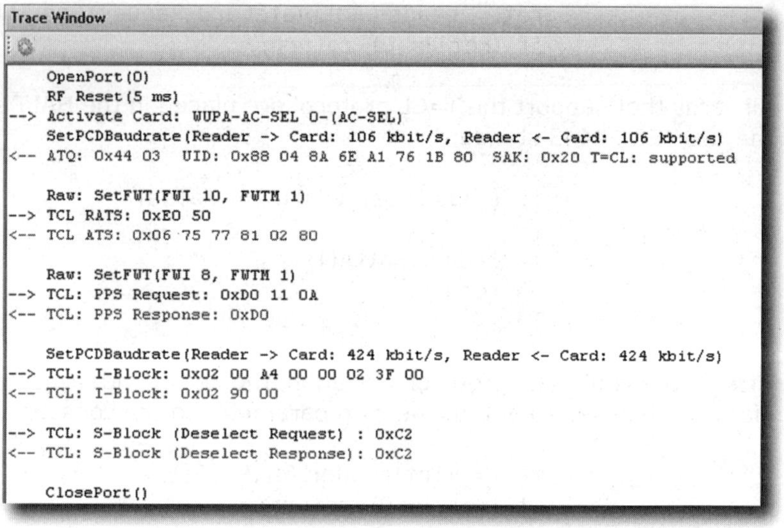

8 CARDS AND TAGS IN APPLICATION

Figure 8.86.
Activating a DES-Fire EV1 card, sending the ISO/IEC 7816 Select Master File command, and deactivating the card.

```
Trace Window

    OpenPort(0)
    RF Reset(5 ms)
--> Activate Card: WUPA-AC-SEL 0-(AC-SEL)
    SetPCDBaudrate(Reader -> Card: 106 kbit/s, Reader <- Card: 106 kbit/s)
<-- ATQ: 0x44 03   UID: 0x88 04 8A 6E A1 76 1B 80   SAK: 0x20 T=CL: supported

    Raw: SetFWT(FWI 10, FWTM 1)
--> TCL RATS: 0xE0 50
<-- TCL ATS: 0x06 75 77 81 02 80

    Raw: SetFWT(FWI 8, FWTM 1)
--> TCL: PPS Request: 0xD0 11 0A
<-- TCL: PPS Response: 0xD0

    SetPCDBaudrate(Reader -> Card: 424 kbit/s, Reader <- Card: 424 kbit/s)
--> TCL: I-Block: 0x02 00 A4 00 00 02 3F 00
<-- TCL: I-Block: 0x02 90 00

--> TCL: S-Block (Deselect Request) : 0xC2
<-- TCL: S-Block (Deselect Response): 0xC2

    ClosePort()
```

8.6.4.2 Multi-Card Activation

If the reader application supports multi-card activation, it will activate several T=CL cards sequentially. With the RATS command, every card is assigned a logical card number (CID – Card Identifier) between 1 and 14 for the duration of activation.

An active card responds only to a T=CL block (I-, R- and S-Frames) if the block has the same value as assigned by the RATS command. All other blocks and ISO/IEC 14443 Part 3 commands (REQA, SELECT, etc.) are ignored. If the card doesn't support a CID, it ignores any block containing a CID.

Multi-card activation allows the reader to communicate with several active T=CL cards in a row, without taking extra time for additional card activation and deactivation. The T=CL protocol can be used to manage up to 14 physical card instances simultaneously, although a commercial reader is usually only able to supply a few cards with energy. The Elektor RFID Reader can operate 2 to 3 cards at once.

Listing 8.30 shows the basic program flow of a multi-card application.

Listing 8.30. *TCL_MultiCardActivation.cs.*

```
using System;
using GS.ISO14443_Reader;
using GS.ElektorRfidReader;
using GS.TclProtocol;
namespace GS.ElektorRfidReaderExample
{
    class TclMultiCard
    {
        public static void Main(MainForm script, string[] args)
        {
```

THE T=CL TRANSMISSION PROTOCOL 8.6

To manage several physical card instances, it is necessary to create an array of `TCL_Protocol` objects.

```
ElektorISO14443Reader reader = new ElektorISO14443Reader();
TCL_Protocol[] tcl = new TCL_Protocol[15];
```

For every element of the `tcl` array, starting with index 1, a `TCL_Protocol` object is created and a corresponding CID in the range 1 – 14 is assigned.

```
try
{
   for (int i =1; i < tcl.Length; i++)
   {
      tcl[i] = new TCL_Protocol(reader.Raw);
      tcl[i].PcdParameter.CID = (byte)i;
      tcl[i].EnableTclTrace = true;
   }
   int ret;
   byte cid = 1;
   byte cardCount = 0;
```

Before running the script, it is necessary to place all the cards within the reader's field. An RF reset places all cards in the defined IDLE state.

```
reader.OpenPort();
reader.RFReset(5);
```

Activation of all cards in the reader field takes place within a `do-while` loop. These cards must support CIDs. Cards that don't support this criterion, e.g. the MIFARE Ultralight card, are placed in the HALT state.

```
do
{
   ret = reader.ActivateCard(RequestCmd.REQA);
   if (ret != 0) break;
   if (reader.IsTCLCard == false)
   {
      reader.HaltA();
      continue;
   }
   ret = (int)tcl[cid].RATS();
   if (ret == 0)
   {
      if (tcl[cid].PiccParameter.CidSupported == Supported.NO)
      {
         reader.HaltA();
         continue;
      }
```

353

```
                cid++;
                cardCount++;
            }
        }
        while (cid < tcl.Length);
```

We then check to see if at least one T=CL card was activated.

```
            if (cardCount > 0)
            {
```

In the next step, all active T=CL cards are sent an APDU command.

```
                for (byte i = 1; i <= cardCount; i++)
                { // Select Master File
                    tcl[i].Exchange("00 A4 00 00 02 3F 00");
                }
```

Finally, all cards are deactivated sequentially.

```
                for (byte i = 1; i <= cardCount; i++)
                {
                    tcl[i].Deselect();
                }
            }
        }
        catch (Exception ex) { Console.WriteLine(ex.Message.ToString() +
                                    Environment.NewLine); }
        finally { reader.ClosePort(); }
    }
  }
}
```

Figure 8.87 shows the example command sequence for multi-card activation of two cards. After successfully activating a card using the REQA, Anti-collision and Select commands, the T=CL protocol activation occurs. For this, the reader sends a RATS command to the card with a CID value of 1. This responds to the ATS. Then, the card responds only to blocks that have a CID of 1. One may then activate a second card without any data collision issues. Thie second card is assigned a CID value of 2.

At this point, both cards in the field are active. In the next step, first Card 1 (CID=1) then Card 2 (CID=2) are sent the Select Master File APDU command. Both cards are selected, or addressed, using the CID. The respective blocks in the card response also contain the CID. Card 1 returns an error code with a value of 0x6A 82 in the INF field. This error code means that the card could not select a file with the File ID 0x3F 00. The second card can successfully carry out the command and thus responds with an error code of 0x90 00. Finally, both cards are deactivated. In this case, the S-Frames in the Deselect Request and Deselect Response also contain the CID.

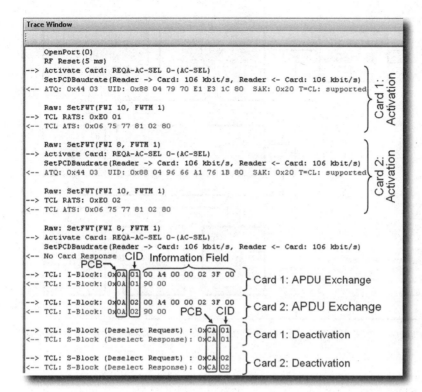

Figure 8.87.
Example of multi-card activation.

8.7 The MIFARE DESFire EV1 Card

The MIFARE DESFire EV1 chip is based on a contactless smartcard controller with a card operating system (COS). The DESFire operating system can be divided into the following functional areas:

- Communication via the contactless interface.
- Command scheduler.
- File management system.
- Preparation of cryptographic algorithms and functions.

Transfer of the card commands is done with the T=CL protocol. The comprehensive card instruction set supports commands for the management of applications and the associated data and key files. When creating an application, it is determined whether selecting mutual authentication is required or not. Depending on configuration, the data transfer can take place encrypted or unencrypted. The data integrity is assured either via a CRC checksum or a cryptographic checksum (MAC – Message Authentication Code).

8 CARDS AND TAGS IN APPLICATION

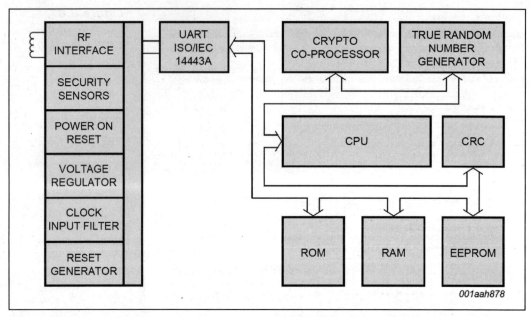

Figure 8.88. Block diagram of a MIFARE DESFire EV1 card (source: NXP Semiconductors).

The cryptographic operations are calculated with the help of a special cryptographic coprocessor. The DESFire EV1 card supports the DES algorithm, the Triple-DES algorithm (both 2- and 3-key variants), as well as the AES algorithm. The encryption algorithm for authentication, data encryption and MAC calculation is also defined when creating the application.

Because the DESFire card's specifications are not public (NDA), this section will present just the basics of card communication and the functional principles of the file system, without any cryptographic functions.

NXP offers the MIFARE DESFire EV1 in the following variants:

- MIFARE DESFire EV1 2K (MF3 IC D21)
- MIFARE DESFire EV1 4K (MF3 IC D41)
- MIFARE DESFire EV1 8K (MF3 IC D81)

These differ only in EEPROM memory size. All other features are identical. Using the practical examples in this section, one can become fully familiar with all three chip types.

8.7.1 MIFARE DESFire EV1 Commands

The MIFARE DESFire EV1 card supports two different command formats:

- DESFire Native APDU structure.
- ISO/IEC 7816 APDU structure.

To assist with differentiation, commands encoded in the proprietary DESFire Native APDU structure are referred to as DESFire Native commands in the following. The term 'APDU' is reserved for the ISO/IEC 7816-3 APDU command structure.

The DESFire Native commands can be wrapped in an APDU command structure. These are referred to as ISO/IEC 7816-wrapped APDUs. This method was developed especially for PC/SC readers that support only this command structure (see Section 11.4.4.7).

ISO/IEC 7816 APDUs

Because the DESFire EV1 card optionally supports the ISO/IEC 7816 standard file system, the card supports the following ISO/IEC 7816-4 standard APDU commands (see Section 8.8.3) for authentication and for reading and writing data: Select File, Read Binary, Update Records, Append Record, Get Challenge, Internal Authenticate and External Authenticate.

8.7.2 DESFire Native Command Structure

The DESFire Native structure is much more efficient than the APDU structure and is designed for time-critical ticketing applications. The minimum length of a MIFARE DESFire Native command is one byte, while for ISO/IEC 7816-3 APDUs, it is four bytes. The structure of the DESFire Native commands to the card differs from the card responses.

8.7.2.1 Card Command Structure

A DESFire Native command consists of at least one command byte (CMD), in which the card command is encoded. Optionally, a variable number of parameter bytes may follow, which the respective command specifies in detail. This is followed by an optional data field.

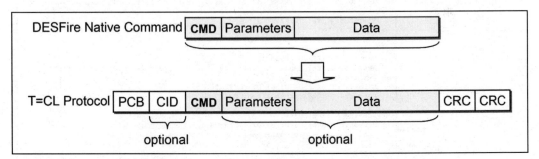

Figure 8.89. The MIFARE DESFire Native command's data structure.

8.7.2.2 Card Response Structure

The card response consists of a one-byte status or error code and an optional data field of variable length. If the previous command was successfully processed by the card, the card returns a value of 0x00 in the Status byte. All other values, with the exception of 0xAF (Additional Frame), are error codes.

8 CARDS AND TAGS IN APPLICATION

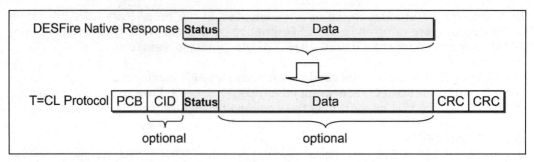

Figure 8.90. *The MIFARE DESFire Native response's data structure.*

> **Note** Both the command to the card and the card's response have no specified length. Even the underlying T=CL protocol has no length field. The size of a contactless card's frame is defined only by Start of Frame and End of Frame in the physical layer (ISO/IEC 14443 Part 3).

8.7.2.3 DESFire Block Chaining

Although the T=CL protocol already offers the ability to chain (see Section 8.6.2.2), the DESFire card offers a special application-level method of block chaining.

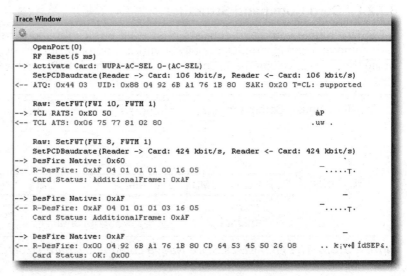

Figure 8.91. *The principle of DESFire chaining.*

Figure 8.91 shows the principle of DESFire block chaining. In this example, the Get Version command (command byte 0x60) is sent to the card. In the first card response, the card returns the Additional Frame (0xAF) status and 7 data bytes.

A reader can request the next available byte with the 0xAF command byte. In the second card response, the card also responds with the Additional Frame Status byte and further data bytes. For this reason, the reader once again sends the 0xAF byte to the card. In the third frame, the card responds with the OK status (0x00) and the remaining data bytes.

This method of block chaining at the application level has the advantage that the DESFire card can be integrated in older reader systems that don't support T=CL chaining. Further, it is not imperative that the reader reads all data from the card. For example, if the information in the card's first frame is all that is needed, the reader can end the command. This is done by the reader simply sending a different command to the card.

8.7.3 The DESFire File System

The MIFARE DESFire EV1 card's file system can manage up to 28 card applications with a maximum of 32 files per application. The big difference from a PC file system is that it can provide access to a card application and assign individual access rights to each data file. The access rights are decided when an application is created. All files on a MIFARE DESFire EV1 card are stored in the EEPROM memory.

8.7.3.1 File Types

A DESFire EV1 card has a directory structure similar to that on a PC file system. Just like a PC drive, each card has a root directory, known here as a master file. In the master file (MF), there is at least one application directory, known in smartcard technology as a dedicated file (DF). All data files in a DESFire application, the so-called elementary files (EF), are found within the application directories (DF).

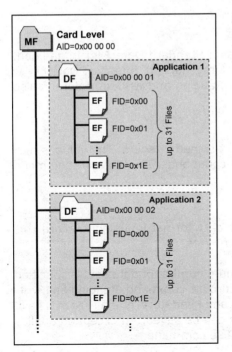

Figure 8.92.
MIFARE DESFire card directory structure.

Unlike in other operating systems, it is not possible to create any further subdirectories within applications on a MIFARE DESFire card. This does not, however, represent any limitation for your typical DESFire applications.

8.7.3.2   Data File Structure

Like most card operating systems, the MIFARE DESFire card supports various data file structures. When creating a new file, one may choose freely between a Standard Data File, Backup Data File, Value File, Linear Record File or a Cyclic Record File.

With the exception of the standard data file, all EF data structures have an automatic rollback mechanism (backup management) in case of a power interruption during writing to a file (see Section 8.4.9.5). Only by executing the Commit Transaction command will the write instruction be valid for a Backup Data File, Value File, Linear Record File or Cyclic Record File. If Commit Transaction has not yet been executed, all write operations on these data types may be cancelled using the Abort Transaction command. In this case, all of the original data is restored.

Standard Data File
Standard Data Files correspond with the ISO/IEC 7816-4 standard transparent data structure. Because the Standard Data File has no internal data structure, the data is simply saved byte-wise in sequence. The interpretation of the data is incumbent on the reader application. The file size is set by CreateStdDataFile when creating a Standard Data File. The data from a Standard Data File is read using the Read Data command and written with the Write Data command.

Backup Data File
A Backup Data File has the same data structure as a Standard Data File. In addition, the Backup Data File supports an automatic rollback mechanism. A Backup Data File is created with the CreateBackupDataFile command. Reading and writing of data is also done using Read Data and Write Data.

Value File
The Value File data structure is used to store 32-bit signed integers. The value is manipulated using the Credit, Limited Credit and Debit commands. The GetValue command reads the value.

Linear Record File
Linear Record Files are used to store records. The data format in a single record corresponds to a Standard Data file, i.e. a sequence of bytes.

The maximum data set length as well as the maximum number of data sets is specified when the file is created, using the CreateLinearRecordFile command. Reading and writing of individual records is done using Write Record and Read Record. If all records have been written to a Linear Record File, no more data can be saved. An exception applies when all records are erased using the ClearRecordFile command.

Cyclic Record File
The Cyclic Record File is structured in basically the same manner as a Linear Record File. With a Cyclic Record File, the oldest record is always overwritten if the file has already been fully written. This structure is therefore suitable for logs.

8.7.3.3 Directory Names

The MIFARE DESFire EV1 card supports three different directory and file name labels: DESFire Application Identifier (AID), ISO/IEC 7816-4 File Identifier (FID), and ISO/IEC 7816-5 Application Identifier.

Table 8.33. Example of Application Directory Encoding

DESFire AID	ISO/IEC 7816-4 FID	ISO/IEC 7816-4 AID
0x11 22 33	0x11 22	0x11 22 33 44 55 66 77 88

Table 8.34. Encoding of the Root Directory

DESFire AID	ISO/IEC 7816-4 FID	ISO/IEC 7816-4 AID
0x00 00 00	0x3F 00	–

DESFire Application Identifier (AID)
The 3-byte DESFire Application Identifier is used to identify a DESFire application. The root directory (MF) of a DESFire card is selected with a DESFire AID of 0x00 00 00. All other AIDs can be freely assigned.

ISO/IEC 7816-4 File Identifier (FID)
The ISO/IEC 7816-4 defines a 2-byte File Identifier for all files, including directories. The value of the Master File is always 0x3F 00. A value of 0xFF FF is reserved for future applications. Furthermore, the ISO/IEC 7816 and other smartcard standards define application-specific values.

ISO/IEC 7816-5 Application Identifier
The Application Identifier (AID) specified by ISO/IEC 7816-5 is used to uniquely identify a smartcard application and it consists of two parts. The first part, a 5-byte Registered Application Provider Identifier (RID), is assigned either by a national or international authority and is unique. The upper nibble of the first byte of an RID is always 0xA for an international RID and 0xD for a national RID. The second part of the AID consists of a 7-byte Proprietary Applicaton Identifier Extension (PIX). The PIX is freely assigned by the creator of the card application.

The ISO/IEC 7816 FID and ISO/IEC 7816 AIDs are thus only required when the DESFire EV1 is emulating an ISO/IEC 7816-4 standard file system. In this case, one may select the application using either the DESFire Select Application command or the ISO/IEC 7816 Select command.

8.7.3.4 File Names

The files on a MIFARE DESFire EV1 card are selected using their file identifiers (FID). The DESFire-specific file identifier is one-byte long. Optionally, one may assign a 2-byte-long ISO/IEC 7816-4 file identifier when creating the file.

Table 8.35. File Name Examples

DESFire AID	ISO/IEC 7816-4 FID
0x01	0x00 01

8.7.4 Data Structure

In the field of smartcards, it is common that the data in a file is stored in TLV format. The first element of the TLV structure, the tag (T), is used to identify data objects. The second element, length (L), encodes the number of data bytes (V – Value) to follow. A tag consists of 1 or 2 bytes and the length information of between 1 and 4 bytes. For the identification of tags, there are different application-specific standards.

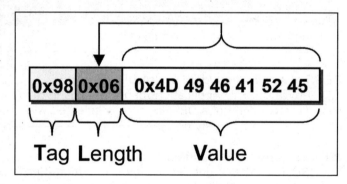

Figure 8.93. Representation of the string, "MIFARE", in TLV format. The tag value has been freely selected.

8.7.5 Elektor RFID Reader Library: MIFARE DESFire EV1

The TCL_Protocol class has 4 overloaded ExchangeDesFireNative() methods. These all return an object of type DesFireResponse. Apart from that, the ExchangeDesFireNative() methods do not differ from the Exchange() methods described in Section 8.6.3. The DesFireResponse helper class allows a separate evaluation of the card statuses, the card data and the reader statuses.

Table 8.36. The Main DesFireNative Exchange() methods.

Method	Description
DesFireResponse ExchangeDesFireNative(byte[] sndBuffer, int sndLength)	The DESFire command (including parameters and data) is passed in the sndBuffer parameter.

Method (continued)	Description (continued)
`DesFireResponse ExchangeDesFireNative( string desFireCommand)`	The DESFire command (including parameters and data) is passed as a hex-formatted string.
`DesFireResponse ExchangeDesFireNative( byte[] sndBuffer, int sndLength, byte? expectedCardStatus)`	These two variants allow for error handling via exceptions. The expected card status byte is passed in the `expectedCardStatus` parameter. Should the card return another status byte, a `DesFireException` is triggered. Any error in the T=CL protocol also triggers an exception. If *null* is passed in the `expectedCardStatus` parameter, all internal error reporting is turned off and no exception is generated in the event of an error.
`DesFireResponse ExchangeDesFireNative( string desFireCommand, byte? expectedCardStatus)`	

Table 8.37. *The properties of the DesFireResponse class.*

Property	R/W	Description
`byte? CardStatus`	R/W	DESFire EV1 card error code.
`byte[] Data`	R/W	Response data from the card without error code.
`int RespLength`	R/W	Length of the card response (status byte + optional data).
`int ReaderStatus`	R/W	Reader firmware error code (see Table 8.11).

8.7.6 Example Programs

8.7.6.1 Creating a DESFire Application

In this example, a DESFire application with a Standard Data File is created. The example assumes that it is a brand new MIFARE DESFire EV1. In this case, one cannot produce an application without prior authentication. To delete the application, authentication with the Card Master Key is required.

Listing 8.31. *MifareDesFireCreateApplication.cs.*

```
using System;
using GS.ISO14443_Reader;
using GS.ElektorRfidReader;
using GS.TclProtocol;
using GS.DesFireUtil;

namespace GS.ElektorRfidReaderExample
{
```

8 CARDS AND TAGS IN APPLICATION

```
        class MifareDesFireCreateApplication
        {
            public static void Main(MainForm script, string[] args)
            {
                ElektorISO14443Reader reader = new ElektorISO14443Reader();

    reader.Tcl.EnableTclTrace = false;
    reader.Tcl.EnableApduTrace = false;
    reader.Tcl.EnableDesFireTrace = true;

    try
    {
        reader.OpenPort();
        reader.RFReset(5);

        reader.ActivateCard(RequestCmd.WUPA);
        reader.Tcl.RATS();
        reader.Tcl.PPS(DivisorInteger.Baud_424K, DivisorInteger.Baud_424K);
```

In the first step, the application directory is created with the DESFire Create Application command. To allow the card to be used for further experimentation, the two optional ISO/IEC 7816 directory names, FID and AID, are specified. The access configuration was chosen so that selection of an application does not require authentication. Please note that the byte order is not the same for all parameters. In most cases, however, the least-significant parameter byte is sent first (little-endian). An example of an exception is the ISO/IEC 7816 AID.

```
        script.TraceWriteLine(
            "\n    Create Application with DESFire: AID = 0x112233; " +
                "\n     ISO/IEC 7816 AID = 0x1122334455667788");
            reader.Tcl.ExchangeDesFireNative(
            "CA " +                    // DESFire Cmd Code: Create Application
            "33 22 11 " +              // DESFire Native AID = 0x112233
                "0F 22 " +             // Key Settings = Free Access
                "22 11 " +             // ISO/IEC 7816 FID = 0x1122
            "1122334455667788",        // ISO/IEC 7816 AID = 0x1122334455667788
            00);                       // Expected card status 0x00 (OK)
```

After creating the application, it is not implicitly selected. It is therefore necessary to explicitly select the application using the DESFire Select Application command.

```
        script.TraceWriteLine("\n Select Application: AID = 0x112233");
        reader.Tcl.ExchangeDesFireNative(
            "5A " +                    // DESFire Cmd Code: Select Application
            "33 22 11",                // AID = 0x112233
            00);                       // Expected card status 0x00 (OK)
```

In the last step, a Standard Data File of size 100 bytes is created. The security parameters are chosen so that we have unrestricted access to the file. Furthermore, the file can be read and written in an unencrypted fashion.

```
        script.TraceWriteLine(
            "\n    Create Standard Data File File No. = 0x01" +
```

```
                    "\n    ISO/IEC 7816 FID = 0x0001; FileSize = 100 Bytes");
                reader.Tcl.ExchangeDesFireNative(
                    "CD " +                 // DESFire Cmd Code: Create File
                    "01" +                  // DESFire Native File No. = 0x01
                    "01 00 " +              // ISO/IEC 7816 FID = 0x0001
                    "00 " +                 // Communication Settings - Plain
                    "EE EE" +               // Access Rights - all Free
                    "64 00 00",             // File Size = 100 Bytes (0x64)
                    00);                    // Expected card status 0x00 (OK)

                reader.Tcl.Deselect();
            }
            catch (Exception ex)
            {
                Console.WriteLine("Exception from " + ex.Source.ToString() +
                    ".dll: " + ex.Message.ToString() + Environment.NewLine);
            }
            finally
            {
                reader.ClosePort();
            }
        }
    }
}
```

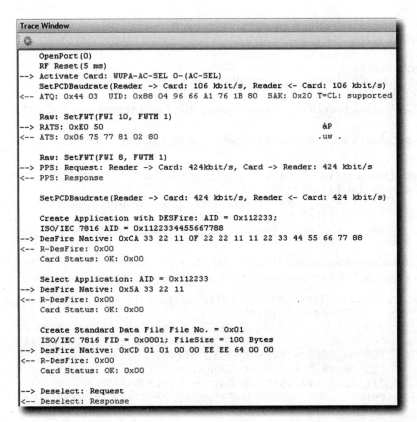

Figure 8.94.
Creating a DESFire application.

8 CARDS AND TAGS IN APPLICATION

8.7.6.2 Standard Data File: Reading and Writing Data

The following example shows how to write data into a MIFARE DESFire standard data file and then read it back.

Listing 8.32. *MifareDesFireStdDataFileWriteRead.cs.*

```
using System;
using GS.Util.Hex;
using GS.ISO14443_Reader;
using GS.ElektorRfidReader;
using GS.TclProtocol;
using GS.DesFireUtil;

namespace GS.ElektorRfidReaderExample
{
   class MifareDesFireStdDataFileWriteRead
   {
      public static void Main(MainForm script, string[] args)
      {
         ElektorISO14443Reader reader = new ElektorISO14443Reader();

         reader.Tcl.EnableTclTrace = false;
         reader.Tcl.EnableDesFireTrace = true;

         try
         {
            reader.OpenPort();

            reader.RFReset(5);
            reader.ActivateCard(RequestCmd.WUPA);
            reader.Tcl.RATS();
            reader.Tcl.PPS(DivisorInteger.Baud_424K, DivisorInteger.Baud_424K);
```

In the first step, it's necessary to select the application using the DESFire Select Application command.

```
            script.TraceWriteLine("\n   Select Application: AID = 0x112233");
            reader.Tcl.ExchangeDesFireNative(
                  "5A " +         // DESFire Cmd Code: Select Application
                  "33 22 11",     // AID = 0x112233
                  00);            // Expected card status 0x00 (OK)
```

The DESFire Write Data command is used to write data to a Standard Data File. Write-Data implicitly selects the file to be written to using the DESFire File Number parameter.

```
            script.TraceWriteLine("   Write Data: File No. = 0x01");
            reader.Tcl.ExchangeDesFireNative(
                  "3D " +              // DESFire Cmd Code: Write Data
                  "01 " +              // DESFire Native File No. = 0x01
                  "00 00 00 " +        // File Offset
                  "05 00 00 " +        // Length = 5 Byte
                  "11 22 33 44 55 ",   // Data = 0x1122334455
                  00);                 // Expected card status 0x00 (OK)
```

366

For demonstration purposes, the data is read back with the DESFire Read Data command, immediately after being written.

```
                script.TraceWriteLine(" Read Data: File No. = 0x01");
                reader.Tcl.ExchangeDesFireNative(
                    "BD " +         // DESFire Cmd Code: Read Data
                    "01" +          // DESFire Native File No. = 0x01
                    "00 00 00" +    // File Offset
                    "0A 00 00",     // Length = 10 Bytes (0x00000A)
                    00);            // Expected card status 0x00 (OK)
                reader.Tcl.Deselect();
            }
            catch (DesFireException ex)
            {
            }
            catch (Exception ex)
            {
                Console.WriteLine("Exception from " + ex.Source.ToString() + ".dll: " +
                    ex.Message.ToString() + Environment.NewLine);
            }
            finally
            {
                reader.ClosePort();
            }
        }
    }
}
```

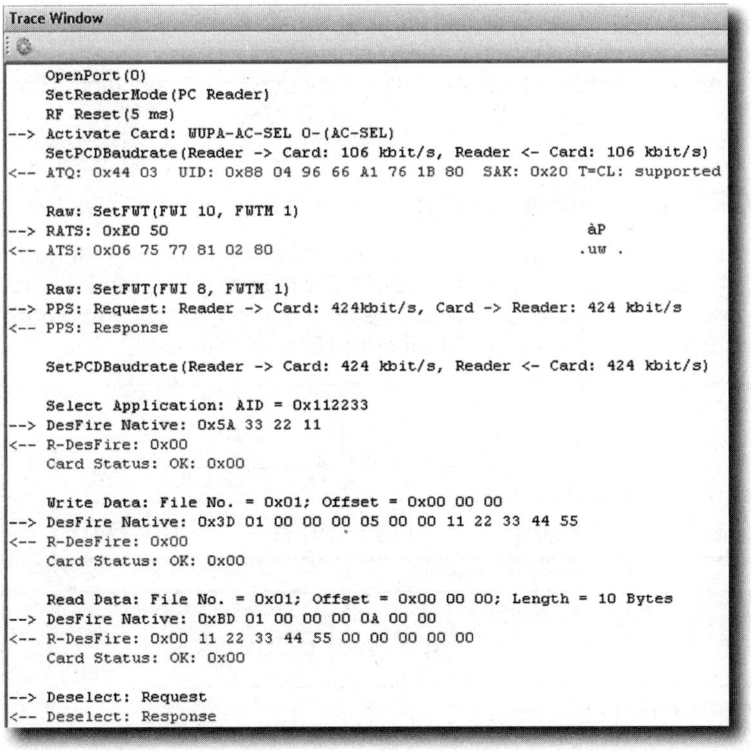

Figure 8.95.
Reading and writing a DESFire EV1 Standard Data File.

```
Trace Window

    OpenPort(0)
    SetReaderMode(PC Reader)
    RF Reset(5 ms)
--> Activate Card: WUPA-AC-SEL 0-(AC-SEL)
    SetPCDBaudrate(Reader -> Card: 106 kbit/s, Reader <- Card: 106 kbit/s)
<-- ATQ: 0x44 03  UID: 0x88 04 96 66 A1 76 1B 80  SAK: 0x20 T=CL: supported

    Raw: SetFWT(FWI 10, FWTM 1)
--> RATS: 0xE0 50                                                      àP
<-- ATS: 0x06 75 77 81 02 80                                           .uw .

    Raw: SetFWT(FWI 8, FWTM 1)
--> PPS: Request: Reader -> Card: 424kbit/s, Card -> Reader: 424 kbit/s
<-- PPS: Response

    SetPCDBaudrate(Reader -> Card: 424 kbit/s, Reader <- Card: 424 kbit/s)

    Select Application: AID = 0x112233
--> DesFire Native: 0x5A 33 22 11
<-- R-DesFire: 0x00
    Card Status: OK: 0x00

    Write Data: File No. = 0x01; Offset = 0x00 00 00
--> DesFire Native: 0x3D 01 00 00 00 05 00 00 11 22 33 44 55
<-- R-DesFire: 0x00
    Card Status: OK: 0x00

    Read Data: File No. = 0x01; Offset = 0x00 00 00; Length = 10 Bytes
--> DesFire Native: 0xBD 01 00 00 00 0A 00 00
<-- R-DesFire: 0x00 11 22 33 44 55 00 00 00 00 00
    Card Status: OK: 0x00

--> Deselect: Request
<-- Deselect: Response
```

8 CARDS AND TAGS IN APPLICATION

8.8 Application Protocol Data Units (APDUs)

At the OSI application layer, communication with microcontroller smartcards is done with the help of Application Protocol Data Units (see Section 8.6). APDUs are divided into Command APDUs (C-APDUs) and Response APDUs (R-APDUs). The C-APDUs are the commands to the card, and the R–APDUs are the card's responses. The general APDU structure is specified in Part 4 of the ISO/IEC 14443 standard. Regardless of whether the T=0, T=1 or T=CL protocol is used, the structure is always the same. Only the T=0 protocol requires an additional protocol-specific APDU command (GET RESPONSE).

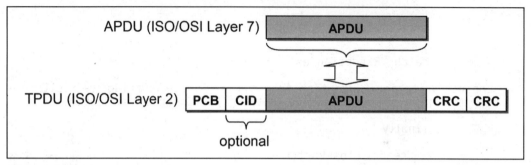

Figure 8.96. In the T=CL protocol, the APDUs are transferred using I-Frames.

The ISO/IEC 7816 standard specifies the encoding of the APDU commands in Parts 4, 7, 8 and 9. Application-specific standards (GSM, EMV, ICAO, Global Platform...) deviate partially from this encoding. In addition, most card operating systems support proprietary commands for card management. Only the APDU command structure always follows the ISO/IEC 7816 standard. The exact APDU command encoding is available from the manufacturer.

8.8.1 Command APDU Data Structure

A Command APDU consists of a 4-byte header and an optional body of variable length. The header is always structured the same and consists of a Class byte (CLA), an Instruction byte (INS) and two Parameter bytes (P1 and P2). The body contains data and length information (Lc and Le), which are not available in all smartcard commands.

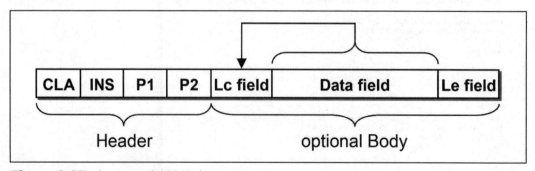

Figure 8.97. Command APDU data structure.

8.8.1.1 Class Byte (CLA)

The CLA byte indicates the instruction set of a smartcard application. For the ISO/IEC 7816 standard commands, the value of the CLA byte is 0x0X. The GSM 11.11 standard for SIM cards use a value of 0xA0.

The ISO/IEC 7816-4 specifies two groups of commands. Bit 8 differentiates between ISO/IEC 7816-compliant (Bit 8=0) and proprietary operating system commands (Bit 8=1). Bit 7 and 6 are, by definition, always 0. Bit 5 indicates whether chaining takes place at the application layer.

Bits 2 and 3 define whether secure messaging is used and what standard is used for this. With secure messaging, data integrity is ensured with a cryptographic checksum (MAC). With additional encryption of the APDU, the confidentiality of the message is ensured.

Bits 1 and 0 encode a logical channel number. If a multi-application card supports logical channels, up to four applications can run within a card session independently and in parallel. The applications are addressed using the logical channel.

8.8.1.2 Instruction Byte (INS)

In the command set, an INS byte serves to encode the individual commands. For historical reasons, the value of the INS byte is always an even number. When smartcards first came out, one could not integrate a charge pump to produce the EEPROM programming voltage onto the chip. With the T=0 protocol, the programming voltage is driven by the least-significant bit in the INS byte. Since generating the necessary EEPROM voltage hasn't been a problem in semiconductor technology for a long time, this bit was assigned a different function in the ISO/IEC 7816 standard. Currently, this bit indicates if the data field is encoded using Basic Encoding Rules – Tag, Length Value (BER–TLV) or not.

8.8.1.3 Parameter Bytes P1 and P2

The command is specified in more detail using the P1 and P2 parameter bytes. These bytes are also used to select command options. In a FILE SELECT command, P1 is used to select between an application, a Dedicated File (DF) or an linear data file (EF – Elementary File). In all three cases, the value of the INS byte is 0xA4.

The READ BINARY command is used to read data from a linear data file (EF – Elementary File). In this command, P1 and P2 define the offset within the file.

8.8.1.4 Coding of Length Fields Lc and Le

If user data is sent to the card, the Lc (Length Command) field encodes the number of data bytes to follow. The Le (Length Expected) field specifies the number of bytes of data expected from the card. The ISO/IEC 7816-4 standard defines two different encoding types for Lc and Le. These are called short APDUs and extended APDUs.

Short APDUs

In short APDUs, one byte is used to encode Lc and Le. Valid values for the Lc byte are between 1 and 255. If no data is sent to the card, there is no Lc byte. The value 0x00 is used as an escape character for distinguishing between Short and Extended APDUs.

In the case of a short APDU, one may request up to 256 bytes from the card. If no data is requested from the card, there is no Le byte. A special case of Le = 0x00 means the maximum, or 256 bytes. Should the file size be smaller than 256 bytes, for example, the card will return only the maximum number of available data bytes in response to a READ BINARY command.

With short APDUs, a maximum of 255 bytes can be written (Lc) and 256 bytes read (Le).

Extended APDUs

With extended APDUs, the length fields, Lc and Le, can be extended to 2 or 3 bytes. The fifth byte of the Command APDU is always an escape character and has the value 0x00. The only exceptions are commands that send no data to the card and request none back either (Case 1: CLA, INS, P1 and P2). In this format, there is no difference between short and extended APDUs. The actual length information is two bytes long. The range of Lc is from 1 to 65,535, and Le ranges from 1 to 65,536, where 65,536 is represented by the value 0x0000.

When a large amount of data is to be read from a passport (e.g. a passport photo), the read speed can be increased considerably by using extended APDUs, as this reduces the number of required APDU sequences. To read 64 KB, one needs 256 short APDUs. When using the extended APDU format, this is reduced to a single APDU.

This doesn't mean that the card sends all 64 kilobytes in one block. With contactless readers and cards, the maximum read buffer size is 256 bytes. Chaining, that is, dividing and transferring the data in blocks, is carried out on the T=CL protocol level. Therefore, on the application level, chaining can be eliminated completely. The support of extended APDUs is optional for both cards and readers.

Case 1 to Case 4

Due to the optional body, there are basically four different cases of C-APDUs. In the standard, these are known as Case 1 to Case 4. Case 2, 3 and 4 are further differentiated by short and extended formats.

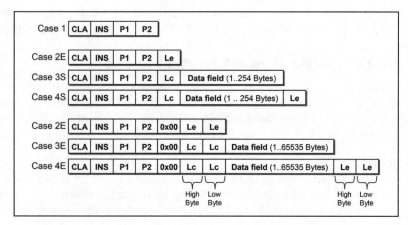

Figure 8.98. The 4 different cases with the short and extended Command APDU2.

8.8.2 Response APDU Data Structure

A response APDU consists of an optional body of variable length and a trailer. The body holds the card response data (Case 2 or Case 4 Command APDU). The data length matches the required number in the Le field. If the Le field requests the maximum amount of data, the card can still only send the amount of data available.

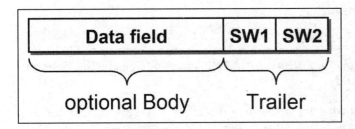

Figure 8.99. Command APDU data structure.

The size of the data field in a Response APDU may be up to 256 bytes in the short format and 65,536 bytes in the extended format. The obligatory trailer contains the two status bytes, SW1 and SW2. These two make up a status word (SW1SW2), which is also known as the return code. A return code of 0x9000 means that the card transmitted the last command without error. In the case of an error, the card responds always with the status word and no data field.

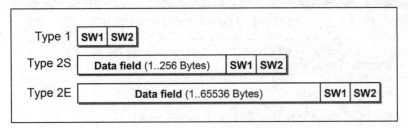

Figure 8.100. The 3 possible response APDU types.

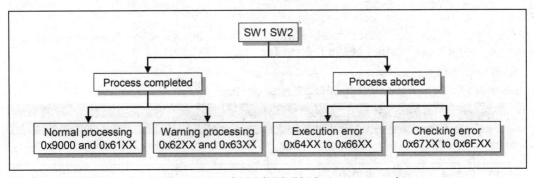

Figure 8.101. The basic structure of ISO/IEC 7816-4 return codes.

8.8.3 Examples of ISO/IEC 7816-Compatible APDUs

In this section, the three APDU commands, SELECT, READ BINARY and UPDATE BINARY are briefly introduced. For the sake of clarity, not all parameters specified in ISO/IEC 7816-4 are handled in detail. The example program in Listing 8.33 shows the proper use of these commands.

8.8.3.1 The SELECT Command

The SELECT command supports either the selection of a directory (DF) or of a file (EF). After successful execution of a SELECT command, the selected directory or file is valid for the file commands that follow (e.g. READ BINARY). A previously selected directory or file is automatically deselected.

Parameter byte P1 determines whether a file or a directory was selected (DF or EF). Only if P2 has a value of 0x00 is information returned in the Response APDU data field's DF/EF (File Control Parameter). The File Control Parameters (FCP) encodes, for example, the file size, the file type, the file identifier and the file access rights.

A smartcard application is always selected using the DF name (P1 = 0x04). Usually, the DF name is a registered Application Identifier (AID).

Table 8.38. Command APDU: SELECT Application

CLA	INS	P1	P2	Lc	Data	Le
0x00	0xA4	0x04	1.)	2.)	Application Identifier (AID)	3.)

[1] P2 = 0x00 – FCI should be returned.   P2 = 0x0C – FCI should not be returned.
[2] Lc – Length of the Application Identifier.
[3] Le – Either absent or 0x00.

Example: selecting an application with AID = 0x11 22 33 44 55 66 77 88.

```
--> C-APDU: 0x00 A4 04 00 08 11 22 33 44 55 66 77 88
<-- R-APDU: 0x90 00
```

After a card reset, the master file is always automatically selected. However, two variants exist for the selection of the master file.

Table 8.39. Command APDU: SELECT Master File.

CLA	INS	P1	P2	Lc	Data	Le
0x00	0xA4	0x00	0x00	0x02	0x3F 00 (Master File FID)	3.)
0x00	0xA4	0x00	0x00	–	–	–

Example: selecting a Master File.

```
--> C-APDU: 0x00 A4 00 00 02 3F 00
<-- R-APDU: 0x90 00
```

Selecting a Dedicated File or an Elementary File is done using a two-byte File Identifier. Table 8.44 Command APDU: SELECT File.

Table 8.40. Command APDU: SELECT File

CLA	INS	P1	P2	Lc	Data	Le
0x00	0xA4	1.)	0x00	0x02	FID	–

[1] P2 = 0x00 – MF, DF or EF. P2 = 0x02 – EF under the current DF. P2 = 0x02 – EF under the current DF.

Example: selecting an elementary file with FID = 0x00 01.

```
--> C-APDU: 0x00 A4 00 00 02 00 01
<-- R-APDU: 0x90 00
```

The SELECT command's return codes are listed in Table 8.41.

Table 8.41. Response APDU: SELECT Command.

Data	SW1	SW2	Description
–	0x62	0x83	The selected file is reversibly locked.
–	0x6A	0x80	Incorrect parameter in the data field.
–	0x6A	0x81	The feature is not supported by the card.
–	0x6A	0x82	Application or file not found.
–	0x6A	0x86	Incorrect P1/P2 parameter.
–	0x6A	0x87	Lc inconsistent with P1/P2 or with com
Optional: FCI	0x90	0x00	The command was executed successfully.

8.8.3.2 The READ BINARY Command

The READ BINARY command is used to read data from an Elementary File with a transparent data structure. The number of data bytes to be read is specified in the Le field (Length Expected). As previously mentioned, Le = 0x00 requests the maximum number of available bytes. Since the data field in a short format Response APDU is a maximum of 256 bytes in size, an offset is required to read larger files.

The READ BINARY command offers the ability to implicity select a file using a 5-bit long Short File Identifier (SFID). The value of the Short File Identifier is identical to the lower 5 bits of the File Identifier.

Table 8.42. Command APDU: READ BINARY.

CLA	INS	P1	P2	Lc	Data	Le
0x00	0xB0	1.)	1.)	2.)	0x3F 00 (Master File FID)	3.)

[1] See Table 8.47.
[2] Encodes the number of data bytes expected from the card.

When Bit 7 of P1 is '0', P1 and P2 encode a 15-bit offset (0 – 32,767).

If the value of P1, Bit 7 is 1, then Bits 0 to 4 in P1 encode the SFID. Bits 5 and 6 of P1 have no special meaning and should be set to 0 (r.f.u.). In this case, P2 encodes an 8-bit offset (0 – 255).

Table 8.43. Parameter bytes P1 and P2.

Bit	P1								P2						
	7	6	5	4	3	2	1	0	7	6	5	4	3	2	1	0
	0	0	0	0	0	0	0	0	0	0	0	0	0	0	0	0
	1	0	0	S	5	5	5	5	0	0	0	0	0	0	0	0

The following two examples illustrate how the READ BINARY command works. 10 bytes are read from offset 5 in the EF with FD = 0x00 01. In the file, the following fictional data is stored: 0x00 01 02 03 04 05 06 07 08 09.

Variant 1: Without Short File Identifier

```
--> C-APDU: 0x00 A4 00 00 02 00 01  // Select File
<-- R-APDU: 0x90 00
--> C-APDU: 0x00 B0 00 05 0A        // Read Binary
<-- R-APDU: 0x00 01 02 03 04 05 06 07 08 09 90 00
```

Variant 2: With Short File Identifier

```
--> C-APDU: 0x00 B0 81 05 0A        // Read Binary
<-- R-APDU: 0x00 01 02 03 04 05 06 07 08 09 90 00
```

Table 8.44. Response APDU: READ BINARY command.

Data	SW1	SW2	Description
–	0x62	0x83	The returned data may be corrupt.
–	0x62	0x80	The end of file was reached prematurely.
–	0x67	0x81	Incorrect length parameter Lc or Le is faulty.
–	0x69	0x82	The command is incompatible with the file structure.
–	0x69	0x86	Command not allowed (no EF selected).
–	0x69	0x87	The access conditions are not met.
–	0x6A		Directory or file not found.
–	0x6B		Incorrect P1 and/or P2 parameter.
File data	0x90	0x00	The command completed successfully.

8.8.3.3 The Update Binary Command

The UPDATE BINARY command writes data to an Elementary File with a transparent data structure. The two parameter bytes, P1 and P2, are coded as shown in Table 8.47 and thus do not differ from the READ BINARY command.

Table 8.45. Command APDU: READ BINARY.

CLA	INS	P1	P2	Lc	Data	Le
0x00	0xD6	1.)	1.)	2.)	Data	–

[1] See Table 8.47. [2] Encodes the number of data bytes to be written.

Table 8.46. Response APDU: UPDATE BINARY command.

Data	SW1	SW2	Description
–	0x65	0x81	EEPROM error.
–	0x67	0x00	Incorrect length parameter Lc or Le is faulty.
–	0x69	0x81	The command is incompatible with the file structure.
–	0x69	0x82	The access conditions were not met.

Data (cont.)	SW1	SW2	Description (continued)
–	0x6A	0x82	Directory or file not found.
–	0x6B	0x00	Incorrect parameter P1 and/or P2.
–	0x90	0x00	The command was executed successfully.

8.8.4 Elektor RFID Reader Library: APDU

Three classes, `CmdApdu`, `RespApdu` and `ApduException` are encapsulated in the `GS.Apdu.dll` assembly. These helper classes simplify working with APDUs and support both the short and the extended APDU format. These classes are not Elektor RFID Reader-specific and are therefore reused for the PC/SC reader examples.

The `TCL_Protocol` class has additional APDU format versions of the `Exchange()` method. These methods return an object of type `RespApdu`. The `RespApdu` class allows for separate evaluation of the data bytes, the status word (SW1SW2) and the reader error codes.

Table 8.47. TCL_Protocol: Selected APDU Exchange() methods.

Method	Description
`RespApdu Exchange(` `byte[] sndBuffer);`	The command APDU is passed as a byte array.
`RespApdu Exchange(` `string cmdApdu)`	The command APDU is passed as a string in hexadecimal format.
`RespApdu Exchange(` `CmdApdu cmdApdu)`	In this variant, the command APDU is passed as a `CmdApdu` object.

`RespApdu Exchange(`
`byte[] sndBuffer`
`ushort? expectedSW1SW2)`
	The status word expected from the card is passed in the `expectedSW1SW2` parameter. Should the card return a different status word, an `ApduException` is triggered. Any T=CL protocol error also triggers an exception. If `null` is passed in the `expectedSW1SW2` parameter, all internal error reporting is turned off and no exception is generated in the event of an error.

`RespApdu Exchange(`
`string cmdApdu,`
`ushort? expectedSW1SW2)`	

`RespApdu Exchange(`
`CmdApdu cmdApdu,`
`ushort? expectedSW1SW2)`	

Three `Exchange()` methods support error handling using the ApduException class. This method of error handling is much clearer than when one evaluates and responds to the SW1SW2 status word and the `Exchange()` method error code after every method call.

8.8 APPLICATION PROTOCOL DATA UNITS (APDUS)

Table 8.48. Properties of the CmdApdu class.

Property	R/W	Default	Description
`byte CLA`	R/W	0x00	Class byte
`byte INS`	R/W	0x00	Instruction byte
`byte P1`	R/W	0x00	Parameter Byte 1
`byte P2`	R/W	0x00	Parameter Byte 2
`int? Lc`	R/W	*null*	Length of the data field (Length Command). Lc is absent, the value is zero. When setting the data property, Le is updated.
`byte[] Data`	R/W	*null*	The data field. If the data field is absent, the value is *null*.
`int? Le`	R/W	*null*	Expected card data length (Length Expected). A value of *null* indicates that no Le field is present.

Table 8.49. A selection of CmdApdu contructors.

Method	Description
`CmdApdu()`	Default constructor.
`CmdApdu(string cmdApduString)`	Initializes a new instance of the `CmdApdu` class.
`CmdApdu(byte[] cmdApduBuffer)`	
`CmdApdu(byte cla, byte ins, byte p1, byte p2, byte[] data, int? le)`	
`CmdApdu(CmdApdu cmdApdu)`	Copy constructor. Creates an independent, low copy of the object.

Table 8.50. The CmdApdu methods.

Method	Description
`void SetValue(string cmdApduString);`	Initializes an instance of the `CmdApdu` class.
`void SetValue(byte[] cmdApduBuffer);`	Initializes an instance of the `CmdApdu` class.

Method (cont.)	Description (cont.)
`byte[] GetBytes();`	Returns a byte array of the command APDU.
`bool Equals(object obj);`	Determines whether this instance and another `CmdApdu` object have the same value. Alternatively, the overloaded operators, == and != may be used.
`string ToString();`	Converts the bytes of a `CmdApdu` object into a string.
`object Clone();`	Creates an independent, deep copy of the object.

Table 8.51. Properties of the RepApdu class.

Property	R/W	Description
`byte[] Data`	R	Data from the card response without status word SW1SW2.
`int RespLength`	R	Length of card response (optional data + SW1SW2).
`byte? SW1`	R	Status byte SW1.
`byte? SW2`	R	Statys byte SW2.
`ushort? SW1SW2`	R	Status word SW1SW2.
`int ReaderStatus`	R	Reader firmware error code (see Table 8.11).

8.8.5 Accessing an ISO/IEC 7816 File System

Figure 8.102 shows the basic APDU command sequence for reading a linear data file (EF).

In most applications, the reader system must be identified by the card. Whether authentication is required prior to selecting the application (DF), and/or prior to selecting a file (EF), depends on the particular application.

Table 8.52 gives a brief overview of the main ISO/IEC 7816-compatible identification and authentication commands. A MIFARE DESFire EV1 card supports Get Challenge, Internal Authenticate and External Authenticate.

If a card does not support mutual authentication, mutual authentication is achieved using the two commands Internal Authenticate and External Authenticate. In this case, these two commands must be executed in succession. The Mutual Authenticate command uses an optimized command sequence, but operates on the same principle.

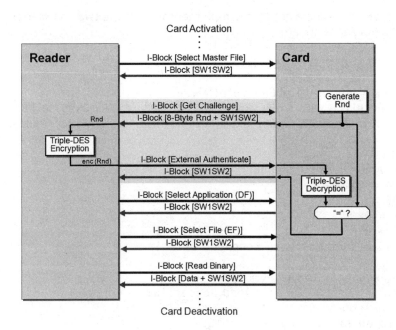

Figure 8.102.
Example of an APDU command sequence for accessing a file (EF).

Table 8.52. *The main ISO/IEC 7816-compatible authentication commands.*

Command	INS	Description
Verify	0x20	Usually checks a PIN (Personal Identification Number).
Get Challenge	0x84	Request a random number for External authenticate or Mutual Authenticate.
Internal Authenticate	0x88	Unilateral authentication of the card with respect to the reader.
External Authenticate	0x82	Unilateral authentication of the reader with respect to the card.
Mutual Authenticate	0x82	Mutual authentication between card and reader.

Figure 8.102 shows the sequence of the External Authenticate commands. In the first step, the reader requests an 8-byte random number using the Get Challenge command. The card returns a random number in the response APDU. The reader has now proven its authenticity by returning the encrypted random number to the card in the External Authenticate command's data field. In the last step, the card decrypts this encrypted random number and compares the result to the random number it generated earlier. If they're identical, the reader is authenticated. In this case, the card returns a status word of 0x9000, or an error code. If the Reader is authenticated, the card enables access to the files in the application directory.

8 CARDS AND TAGS IN APPLICATION

The following code snippet shows the algorithmic implementation of the two commands, Get Challenge and External Authenticate.

```
// Get Challenge
RespApdu respApdu = reader.Exchange("00 84 00 00 08", 0x9000);

TripleDESCrypto tdes = new TripleDESCrypto();
tdes.DesKeyA = new byte[] {0xFF, 0xEE, 0xDD, 0xCC, 0xBB, 0xAA, 0x99, 0x88};
tdes.DesKeyB = new byte[] {0x77, 0x66, 0x55, 0x44, 0x33, 0x22, 0x11, 0x00};

// External Authenticate
CmdApdu extAuthCmdApdu = new CmdApdu();
extAuthCmdApdu.CLA = 0x00;
extAuthCmdApdu.INS = 0x82;
extAuthCmdApdu.P1 = 0x00;
extAuthCmdApdu.P2 = 0x83; // Auth. DF use Key No. 3
extAuthCmdApdu.Data = tdes.Encrypt( respApdu.Data );

respApdu = reader.Exchange( extAuthCmdApdu );
if (respApdu.SW1SW2 == 0x9000 && respApdu.ReaderStatus == 0x00)
{
    // The reader is authenticated.
}
else
{
    // The reader is not authenticated.
}
```

8.8.5.1 Example Program

The principle of write and read access to an ISO/IEC 7816 file system is illustrated in a short example program. Furthermore, the different options for the overloaded `Exchange()` method and the correct use of the `CmdApdu` and `RespApdu` classes are presented.

The example is best understood with the MIFARE DESFire EV1 card, which was already used for the example program in Section 8.7.6. The DESFire example program in Listing 8.31 creates the required directory and file structure (see Figure 8.103). As mentioned several times already, it is not legally possible to describe the MIFARE DESFire EV1 cryptographic functions in detail here. The file system will therefore be available for unrestricted access.

The required APDU command sequence for writing and reading data (FID = 0x0001) corresponds with the sequence in Figure 8.102, but without authentication. All of the APDUs used in this example were presented in detail in Section 8.8.3.

Listing 8.33. *MIFAREDesFireISO7816APDUs.cs.*

```
using System;
using GS.Util.Hex;
using GS.ISO14443_Reader;
```

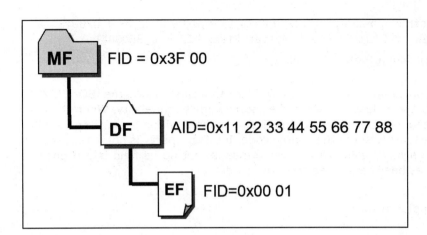

Figure 8.103.
Listing 8.31 creates an application (DF), which contains a single file (EF).

```
using GS.ElektorRfidReader;
using GS.TclProtocol;
using GS.Apdu;

namespace GS.ElektorRfidReaderExample
{
   class MIFAREDesFireISO7816APDUs
   {
    public static void Main(MainForm script, string[] args)
   {
   ElektorISO14443Reader reader = new ElektorISO14443Reader();
   reader.Tcl.EnableTclTrace = false;
   reader.Tcl.EnableApduTrace = true;

   try
   {
      reader.OpenPort();
      reader.RFReset(5);

      reader.ActivateCard(RequestCmd.WUPA);
      reader.Tcl.RATS();
      reader.Tcl.PPS(DivisorInteger.Baud_424K, DivisorInteger.Baud_424K);
```

Immediately after card activation, the master file (MF) is implicitly selected, so it's not required that the MF be explicitly selected, but it is customary to do so. In the simplest case, you can write the command APDU to a byte array.

```
//----------------------------------------------------------------
// Select Master File
//----------------------------------------------------------------
// CLA = 0x00, INS = 0xA4, P1 = 0x00, P2 = 0x00, Lc = 2(data length),
// Data = 0x3F 00, Le -
```

8 CARDS AND TAGS IN APPLICATION

```
byte[] selectMF = new byte[]{0x00,0xA4,0x00,0x00,0x02,0x3F,0x00};
script.TraceWriteLine(" Select Master File: FID = 0x3F 00");
reader.Tcl.Exchange(selectMF, 0x9000);
```

The Select Application command selects the DESFire application using the ISO/IEC 7816 AID. Use of the `CmdApdu` class increases the source code's readability. When the data property is initiatlized, the Lc field is automatically updated within the `CmdApdu` class. Thus, the Lc is not explicitly set in the source code. If this property is not initialized, the Le field will contain a *null* value. A null value means that no Le field is present and therefore no data has been requested from the card.

```
//--------------------------------------------------------------
// Select Application
//--------------------------------------------------------------
CmdApdu selectApplication = new CmdApdu();
selectApplication.CLA = 0x00;
selectApplication.INS = 0xA4;
selectApplication.P1 = 0x04;
selectApplication.P2 = 0x00;
selectApplication.Data = new byte[] {0x11,0x22,0x33,0x44,
                                     0x55,0x66,0x77,0x88};
script.TraceWriteLine("Select Application: AID=0x1122334455667788");
reader.Tcl.Exchange(selectApplication, 0x9000);
```

In the next step, the file (EF) is selected using FID = 0x0001. The `HexEncoding.GetBytes()` method from the `GS.Util.Hex` namespace converts a string into a byte array. The string data is interpreted as hexadecimal numbers. For improved readability, white spaces may be used, as they are ignored by the conversion process.

```
//--------------------------------------------------------------
// Select File
//--------------------------------------------------------------
CmdApdu selectFile = new CmdApdu();
selectFile.CLA = 0x00;
selectFile.INS = 0xA4;
selectFile.P1 = 0x00;
selectFile.P2 = 0x00;
selectFile.Data = HexEncoding.GetBytes("00 01"); //FID = 0x0001
script.TraceWriteLine(" Select File: FID = 0x0001");
reader.Tcl.Exchange(selectFile, 0x9000);
```

Using the Update Binary command, an ASCII value of "12345678" (hex: 0x31 32 33 34 35 36 37 38) is written to the file, at offset 0x0000. The command APDU can simply be passed as a hexadecimal string. This overloaded `Exchange()` method also supports automatic calculation of the Lc field. Instead of the Le value, the # escape character must be inserted in the correct position. Furthermore, it is necessary to include the data field in parentheses.

```
//-----------------------------------------------------------------
// Update Binary
//-----------------------------------------------------------------
// CLA = 0x00, INS = 0xD6, P1 = 0x00, P2 = 0x00 (offset),
// Lc = 8 (data length), Data = 0x31 32 33 34 35 36 37 38, Le -
script.TraceWriteLine("    Update Binary:");

reader.Tcl.Exchange("00 D6 00 00 # (3132333435363738)", 0x9000);
```

The next Update Binary command writes "Hello World" to the same file, this time at offset 0x0008. The string value must be enclosed in quotation marks. Because the cmdApduString in the Exchange() method is of the string type (see Table 8.53), it is necessary to escape the quotation marks using the \" escape sequence. The string is also enclosed in parentheses, as the Le field is also automatically calculated.

```
//-----------------------------------------------------------------
// Update Binary
//-----------------------------------------------------------------
// CLA = 0x00, INS = 0xD6, P1 = 0x00, P2 = 0x08 (offset), Lc = 11
// data = 0x48 65 6C 6C 6F 20 57 6F 72 6C 64 , Le -
script.TraceWriteLine("    Update Binary:");

reader.Tcl.Exchange("00 D6 00 08 # (\"Hello World\")", 0x9000);
```

The Read Binary command with Le = 0x00 reads the entire file contents (100 bytes).

```
//-----------------------------------------------------------------
// Read Binary
/------------------------------------------------------------------
// CLA = 0x00, INS = 0xB0, P1 = 0x00, P2 = 0x00 (offset), Lc = -,
// data = - , Le = 0x00
script.TraceWriteLine("    Read Binary:");

RespApdu respApdu2 = reader.Tcl.Exchange("00 B0 00 00 00", 0x9000);
```

In the last example, the previously stored string is read (offset = 0x0008 and Le = 0x10) and output to the console. The RespApdu class simplifies access to the response Apdu's data field. It is also possible to query the value of the status word as well as the reader error code. The .NET method, Encoding.ASCII.GetSTring() from the System.Text namespace converts a byte array into a string.

```
//-----------------------------------------------------------------
// Read Binary
//-----------------------------------------------------------------
// CLA = 0x00, INS = 0xB0, P1 = 0x00, P2 = 0x08 (offset), Lc = -,
// data = - , Le = 0x00
script.TraceWriteLine("    Read Binary:");

RespApdu respApdu1 = reader.Tcl.Exchange("00 B0 00 08 10");
```

8 CARDS AND TAGS IN APPLICATION

```
if (respApdu1.SW1SW2 == 0x9000 && respApdu1.ReaderStatus == 0x00)
{
string cardText;
cardText = System.Text.Encoding.ASCII.GetString(respApdu1.Data);
Console.WriteLine("Card Text: " + cardText);
}
reader.Tcl.Deselect();
}
```

Figure 8.104.
Writing and reading to and from a file (EF).

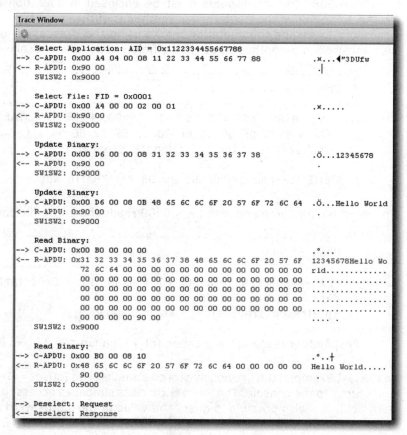

9
Elektor RFID Projects

Gerhard H. Schalk

9.1 Programming the MF RC522 Reader IC

The MF RC522 reader IC is programmed and configured using a total of 48 special function registers (SFRs). A host computer can control the MF RC522 registers via RS-232, I^2C or SPI interface. Selection of the serial interface is done with a voltage at the input pins I2C and EA (hardwired). In the Elektor RFID Reader (Figure 9.1), connection to the P89LPC396 microcontroller is done via I^2C.

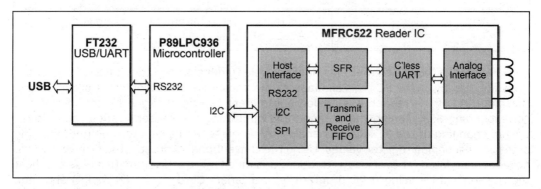

Figure 9.1. Elektor RFID Reader simplified block diagram.

The Elektor RFID Reader's firmware, that is, the P89LPC936 microcontroller program, supports the two stand-alone modes of operation, Terminal mode and Access Control mode, as well as a PC Reader mode. For performance reasons, the complex MF RC522 control software is implemented entirely in the reader firmware.

In PC Reader mode, the communication with the card is always initiated by the PC. The PC sends a firmware command to the microcontroller via the FT232R USB-to-RS-232 converter. This calls the corresponding MF RC522 function. The MF RC522 registers are read and written via I^2C. Writing and reading of card data always takes place through the 64-byte FIFO (first-in-first-out) buffer. By repeatedly writing to or reading from FIFODataReg, data is transferred to or from the FIFO buffer. The generation of protocol frames, including CRC calculation, is carried out entirely by the contactless UART. RF modulation of the serial bit stream ultimately takes place in the analog interface. The demodulation and decoding of the card's response also takes place in the analog section. The received data bytes are automatically written to the FIFO buffer by the contactless UART. Because only one FIFO is available for both reading and writing, the card command is overwritten. Since all communication is based on the master-slave principle, overwriting the command data presents no problems. The entire procedure for contactless communication is controlled solely by the MF RC522's control, configuration and status registers.

9.1.1 Elektor RFID Reader Library: MF RC522

The `RC522.ReadSFR()` and `RC522.WriteSFR()` methods in the Elektor RFID Reader library enable direct access to the MF RC522 registers. In this case, the reader firmware's function is reduced to that of a special USB-to-I2C converter for controlling the MF RC522 UART.

Table 9.1. Methods available for direct control of the MF RC522 SFRs.

Method	Description
`byte ReadSFR(byte sfrAddr);`	Returns the value of the given MF RC522 special function register (SFR).
`void WriteSFR(byte sfrAddr, byte sfrValue)`	Writes the `sfrValue` parameter to the given MF RC522 SFR.

With the help of these two methods, it is possible to develop and test parts of the reader firmware directly on the PC. This eliminates time-consuming microcontroller programming (flashing). All of the reader IC's analog parameters can be changed by configuring the corresponding SFR. Tuning the antenna and optimizing the reader's range requires repeated changing of the MF RC522's analog parameters, and thus the repeated updating of the corresponding register values. Using the conventional approach, the firmware would have to be updated for every new iteration. It is much more efficient in this case to adjust the register values using a PC program (e.g. a Smart Card Magic.Net script) and then perform the RF measurements. It's only necessary to load a modified version of the reader firmware into flash memory after successful analog performance optimization.

9.1.2 Program Examples

9.1.2.1 Changing the RF Parameter Configuration

This example shows how one overwrites the modulator and demodulator default values, for test purposes. The default values are automatically restored from the reader firmware when power is supplied to the RFID Reader.

> **Note** The MF RC522 offers the possibility of connecting different antenna configurations. Therefore, you should not change the Elektor RFID Reader MF RC522 SFRs that define the output driver circuit. Otherwise, in extreme cases, this could lead to a short circuit that could destroy the reader IC or other circuit components.

Listing 9.1. *Excerpt from RC522RfParameterConfig.cs. (from www.smartcard-magic. net/smart-card-magic-net/)*

```
using System;
using GS.Util.Hex;
using GS.ElektorReader;
using GS.ISO14443_Reader;

namespace GS.ElektorRfidReader
{
   class RC522RfParameterConfig
   {
    const byte RxThresholdReg = 0x18; // selects threshold for the bit decoder
    const byte DemodReg = 0x19;       // defines demodulator settings
    const byte ModWidthReg = 0x24;    // controls the ModWidth setting
    const byte RFCfgReg = 0x26;       // configures the receiver gain

      public static void Main(MainForm script, string[] args)
      {
        ElektorISO14443Reader reader = new ElektorISO14443Reader();

        try
        {
           reader.OpenPort();

           // RF parameter configuration
          reader.RC522.WriteSFR(RxThresholdReg, 0x55); // Bit decoder threshold
                                                       // MinLevel, CollLevel
          reader.RC522.WriteSFR(DemodReg, 0x4D);       // Demodulator Setting
          reader.RC522.WriteSFR(ModWidthReg, 0x26);    // Modulation width
          reader.RC522.WriteSFR(RFCfgReg, 0x59);       // Receiver Gain

           int ret;
           Console.WriteLine("Wait for new Card...");
           do
           {
             ret = reader.ActivateCard(RequestCmd.REQA);
           }
           while((ReaderStatus)ret != ReaderStatus.OK);
```

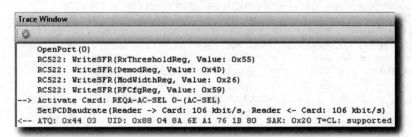

Figure 9.2. *Any of the MF RC522 parameters can be changed using a script.*

9.1.2.2 MF RC522 SFR Programming — Card Activation

The programming of the MF RC522 is presented with a card activation example. In this implementation, the bit frame anti-collision algorithm is omitted for clarity. Should more than one card enter the reader field, the anti-collision command will not support collision resolution, and an error code will be returned. Furthermore, the maximum card response time (frame waiting time) is not monitored. Because direct control of the MF RC522 using PC software is not especially fast, the card response can be read from the FIFO immediately after sending the card command. This presents no problem with any of the ISO/IEC 14443 Part 3 commands (REQA, Anti-collision and Select), as the card responds within the Frame Delay Time (FDT ≈ 90 μs). When exchanging T=CL frames, the command delays between reader and PC might not be long enough. In this case you can simply add an additional delay (for example, with the `Thread.Sleep()` method). The Elektor RFID Reader firmware has all error handing implemented.

> **Note** The Elektor reader firmware supports both bit frame anti-collision and monitoring of the Frame Waiting Time (FWT).

Listing 9.2. *MFRC522_SFR_Programming.cs.*

```
using System;
using System.Threading;
using GS.ElektorReader;
using GS.Util.Hex;

namespace GS.ElektorReader
{
    class RC522RegProg
```

In this example, all of the registers are accessed using symbolic names.

```
        const byte CommandReg =      0x01; // starts and stops command execution
        const byte ComIEnReg =       0x02; // enable and disable IRQ request control bits
        const byte DivIEnReg =       0x03; // enable and disable IRQ request control bits
        const byte ComIrqReg =       0x04; // interrupt request bits
        const byte DivIrqReg =       0x05; // interrupt request bits
        const byte ErrorReg =        0x06; // error bits showing the error status of the
                                           // last command executed

        const byte Status1Reg =      0x07; // communication status bits
        const byte Status2Reg =      0x08; // receiver and transmitter status bits
        const byte FIFODataReg =     0x09; // input and output of 64 byte FIFO buffer
        const byte FIFOLevelReg =    0x0A; // number of bytes stored in the FIFO buffer
        const byte WaterLevelReg =   0x0B; // level for FIFO underflow & overflow warning
        const byte ControlReg =      0x0C; // miscellaneous control registers
        const byte BitFramingReg =   0x0D; // adjustments for bit-oriented frames
        const byte CollReg =         0x0E; // bit position of the first bit-collision
                                           // detected on the RF interface

        const byte ModeReg =         0x11; // modes for transmitting and receiving
        const byte TxModeReg =       0x12; // transmission data rate and framing
        const byte RxModeReg =       0x13; // reception data rate and framing
        const byte TxControlReg =    0x14; // antenna driver pins TX1 and TX2
        const byte TxASKReg =        0x15; // transmission modulation
        const byte TxSelReg =        0x16; // selects the sources for the antenna driver
        const byte RxSelReg =        0x17; // selects internal receiver settings
        const byte RxThresholdReg =  0x18; // selects thresholds for the bit decoder
```

```
const byte DemodReg        = 0x19; // defines demodulator settings
const byte MfTxReg         = 0x1C; // MIFARE communication transmit parameters
const byte MfRxReg         = 0x1D; // MIFARE communication receive parameters
const byte SerialSpeedReg  = 0x1F; // speed of the serial UART interface

const byte CRCResultReg    = 0x21; // MSB and LSB values of the CRC calculation
const byte ModWidthReg     = 0x24; // controls the ModWidth setting
const byte RFCfgReg        = 0x26; // configures the receiver gain
const byte GsNReg          = 0x27; // defines the conductance of the antenna
                                   // driver pins TX1 and TX2 for modulation
const byte CWGsPReg        = 0x28; // defines the conductance of the p-driver
                                   // output during periods of no modulation
const byte ModGsPReg       = 0x29; // defines the conductance of the p-driver
                                   // output during periods of no modulation
const byte TModeReg        = 0x2A; // defines settings for the internal timer
onst byte TPrescalerReg    = 0x2B; // prescaler for the internal timer
const byte TReloadReg      = 0x2C; // defines the 16-bit timer reload value
const byte TCounterValReg  = 0x2E; // shows the 16-bit timer value

const byte TestSel1Reg     = 0x31; // general test signal configuration
const byte TestSel2Reg     = 0x32; // general test signal config. and PRBS control
const byte TestPinEnReg    = 0x33; // enables pin output driver on pins D1 to D7
const byte TestPinValueReg = 0x34; // values for D1..D7 when it is used as an I/O
const byte TestBusReg      = 0x35; // shows the status of the internal test bus
const byte AutoTestReg     = 0x36; // controls the digital self test
const byte VersionReg      = 0x37; // shows the software version
const byte AnalogTestReg   = 0x38; // controls the pins AUX1 and AUX2
const byte TestDAC1Reg     = 0x39; // defines the test value for TestDAC1
const byte TestDAC2Reg     = 0x3A; // defines the test value for TestDAC2
const byte TestADCReg      = 0x3B; // shows the value of ADC I and Q channels
```

To allow any method to call `script.TraceWrite()`, `script.TraceWriteLine()` and `script.Break()` without explicitly passing a MainForm object, it is necessary to declare a static class variable of type `MainForm`. Initialization of the `script` class variable takes place in the `Main()` method.

```
static MainForm script;
ElektorISO14443Reader reader;

public RC522RegProg()
{
    reader = new ElektorISO14443Reader();
}

public static void Main(MainForm script, string[] args)
{
    RC522RegProg readerScript = new RC522RegProg();
    RC522RegProg.script = script;
    readerScript.Run();
}
```

The `MFRC522_Init()` method initializes the MF RC522. The reader firmware carries out an identical initialization. Before actually writing the parameters, a chip software reset ensures that all SFRs are loaded with their default values. For example, the baud rate is automatically set to 106 Kbit/s. Therefore, the initialization routine need only write to the configuration registers whose values need to differ from the defaults.

```
// ------------------------------------------------------------
// MFRC522 Configuration
// ------------------------------------------------------------
void MFRC522_Init()
{
    script.TraceWriteLine(" MFRC522_Init()");
    reader.RC522.WriteSFR(CommandReg, 0x0F); // Softreset
    // Please do not modify the antenna driver configuration!
    reader.RC522.WriteSFR(TxControlReg, 0x83);
    reader.RC522.WriteSFR(GsNReg, 0xF4);
    reader.RC522.WriteSFR(CWGsPReg, 0x3F);
    reader.RC522.WriteSFR(ModGsPReg, 0x11);

    // RF parameters configuration
    reader.RC522.WriteSFR(TxASKReg, 0x40);        // Force100ASK
    reader.RC522.WriteSFR(RxThresholdReg, 0x55);  // Bit decoder thresholds
    reader.RC522.WriteSFR(DemodReg, 0x4D);        // Demodulator Setting
    reader.RC522.WriteSFR(ModWidthReg, 0x26);     // Modulation width
    reader.RC522.WriteSFR(RFCfgReg, 0x59);        // Receiver Gain
}
```

An RF reset is performed by turning off the 13.56 MHz carrier frequency and then turning it back on after a delay.

```
// ------------------------------------------------------------
// RF Reset
// ------------------------------------------------------------
void RFReset(byte ms)
{
    script.TraceWriteLine(string.Format( "     RFReset({0})", ms));
    reader.RC522.WriteSFR(TxControlReg, 0x80);  // Tx Off - Rf off
    Thread.Sleep(ms);
    reader.RC522.WriteSFR(TxControlReg, 0x83);  // TX on- Rf-on
    Thread.Sleep(5);
}
```

The `Request()` method sends a REQA or WUPA command. The two-byte card response, ATQA, is returned in the parameter, `atqA`.

```
// ------------------------------------------------------------
// ReqA (cmd = 0x26) or WupA (cmd = 0x52
// ------------------------------------------------------------
byte Request(byte cmd, out byte[] atqA)
{
    if(cmd == 0x26) script.TraceWriteLine("--> REQA: 0x26");
    else if (cmd == 0x52) script.TraceWriteLine("\n--> WUPA: 0x52");
```

The REQA and WUPA commands are always sent without a CRC.

```
    reader.RC522.WriteSFR(TxModeReg, 0x00); // Disable CRC
    reader.RC522.WriteSFR(RxModeReg, 0x00); // Disable CRC
```

Before any write to the FIFO, the FIFO pointer (`FIFOLevelReg`) is reset (flush FIFO). After writing command bytes, the MF RC522 Transceive command is activated using `CommandReg`. Transmission is triggered by setting the `StartSend` bit (Bit 7) of

`BitFramingReg`. This SFR is also used to configure the transmission frame (short frame, standard frame and bit-oriented anti-collision frame). REQA and WUPA are always transmitted in a short frame, i.e. in only 7 bits.

```
// Send REQA or WUPA - 7-bit short frame
reader.RC522.WriteSFR(FIFOLevelReg, 0x80); // flush FIFO
reader.RC522.WriteSFR(FIFODataReg, cmd);   // Write cmd to FIFO
reader.RC522.WriteSFR(CommandReg, 0x0C);   // Start Transceive
reader.RC522.WriteSFR(BitFramingReg, 0x87); // Start Send, 7-bit frame
```

Since the card responds within the FDT ($\approx$90 µs), one can read the error status (`ErrorReg`) and the number of received bytes (`FIFOLevelReg`) immediately. Should there be no card in the field, both values will be 0.

```
// Read Card Response (ATQA)
byte errorValue = reader.RC522.ReadSFR(ErrorReg);    // read error status
byte fifoLevel  = reader.RC522.ReadSFR(FIFOLevelReg); // read FIFO level

// Verify if we have received the two ATQA bytes without an error
if( ( (errorValue & ~0x08) != 0) || (fifoLevel != 2) )
{
    script.TraceWriteLine(" <-- No Card Response ");
    atqA = null;
    return errorValue;
}
```

If the card response is correct, `FIFOLevelReg` will be equal to 2. By reading `FIFODataReg` SFR twice in succession, the two ATQA bytes can be retreived.

```
        atqA = new byte[2];
        atqA[0] = reader.RC522.ReadSFR(FIFODataReg); // Read ATQA LSB
        atqA[1] = reader.RC522.ReadSFR(FIFODataReg); // Read ATQA MSB
        script.TraceWriteLine( string.Format( "<-- ATQA: 0x{0:X02} {1:X02}",
                                              atqA[0], atqA[1] ) );

        return errorValue;
    }
```

By dispensing with the anti-collision implementation, all of other commands are transferred within a standard frame, i.e. 8 bits. The `MFRC522_Transceive()` method handles sending and receiving standard frames. `bEnableCRC` controls whether the reader appends a 16-bit CRC at the end of a standard frame or not.

```
// ------------------------------------------------------------
// MFRC522_Transceive
// ------------------------------------------------------------
byte MFRC522_Transceive(byte[] sndData, out byte[] recData, bool bEnableCRC)
{
    return MFRC522_Transceive(sndData, out recData, bEnableCRC, "");
}
```

This variant of the `MFRC522_Transceive()` method also supports custom debug output to the Trace window (`sndInfo` parameter).

```csharp
// -------------------------------------------------------------
// MFRC522_Transceive
// -------------------------------------------------------------
byte MFRC522_Transceive(byte[] sndData, out byte[] recData, bool bEnableCRC,
                       string sndInfo)
{
    byte sfrValue;
    script.TraceWriteLine(HexFormatting.DumpHex("--> " + sndInfo + "0x",
                                        sndData, sndData.Length, 16));
    if(bEnableCRC)
    {
        // enable CRC generation during data transmission
        sfrValue = reader.RC522.ReadSFR(TxModeReg);
        sfrValue |= 0x80;      // TxCRCEn = 1
        reader.RC522.WriteSFR(TxModeReg, sfrValue);

        // enable CRC generation during data reception
        sfrValue = reader.RC522.ReadSFR(RxModeReg);
        sfrValue |= 0x80;      // RxCRCEn = 1
        reader.RC522.WriteSFR(RxModeReg, sfrValue);
    }
    else
    {
        // disable CRC generation during data transmission
        sfrValue = reader.RC522.ReadSFR(TxModeReg);
        sfrValue &= 0x7F;      // TxCRCEn = 0
        reader.RC522.WriteSFR(TxModeReg, sfrValue);

        // disable CRC generation during data reception
        sfrValue = reader.RC522.ReadSFR(RxModeReg);
        sfrValue &= 0x7F;      // RxCRCEn = 0
        reader.RC522.WriteSFR(RxModeReg, sfrValue);
    }
```

In the same way, data bytes are sent using the `Request()` method.

```csharp
    // Write sndData to FIFO and start transmission
    reader.RC522.WriteSFR(FIFOLevelReg, 0x80);    // flush FIFO
    for(int i = 0; i < sndData.Length; i++)
    {
        reader.RC522.WriteSFR(FIFODataReg, sndData[i]); //Write cmd into FIFO
    }
    reader.RC522.WriteSFR(CommandReg, 0x0C);       // Start Transceive
    reader.RC522.WriteSFR(BitFramingReg, 0x80);    // StartSend, Standard Frame
```

Reading card data is also done in the same way as the `Request()` method.

```csharp
    byte errorValue, fifoLevel;
    errorValue = reader.RC522.ReadSFR(ErrorReg);     // read error flags
    fifoLevel = reader.RC522.ReadSFR(FIFOLevelReg); // read FIFO level

    if( (errorValue != 0) || (fifoLevel == 0) )
    {
        script.TraceWriteLine(" <-- No Card Response ");
        recData = null;
    }
    else
    {
```

```csharp
            recData = new byte[fifoLevel];
            for(int i = 0; i < fifoLevel; i++)
            {
                // read data form FIFO
                recData[i] = reader.RC522.ReadSFR(FIFODataReg);
            }
            script.TraceWriteLine(HexFormatting.DumpHex("<-- 0x",recData,
                                                    recData.Length, 16));
        }
        return errorValue;
    }
```

The `Run()` method starts the script's main program. After opening the reader ports, the MF RC522 IC is initialized and the carrier frequency is turned off for 5 ms (RF reset).

```csharp
    // --------------------------------------------------------
    // Run Script
    // --------------------------------------------------------
    public void Run()
    {
        try
        {
            reader.OpenPort();

            MFRC522_Init();
            RFReset(5);
```

After RF reset, we must ensure that the RF carrier is active for at least 5 ms before a REQA command (0x26) is sent to the card. In our situation, the 5 ms is easily ensured due to the PC control of the MF RC522. Then we check if the card supports bit frame anti-collision.

```csharp
    // Request A
    byte[] atqA;
    byte retValue = Request(0x26, out atqA);
    if( (retValue != 0) || (atqA == null) ) return;
    if( (atqA[0] & 0x1F) == 0)
    {
    script.TraceWriteLine("Warning: Card does not support bit frame anticollision!");
    return;
    }
```

All other commands are implemented using the `MFRC522 _ Transceive()` method. Even if the reader doesn't support the anti-collision algorithm, as in this example, it is still necessary to send the Anti-collision Cascade Level 1 command (0x93 20) to the card. This command is always sent without a CRC. This command consists of the command byte, 0x93, and the NVB byte (Number of Valid Bits). The Anti-collision and Select commands don't have different command bytes, but they do differ in the number of UID bytes / bits transferred. NVB defines the length of the command including the instruction bytes and NVB byte. The upper nibble encodes the number of bytes and the lower nibble the number of bits. So, a value of 0x20 would mean 2 bytes and 0 bits. A number of bits other than 0 is used during the anti-collision algorithm. In this case, the reader sends the command byte, the NVB byte and a portion of the UID bits in an anti-collision frame.

When a card appears in the reader field, it responds with a 4-byte UID and a 1-byte checksum, or BCC (block check character). This is generated by XORing all 4 bytes together.

```
// AntiCollision Cascade Level 1
byte[] baCmdAcCL1 = {0x93, 0x20};
byte[] baUidCL1;
retValue = MFRC522_Transceive(baCmdAcCL1,out baUidCL1, false,
                        "AntiCollision Cascade Level 1: ");
if(retValue != 0) return;
```

The card is selected using the Select Cascade Level 1 command. This consists of the command byte, 0x93, the NVB byte (Number of Valid Bits), the 4-byte UID and the BCC. That's a total of 7 bytes. Thus the NVB byte will be equal to 0x70 (7 bytes, 0 bits).

```
// AntiCollision Cascade Level 1
byte[] baCmdSelectCL1 = new byte[7];
byte[] baSak;

baCmdSelectCL1[0] = 0x93;          // SEL
baCmdSelectCL1[1] = 0x70;          // NVB
baCmdSelectCL1[2] = baUidCL1[0];   // uid0
baCmdSelectCL1[3] = baUidCL1[1];   // uid1
baCmdSelectCL1[4] = baUidCL1[2];   // uid2
baCmdSelectCL1[5] = baUidCL1[3];   // uid3
baCmdSelectCL1[6] = baUidCL1[4];   // BCC

retValue = MFRC522_Transceive(baCmdSelectCL1,out baSak, true,
                        "AntiCollision Cascade Level 1: ");
if(retValue != 0) return;
```

SAK byte Bit 2 indicates whether the reader has received the complete UID or not. If the UID is incomplete, the reader sends the Anti-collision Cascade Level 2 and Select Cascade Level 2 commands to get the remaining UID bytes of a double size UID. For simplicity, this example does not support triple size UID cards.

```
if( (baSak[0] & 0x04) == 0x04 )
{
    // AntiCollision Cascade Level 2
    byte[] baCmdAcCL2 = {0x95, 0x20};
    byte[] baUidCL2;
    retValue = MFRC522_Transceive(baCmdAcCL2,out baUidCL2, false,
                            "AntiCollision Cascade Level 2: ");
    if(retValue != 0) return;

    // AntiCollision Cascade Level 2
    byte[] baCmdSelectCL2 = new byte[7];
    baCmdSelectCL2[0] = 0x95;          // SEL
    baCmdSelectCL2[1] = 0x70;          // NVB
    baCmdSelectCL2[2] = baUidCL2[0];   // uid0
    baCmdSelectCL2[3] = baUidCL2[1];   // uid1
    baCmdSelectCL2[4] = baUidCL2[2];   // uid2
    baCmdSelectCL2[5] = baUidCL2[3];   // uid3
    baCmdSelectCL2[6] = baUidCL2[4];   // BCC

    retValue = MFRC522_Transceive(baCmdSelectCL2,out baSak, true,
                            "AntiCollision Cascade Level 2: ");
    if(retValue != 0) return;
}
```

If the card supports the T=CL protocol, the reader sends a RATS command and the card responds with the ATS.

```
                // Does the card support the T=CL protocol?
                if( (baSak[0] & 0x20) == 0x20)
                {
                   // Request for Answer to Select (RATS)
                   byte[] baCmdRats = {0xE0, 0x50};
                   byte[] baATS;

                  retValue = MFRC522_Transceive(baCmdRats,out baATS,true,"RATS: ");
                  if(retValue != 0) return;
                }
            }
            catch (Exception ex)
            {
                script.TraceWriteLine("Exception from " + ex.Source.ToString() +
                        ".dll: " + ex.Message.ToString() + Environment.NewLine);
            }
            finally
            {
                reader.ClosePort();
            }
        }
    }
}
```

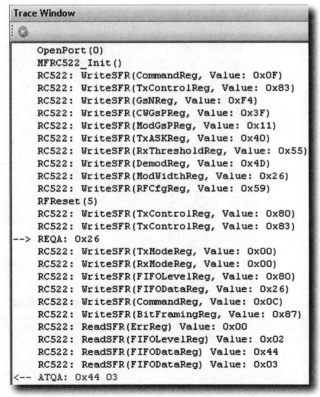

Figure 9.3.
Direct programming of the MF RC522 Special Function Registers.

9.2 RFID Access Control Systems

In modern buildings, RFID access control systems are the first choice when safety, convenience and flexibility are desired. RFID chips can be used as keys in the form of a card or a tag on a keychain. These systems offer significant advantages over mechanical locking systems. The logging of access and access attempts allows the monitoring of who has entered a particular room and when. The card's access privileges can be limited, changed or revoked at any time. Electronic access control systems can be divided into online and offline systems.

9.2.1 Online Systems

The RFID reader is permanently connected to a central computer. Upon a new arrival, the RFID reader reads the card data and sends it to the central computer. By consulting a database, the computer decides whether the cardholder has access rights. In addition to access control, online systems also make employee time and attendance logging possible.

9.2.2 Offline Systems

The RFID reader must decide on its own whether the person has the right to enter the room or not. These systems are used primarily where many rooms are accessed by only a few people.

Figure 9.4.
Example of offline access control.

9.2.3 Elektor RFID Reader as Access Control System

The Elektor RFID Reader is especially suitable to realize of an offline access control system. A gate and door opener can be controlled either directly via the open collector output (T3) or via an additional relay. In the circuit diagram, the output is fitted with a BC517. LED D7 indicates the switch status. Since the Elektor RFID Reader provides the ability to connect directly to an LCD, additional user information can be displayed.

This project can also serve as a basis for similar tasks. For example, the Elektor RFID reader could authorize access to different hard drives in a family computer, or control the TV set.

The Access Control mode is an integral part of the Elektor RFID Reader firmware, and can thus be used even by readers of this book who have no way to program the LPC936 microcontroller.

9.2.3.1 Functional Description

Just like many older commercial access control systems, the card's serial number (UID) serves as sole proof of access authorization.

As soon as the reader detects a new card, the card is activated and the UID is read and compared with a list of card serial numbers stored in the reader. If the number is in the list, switching output T3 is activated. The switching output remains active until the card is removed from the reader.

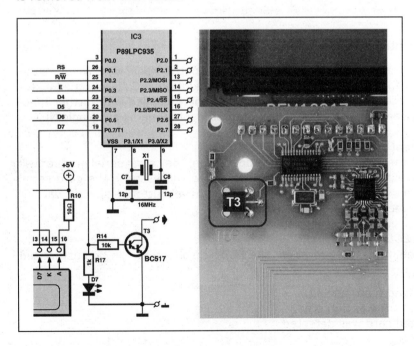

Figure 9.5.
Open collector switching output T3. The BC517 can deliver a maximum of 40 V / 0.4 A / 0.625 W, according to the datasheet.

The Elektor RFID Reader's firmware supports the following functions:
- Management of up to ten cards.
- Predefined text for LCD (Default Text).
- User-configurable text for LCD (User Text).
- Permanent activation and deactivation of Access Control mode.

> **Note** The card serial number (UID) being the only proof of credentials is not good enough for commercial applications. An attacker need merely read and store the UID of a valid card, then use a card emulator, which allows for the generation of any desired card serial number, among other things. Once the attacker has a card emulator with a valid serial number, he is able to compromise the system.
>
> Mutual authentication is an effective defense against this type of attack.

Any ISO/IEC 14443-3 Type A-compatible card or NFC mobile phone can be used. The Access Control mode also works without an LCD connected.

Figure 9.6. *The different hard-coded LCD display texts.*

9.2.3.2 Access Control Manager

Access Control mode configuration is done using Access Control Manager, a Windows application developed for the purpose. This application was written in C# using the Elektor RFID Reader library. Access Control Manager can easily be extended with further functions. Access Control Manager's window is also integrated into Smart Card Magic.NET.

Credential Management
If you place a card on the reader and click the 'Add' button, the card's serial number will be added to the reader flash memory. After being saved successfully, the card serial number is displayed in the user interface. Using the 'Remove' button, a specific card or all cards can be removed from the reader's list.

Configuration of the LCD Display Text
For the following three modes of operation, the LCD's display text can be customized at will:

- Waiting for a card (card polling).
- An authorized card was detected.
- An unauthorized card was detected.

For each of these three situations, there are two text boxes, a 'Default Text' button and a 'User Text' button.

The custom text can easily be edited in the corresponding text box. The 'User Text' button saves the text to the reader's flash memory and enables user-defined text output. Using the 'Default Text' button, user-defined text can be deactivated again.

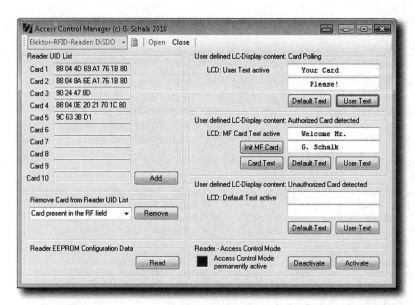

Figure 9.7.
Access Control mode configuration.

Individual text strings can be saved to a MIFARE 1K/4K card for display. The 'User defined LC-Display content: Authorized Card detected' group has two text boxes for the card text, which are written to the card when the 'Init MF Card' button is clicked. Then, by clicking on the 'Card Text' button, display of the card text can be activated.

9.2.3.3 Microcontroller Firmware

The UID list and user-defined LCD text are stored in the (up to) 512-byte large internal data EEPROM on of the P89LPC936 controller. On a brand-new microcontroller, the entire EEPROM is set to 0xFF. Both reading and writing take place via three special function registers (SFRs). Direct memory access, like RAM, is not possible with the EEPROM memory. On the other hand, in contrast to the flash memory, individual bytes can be written to the EEPROM.

Table 9.2. LPC939 EEPROM SFRs.

SFR	Description
Address Register (DEEADR)	EEPROM address bits 7 through 0. Address bit 8 is in the DEECON SFR.
Control Register (DEECON)	Sets different programming modes. The Data EEPROM interrupt flag (EEFI – Bit 7) is set when a read or write operation has completed.
Data Register (DEEDAT)	Used for writing and reading EEPROM data.

Listing 9.3. Excerpt from EEPROM.c.

```c
void eeWrite(unsigned int addr, unsigned char dat)
{
   EA = 0;                                          // disable all Interrupts
   DEECON=(unsigned char)((addr >> 8) & 0x01);      // mode: write byte, set address
   DEEDAT=dat;                                      // write data
   DEEADR=(unsigned char) addr;                     // start EEPROM programming
   EA=1;                                            // enable all Interrupts during write
   while( (DEECON & 0x80) == 0 );                   // wait until programming is complete
}
unsigned char eeRead(unsigned int addr)
{
   DEECON=(unsigned char)((addr >> 8) & 0x01);      // mode: read byte, set adress
   DEEADR=(unsigned char) addr;                     // start read
   while((DEECON & 0x80) == 0);                     // wait until read is complete
   return DEEDAT;                                   // return data
}
```

The `mifare_access_control.h` header file defines the starting addresses of all EEPROM configuration data. In order for the example to work on a P89LPC936 controller with an erased EEPROM (factory-supplied), the value of the DEACTIVE constant is defined as 0xFF.

Listing 9.4. Excerpt from mifare_access_control.h.

```c
#define ACTIVE            0xA5
#define DEACTIVE          0xFF
#define ALLOWED           1
#define NOT_ALLOWED       0
#define EEAddrAccessActFlag          0x100
#define EEAddrUseMFTextFlag          0x101
#define EEAddrUIDData                0x108
#define UIDDataOffset                0x08
#define MAX_UID_LIST_ENTRIES         10
#define EEAddrLCD_CardPolling_Line1  0x158
#define EEAddrLCD_CardPolling_Line2  0x168
#define EEAddrLCD_Auth_Card_Line1    0x178
#define EEAddrLCD_Auth_Card_Line2    0x188
#define EEAddrLCD_Unauth_Card_Line1  0x198
#define EEAddrLCD_Unauth_Card_Line2  0x1A8
#define MAX_LCD_LINE                 16

void AccessControlMode(void);
void ReadEEPROM_LcdText(unsigned int addr, char *lcdText);
short ReadMifareCard_LcdText(unsigned char block, unsigned char *uid,
                                                  char *lcdText);
```

During main program initialization, following every power-on reset, the RFID reader checks whether Access Control mode has been permanently activated.

Listing 9.5. *Excerpt from main.c.*

```c
//------------------------------------------------------------------
// Initialize Software States and Flags
//------------------------------------------------------------------

// Verify the LPC EEPROM "ACCESS CONTROL DEMO Enable" Flag
if (eeRead(EEAddrAccessActFlag) == ACTIVE)
{
    // Access Control Demo is active => Start Access Control Demo
    gbMainState = MainState_ACCESS_CONTROL; // Initialize Main State
}
else
{
    // Access Control Demo is not active => Start Terminal Mode Demo
    gbMainState = MainState_TerminalMode;   // Initialize Main State
}

gbRecState = 0;              // Reset Serial Port Receive State Machine.
gbNewCmdReceived = CLEAR;    // Clear new command received flag.
```

Should Access Control mode be permanently activated, the main program loop will call the `AccessControlMode()` function repeatedly.

Listing 9.6. *Excerpt from AccessControlMode.c.*

```c
void AccessControlMode(void)
{
    unsigned char abUID[8], bSAK;
    unsigned char baATQ[2], bUIDLen;
    unsigned char bAccessStatus;
    unsigned int iEEAddr;
    unsigned char i,j;
    unsigned char baEE_UID_List_Entry[8];
    char szLcdText[17];
    bAccessStatus = NOT_ALLOWED;
    short status;
```

At each call of `AccessControlMode()`, the reader checks if a card is in the field. Should no card be found, the switching output (`OpenerOutput` variable) is deactivated. Subsequently, either the predefined text in the program code or the user-defined text in the P89LPC936 EEPROM is displayed on the LCD.

```c
    Rc522RFReset(5);
    status = ActivateCard(ISO14443_3_REQA,baATQ,abUID,&bUIDLen,&bSAK);
    if(status != STATUS_SUCCESS)
    {
        // No Card in the RF Field...
        OpenerOutput = 0;   // Clear (Door) Opener Output pin
        if(eeRead(EEAddrLCD_CardPolling_Line1) == DEACTIVE)
        {
            // Default LCD text
            lcdPrintf(1,1," ACCESS CTRL ");
            lcdPrintf(2,1," YR CARD PLZ ");
```

```
            }
            else
            {
                // User-defined LCD text
                ReadEEPROM_LcdText(EEAddrLCD_CardPolling_Line1, szLcdText);
                lcdPrintf(1,1,"%s",szLcdText);
                ReadEEPROM_LcdText(EEAddrLCD_CardPolling_Line2, szLcdText);
                lcdPrintf(2,1,"%s",szLcdText);
            }
            return;
        }
```

After a successful card activation, the card's UID is checked against the UID list in the EEPROM. If there's a match, the card is authorized for access.

```
        else
        {
            // Verify if the current actived card is a valid card.
            iEEAddr = EEAddrUIDData;
            for(i=0; i < MAX_UID_LIST_ENTRIES; i++)
            {
                // Read one UID entry form the (LPC936 EEPROM) UID list.
                for(j=0; j < 8; j++) { baEE_UID_List_Entry[j] = eeRead(iEEAddr + j); }

                // Compare the card UID with list entry.
                if(memcmp(abUID,baEE_UID_List_Entry,bUIDLen) == 0)
                {
                    // Match ==> ACCESS allowed.
                    bAccessStatus = ALLOWED;
                    break;
                }
                iEEAddr += UIDDataOffset;
            }
```

An authorized card activates the switch output.

```
        if(bAccessStatus == ALLOWED)
        {
            // ACCESS ALLOWED
            OpenerOutput = 1;      // SET (Door) Opener Output pin
```

Should the display of card-specific LCD text be active, the `ReadMifareCard_LcdText()` function is called. This function reads the text stored on a MIFARE 1K/4K card (Blocks 4 and 5) and displays it on the LCD.

```
            if( eeRead( EEAddrUseMFTextFlag ) == ACTIVE)
            {
                // Mifare Card defined LCD text
                ReadMifareCard_LcdText(4, abUID, szLcdText);
                lcdPrintf(1,1,"%s",szLcdText);
                ReadMifareCard_LcdText(5, abUID, szLcdText);
                lcdPrintf(2,1,"%s",szLcdText);
            }
```

Alternately, either the predefined text in the program memory or the user text in the EEPROM is displayed.

```c
    else
    {
        // Default or user defined LCD text
        if(eeRead(EEAddrLCD_Auth_Card_Line1) == DEACTIVE)
        {
            // Default LCD text
            lcdPrintf(1,1," ACCESS "); // Write to LCD Display
            lcdPrintf(2,1," ALLOWED "); // Write to LCD Display
        }
        else
        {
            // User defined LCD text
            ReadEEPROM_LcdText(EEAddrLCD_Auth_Card_Line1, szLcdText);
            lcdPrintf(1,1,"%s",szLcdText);
            ReadEEPROM_LcdText(EEAddrLCD_Auth_Card_Line2, szLcdText);
            lcdPrintf(2,1,"%s",szLcdText);
        }
    }
}
```

If the card is not authorized for access, the switching output is deactivated and one of the two possible text strings is displayed on the LCD.

```c
    else
    {
        // ACCESS DENIED
        OpenerOutput = 0; // Clear (Door) Opener Output pin

        // Write to LCD Display Line 1
        if(eeRead(EEAddrLCD_Unauth_Card_Line1) == DEACTIVE)
        {
            // Default LCD text
            lcdPrintf(1,1," unauthorized "); // Write to LCD Display
            lcdPrintf(2,1," CARD "); // Write to LCD Display
        }
        else
        {
            // User defined LCD text
            ReadEEPROM_LcdText(EEAddrLCD_Unauth_Card_Line1, szLcdText);
            lcdPrintf(1,1,"%s",szLcdText);
            ReadEEPROM_LcdText(EEAddrLCD_Unauth_Card_Line2, szLcdText);
            lcdPrintf(2,1,"%s",szLcdText);
        }
    }
}
}
```

The `ReadEEPROM_ LcdText()` function reads a C string from the microcontroller's EEPROM and stores it in a 17-byte `char` array. The routine always stores the string terminator as the last element in the array.

9 ELEKTOR RFID PROJECTS

```
void ReadEEPROM_LcdText(unsigned int addr, char *lcdText)
{
   char i;

   for(i = 0; i < 16 ;i++)
   {
      lcdText[i] = eeRead(addr + i);
   }
   lcdText[16] = '\0';
}
```

The `ReadMifareCard _ LcdText()` function reads a MIFARE 1K/4K block and stores the data in a 17-byte `char` array. Because the data will subsequently be interpreted as C strings, the string terminator is stored as the last element of the `char` array.

The MIFARE Authentication is always done with the MIFARE Classic Key A and the key 0xFF FF FF FF FF FF.

```
short ReadMifareCard_LcdText(unsigned char block, unsigned char *uid,
                                                  char *lcdText)
{
   short status;
   unsigned char defaultMifareKey[] = {0xFF, 0xFF, 0xFF, 0xFF, 0xFF, 0xFF};
   status = Authentication(MIFARE_AUTHENT_A, defaultMifareKey, uid, block);
   status |= Read(block, lcdText);

   if(status != STATUS_SUCCESS)
   {
      lcdText[0] = '\0';
      return status;
   }

   lcdText[16] = '\0';
   return status;
}
```

9.2.3.4 Reading and Deleting from the P89LPC936 EEPROM.

For test purposes, you can read the entire contents of the P89LPC936 data EEPROM with the help of a Smart Card Magic.NET script.

Listing 9.7. *P89LPC936DumpDataEEPROM.cs.*

```
using System;
using GS.Util.Hex;
using GS.FTD2XXLibrary;
using GS.ISO14443_Reader;
using GS.ElektorReader;

namespace GS.ElektorReader
{
   class P89LPC936DumpDataEEPROM
```

RFID ACCESS CONTROL SYSTEMS   9.2

```
{
    public static void Main(MainForm script, string[] args)
    {
        ElektorISO14443Reader reader = new ElektorISO14443Reader();

        try
        {
            reader.OpenPort();

            for(int eeAddr = 0x100; eeAddr < 0x200; eeAddr+=16)
            {
                byte[] eeData = new byte[16];
                reader.LPC936.ReadEEPROM(eeAddr, eeData, 0, 16);
                Console.WriteLine(string.Format("0x{0:X04}: ", eeAddr)
                                + HexFormatting.Dump(eeData) );
            }
        }

        catch (Exception ex)
        {
            script.TraceWriteLine(ex.Message.ToString() + Environment.NewLine);
        }

        Finally { reader.ClosePort(); }
    }
}
```

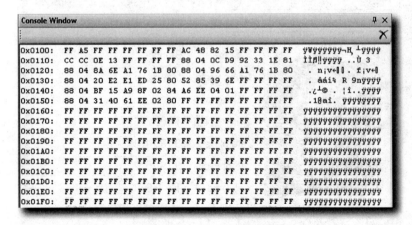

Figure 9.8.
Dumping the P89LPC936 EEPROM content.

The Access Control example uses the EEPROM memory area from 0x100 to 0x1FF. This entire area can be deleted by calling the `ClearEEPROM()` method twice (see the `P89LPC-936ClearDataEEPROM.cs` script).

9.3 An Electronic ID Card

The operation of contactless smart cards can be demonstrated elegantly by implementing a simple electronic ID card. The following project implements an electronic ID card (without security features) using a MIFARE 4K card. A simple Window Form application was created in C# for personalizing and reading the card data. As always, the Mifare-IdCard application and source code can be downloaded from the author's website (*www.smartcard-magic.net*).

A name, surname, nationality, date of birth, date of issue, expiry date, a card ID and an image are stored on the card. The personal data are stored in ASCII format in the first three MIFARE sectors (MIFARE Blocks 1 to 10). Each field, such as the surname, is stored in only one MIFARE block. Therefore, the number of characters is limited to the size of a MIFARE block, i.e. 16 bytes. A JPEG image file with a maximum size of 3,364 bytes can be stored in the remaining MIFARE blocks.

9.3.1 Personalization

The MifareIdCard application consists of two application dialogs: Read Card Data and Card Personalization. The Card Personalization dialog enables the personalization of the MIFARE 4K card. Because the Read Card Data dialog is always opened automatically when the application is started, it is necessary to open the Card Personalization dialog using

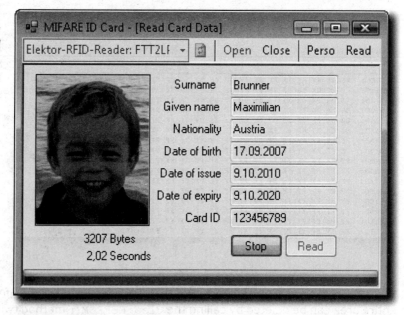

Figure 9.9.
The Windows MifareIdCard application.

the 'Perso' button. The 'Open' button opens the USB port. The personal data can be edited in the corresponding text box. A mouse click in the picture box opens the file dialog for selecting a `.JPG` image file. The 'Write Data' button writes the data to the MIFARE 4K card. The MifareIDCard application assumes that the values of the MIFARE Sector Trailer (Mifare Key A, AC and MIFARE Key B) are the same as those of a brand-new card (MIFARE Key A = Key = `0xFF FF FF FF FF FF`).

9.3.2 Reading the ID Card Data

After opening the reader port with the 'Open' button, the 'Read' button in the Read Card Data application dialog is activated. Once the 'Read' button is pressed, the reader attempts to activate a card by repeatedly sending the REQA command. As soon as the reader detects a card, it is activated. Subsequently, the personal data and the image are read. As long as the card is in the reader field, the data is displayed. When the card is removed, the display is cleared and the reader is readied to read another card. The 'Stop' button stops the entire read cycle.

Due to the MIFARE 4K card's memory architecture, all data must be read in 16-byte blocks. Furthermore, the reader must perform a MIFARE authentication for every sector. Therefore, the reading time required for a 3.2 KB image file is around 2 seconds.

> **Note** For reading and writing the image file, the MifareIdCard application uses the `ByteArrayToMifare()` and `MifareToByteArray()` methods. Both methods are implemented once in the `MifareByteArrayUtil` class and once in the `MifareByteArrayUtilSpeedOptimized` class. The latter speed-optimized version uses the `ReadBlocks()` and `WriteBlocks()` methods to read and write MIFARE blocks (see Section 8.4.8.2).
>
> The `ByteArrayToMifare()` and `MifareToByteArray()` methods are generally applicable and can be reused for other projects. For example, one could capture the measured values of a data logger to a MIFARE 4K card.

9 ELEKTOR RFID PROJECTS

9.4 Launching a Windows Application

The Smart Card Magic.NET script in Listing 9.8 demonstrates how a website can be launched using a MIFARE Ultralight card and then closed again.

Firstly, a web link must be saved to an Ultralight card. The easiest way to do this is with the `MifareUltraLightStoreWebLink.cs` script. This script writes the link (for example, *http://www.mifare.net*) to the card. When creating the link, remember that a maximum of 48 characters can be stored on a MIFARE Ultralight card.

Listing 9.8. *Excerpt from MifareUltraLightOpenWebLink.cs.*

```
using System;
using System.Diagnostics;
using System.Threading;
using GS.Util.Hex;
using GS.ISO14443_Reader;
using GS.ElektorReader;

namespace GS.ElektorRfidReader
{
    public class MifareULOpenWebLink
    {
        ElektorISO14443Reader reader;

        public MifareULOpenWebLink ()
        {
            reader = new ElektorISO14443Reader();
        }

        public static void Main(MainForm script, string[] args)
        {
            MifareULOpenWebLink openWebLink = new MifareULOpenWebLink();
            openWebLink.Run(script);
        }

        void Run(MainForm script)
        {
            try
            {
                reader.OpenPort();
                int ret;
```

The script runs in a continuous loop and therefore must be ended manually using the menu command *Script → Stop*.

```
                while(true)
                {
```

The script is stalled until a card is activated.

```
Console.WriteLine("Wait for new Card...");
do
{
    ret = reader.ActivateCard(RequestCmd.REQA);
    Thread.Sleep(100);
}
while( (ReaderStatus)ret != ReaderStatus.OK );
```

An external application can be started using the `Start()` method of the `Process` class in the `System.Diagnostics` namespace. The path and application name are passed to the `StartInfo.FileName` property. Should Internet Explorer not start correctly on the system, this could be for several reasons. Internet Explorer may not be installed, or it may be installed in a different directory. The example should work with any other web browser without any problems.

The web link read from the card is passed to the `StartInfo.Arguments` property. Should the `WindowStyle` property be assigned the value `ProcessWindowSytle.Maximized`, Internet Explorer will be launched maximized.

```
Console.Clear();
string strWebSite = ReadString(4);
Console.WriteLine("Start: " + strWebSite);

if( !String.IsNullOrEmpty(strWebSite) )
{
Process proc = new Process();
proc.StartInfo.FileName = "C:\\Program Files\\Internet Explorer\\iexplore.exe";
proc.StartInfo.Arguments = strWebSite;
proc.StartInfo.WindowStyle = ProcessWindowStyle.Maximized;
proc.Start();
```

After opening the website, the script is again stalled until the card is removed from the reader.

```
do
{
    byte[] baData = new byte[16];
    ret = reader.Mifare.Read(4, out baData);
}
while((ReaderStatus)ret == ReaderStatus.OK);
```

Once the card is removed from the reader, the `Kill()` method closes the web browser.

```
if( !proc.HasExited )
{
    proc.Kill();
```

9 ELEKTOR RFID PROJECTS

```
                }
                Console.Clear();
            }
        }
    }
```

With a minor modification to the program, one could open any other Windows or console application in the presence of a card and then close it again. For example, the following code opens the file, test.txt, in Notepad.

```
Process proc = new Process();
proc.StartInfo.FileName = "notepad.exe";
proc.StartInfo.Arguments = "c:\\test.txt"
proc.StartInfo.WindowStyle = ProcessWindowStyle.Maximized;
proc.Start();
```

10
Smart Card Reader API Standards

Gerhard H. Schalk

10.1 Introduction

The first API (Application Programming Interface) between a smartcard reader and a PC system was standardized by various organizations in the mid 90s. Until that time, each reader had a proprietary software interface. When developing a PC application, both the card and the reader specifications had to be considered. For this reason, card applications usually supported a small number of card readers – often only a single one. The goal of the various industry standards is to enable the development of PC card applications that are independent of the reader hardware.

CT-API (Card Terminal Application Programming Interface) is an older standard that is widespread in the German-speaking world, but it has no international recognition. The OCF (Open Card Framework) and the PC/SC standard (Personal Computer / Smartcard) standards are, however, recognized internationally.

Figure 10.1. Various types of PC/SC readers.

The PC/SC standard is discussed in detail below, as this standard is supported by virtually all major operating systems. Reader manufacturers offer PC/SC compatible contact, contactless and dual-interface smartcard readers in a wide variety of different designs.

These range from payment terminals, special passport and ID card readers and desktop PC readers to ones in the form of USB adapters. In laptops, integrated PC/SC readers for contact-type cards are already standard. Some laptop manufacturers are already integrating contactless ISO/IEC 14443, ISO/IEC 15693 and NFC readers in their laptops.

10.2 Card Terminal API (CT-API)

The German health service card is based on the MKT (Multifunktionale Karten-Terminals) specification, which is defined by TeleTrusT Deutschland. Due to the thousands of terminals for approximately 80 million health insurance contact-type cards, this standard has achieved appropriate national significance in Germany. The MKT specification consists of 7 parts, and Part 3 specifies the card terminal API (CT-API). This consists merely of a procedurally-built dynamic-link library (DLL) and is thus independent of the programming language.

The reader manufacturers must make the following basic functions available in a CT-API library:

- Open a communication channel with the `CT_init()` function.
- Exchange data with the `CT_data()` function.
- Close the communication channel with the `CT_close()` function.

The CT-API is easily integrated into the PC/SC environment, as a reader manufacturer provides a CT-API Service Provider interface. In this case, it is necessary to install both the PC/SC and CT-API drivers. The MKT specification defines only contact-type cards. Some reader manufacturers, however, provide both a CT-API driver and a PC/SC driver for contactless readers. Long before the PC/SC standard, the MKT specification standardized the use of additional reader functions such as PIN pads, displays and so on.

The MKT specifications are available for free at *http://www.teletrust.de/publikationen/spezifikationen/mkt/*.

10.3 Open Card Framework (OCF)

The Open Card Framework (OCF) is a smart card middleware implementation in the Java programming language, which was specified in 1997 by the Open Card Consortium. This framework supports both contact and contactless cards with an APDU command structure specified according to ISO/IEC 7816. The OCF architecture also supports integration of PC/SC readers.

The Open Card Consortium was founded by Gemplus, Giesecke & Devrient, IBM, Schlumberger, Sun Microsystems, Visa International and other companies. After the publication of Version 1.2 of the specification and a reference implementation by IBM, the OCF Consortium was abandoned in February 2000. Since about 2008, the official website isn't active either. The original source code was made into a SourceForge project *(http://opencard.sourceforge.net/)*.

Although the project is not maintained on the SourceForge website, a range of Java-based open source projects use the framework. An example is the Open Smart Card Development Platform (OpenSCDP) project. The OpenSCDP is a collection of smart cards and PKI (public key infrastructure) development and test tools for the Java programming language. Additional information, documentation and source code can be downloaded for free from *http://www.openscdp.org/ocf/index.html.*

10.4 Personal Computer/Smartcard (PC/SC)

The PC/SC specification builds on the existing ISO/IEC 7816 and EMV (Eurocard, MasterCard, VISA) standards. Essentially, PC/SC extends the standards to defined APIs for drivers, service providers and PC applications. The central role is played by the Resource Manager. This operating system component manages both the reader connected to the PC and the smartcards.

In December, 1997, Version 1.0 of the Interoperability Specification for ICCs and Personal Computer Systems (PC/SC) was published by the PC/SC Working Group. The Working Group consisted of the following companies: Bull, Gemplus, Hewlett-Packard Corporation, IBM Corporation, Microsoft Corporation, Axalto, Siemens Nixdorf, Sun Microsystems, Toshiba, and Verifone. In Version 1.0, only contact-type smartcards and simple readers were supported.

Contactless ISO/IEC 14443 and ISO/IEC 15693 smartcards, asynchronous contact memory smartcards and higher-category integrated readers (for example: keyboards, readers with biometric sensors, displays, keypads, etc.) were first supported by Version 2.0. This extension was published in August 2004. The PC/SC Working Group at this time consisted of the following companies: Apple, Axalto, Gemplus, Infineon Technologies, Ingenico, Microsoft Corporation, Philips Semiconductors and Toshiba. The 10-part specification can be downloaded at no cost from the PC/SC Workgroup website *(http://www.pcscworkgroup.com).*

The PC/SC standard is completely platform- and vendor-independent and can therefore be integrated into any desktop operating system. The PC/SC specification was first implemented in Windows. PC/SC is already standard in all current Microsoft operating systems. In 2000, David Cororan of Purdue University began the PC/SC-Lite project, with the goal of supporting smartcards under UNIX. The first version of MUSCLE (Movement for the Use of SmartCards in a Linux Environment) was published. PC/SC-Lite draws largely from the PC/SC standard and the Microsoft WinSCard API, but doesn't support all of the functionality. The published drivers and source code are on the MUSCLE website *(http://www.musclecard.com).* Currently, PC/SC-Lite is used in the following operating systems: Linux, Solaris, Mac OS X, FreeBSD and HP-UX.

10.4.1 The PC/SC Architecture

In Figure 10.2, the PC/SC hardware and software architecture is shown, and we discuss this in detail below.

10.4.1.1 Integrated Circuit Card (ICC)

The PC/SC standard currently supports contact-type, asynchronous smartcards, contactless ISO/IEC 14443 and ISO/IEC 15693 smartcards, as well as contact and contactless processor cards.

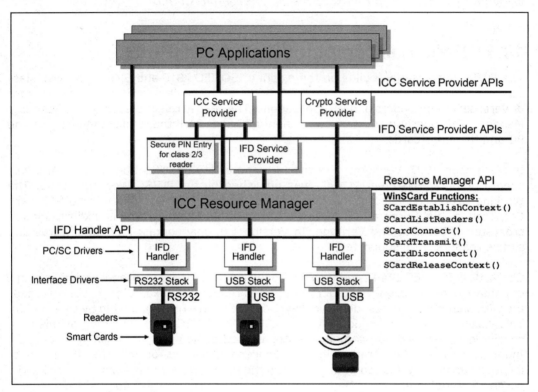

Figure 10.2. *The PC/SC architecture.*

10.4.1.2 Interface Device (IFD)

The term 'Interface Device' (IFD) refers to a smartcard reader that is connected to a PC using a physical interface. Any PC interface can be chosen by the manufacturer. USB, the keyboard interface, RS-232 and PC Card (PCMCIA) are some of the possibilities.

10.4.1.3 Interface Device Handler (IFD Handler)

The IFC software interface, which every reader driver (IFD Handler) must implement, is in Part 3 of the PC/SC specification. The Resource Manager, an underlying layer of software, communicates with the card reader via this interface, without needing to know how the reader is physically connected to the PC.

The vendor-specific IFD handler will usually be shipped together with the interface driver (for example, a USB driver). Windows PC/SC drivers run in Kernel mode, and must necessarily pass the Windows Hardware Quality Lab Tests. If a reader supports the CCID USB device class, then no manufacturer-specific drivers (PC/SC or USB) are required.

Integrated Circuit(s) Cards Interface Device (CCID)
The USB implementers' forum (http://www.usb.org) specifies the Integrated Circuit(s) Card Interface Device class for smartcard readers. Microsoft provides a CCID driver in all current Windows versions automatically. Therefore, the CCID-compliant smartcard reader eliminates both the installation of the PC/SC and the need to install a USB driver.

Contactless Cards
Because contactless cards were not supported in the original V1.0 PC/SC specification, the IFD Handler of a contactless reader always emulates a contact-type chip card. For example, the ATS (Answer to Select) from a contactless ISO/IEC 14443 Type A card will always be converted and mapped to an ATR (Answer to Reset). For contact cards, the ATR plays a similar role as the contactless cards' T=CL ATS. The same procedure is also used in contactless memory cards apply (e.g. MIFARE Ultralight, MIFARE 1K/4K, etc.), although they have neither an ATS nor an ATR.

Regardless of card type, the communication with a contactless card always uses APDUs (Application Protocol Data Units). The PC/SC standard defines special pseudo-APDU commands for the most important proprietary memory card commands, such as the MIFARE Read command.

10.4.1.4 ICC Resource Manager (RM)

The ICC Resource Manager is the central component within the PC/SC architecture. Since Windows 2000, the ICC Resource Manager has been integrated into all Microsoft PC operating systems. The ICC Resource Manager manages all the necessary resources that are required for the integration of smartcards in the operating system. Further, it provides a defined API for PC applications and service providers. The API function names begin with the `SCard` prefix. Resource identification, resource management and transaction support are the three basic tasks of the ICC Resource Manager.

Resource Identification
The ICC Resource Manager manages all installed readers and makes this information available to other applications. A list of available readers, for example, is retrieved using the WinSCard API Function, `SCardListReaders()`.

The ICC Resource Manager also supports the management of card types. In Windows operating systems, the cards and their features are stored in the registry (HKEY_LO-CAL_MACHINE\SOFTWARE\Microsoft\Cryptography\Calais\SmartCards).

Applications such as Windows logon or Secure Mail always communicate with a smartcard via a Crypto Service Provider and must therefore be registered with the operating system. If an application communicates directly with a card via the APDUs, it need not be registered.

Another task performed by Resource Identification is the allocation of card status. When inserting and removing a card, ICC Resource Manager triggers the appropriate event and this is passed to other applications. The current card status can be retrieved using the SCardGetStatus() function.

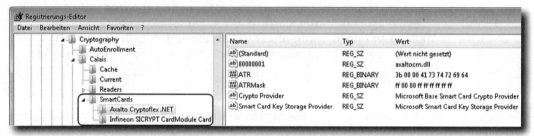

Figure 10.3. *Windows Vista registry.*

Resource Management
Communication between a PC application and the reader or card takes place exclusively via the ICC Resource Manager. Access to the resources (card and reader) can be either exclusive or shared. If the card supports multiple applications (Multi-Application Card), then several PC applications may access the card simultaneously.

Transaction Support
In the case of multiple applications within a session all using the same card, this invariably leads to uncontrolled access. This becomes a problem when several sequential APDU commands are required to complete a transaction. Reading data from a file, for example, requires two APDU commands: SELECT FILE selects the file, then READ BINARY reads the data. During the read operation, should Application A be interrupted by Application B after the FILE SELECT command, B may also select a file. In this case, Application A would read the wrong file.

To avoid such situations, the Resource Manager's API provides the two functions, SCardBeginTransaction() and SCardEndTransaction(). A transaction initiated by SCardBeginTransaction() cannot be interrupted by another application until SCardEndTransaction() is called.

10.4.1.5 Service Provider

Service providers are used to encapsulate complex card and reader functions. The PC/SC standard defines service providers for the card operating system as well as for additional reader components, such as displays, keyboards and biometric sensors (e.g. fingerprint scanners).

The functionality of a service provider is best explained with a small example: The FILE-ACCESS class is part of the ICC OS Service Provider API and offers the following methods: ChangeDir(), Create(), Delete(), Open(), Close(), Read() and Write(). Using these methods, you can manage a card's file system without knowing the required APDU commands.

Version 1.0 of PC/SC supports an ICC Service Provider, which always supports a specific card type. Differences between the various cards and readers are determined using the ATR. A card always responds with the same ATR after a power-on reset. The Resource Manager performs preparation of the static connection between the card and the ICC service provider. For every card type, the ATR value and a link to the service provider is saved in the Windows Registry (see Figure 10.3). This method does not make it possible to connect a single application to a multi-application card using a service provider. For this reason, PC/SC Version 2.0 introduced the Application Doman Service Provider (ADSP). The card issuer provides another software component: the Application Domain Service Provider Locator (ADSPL). With the ADSPL, the Resource Manager provides all the information about the different card applications. The connection between an Application Domain Service Provider and a card application is made dynamically.

10.4.1.6 ICC-Aware Applications

A PC application is referred to as an ICC-aware application in the PC/SC standard. Access to the card takes place either directly via the Resource Manager API or via the ICC Service Provider API.

11
PC/SC Readers

Gerhard H. Schalk

11.1 Contactless Cards

As we've already mentioned, contactless cards were first supported in Version 2.0 of the PC/SC specification. Part 3 of this specification encompasses several documents and is continuously being extended.

The integration of contactless cards is done in the PC/SC driver (IFD Handler). This always emulates a contact microcontroller smartcard, regardless of actual card type. In the PC/SC specification, contactless smartcards and contactless memory cards are treated differently.

11.1.1 Contactless Microcontroller Smartcards

Microcontroller smartcards with a card operating system are referred to in the PC/SC standard as smartcards. Regardless of the physical interface (contact or contactless) and the transmission protocol (T=0, T=1 or T=CL) the exchange of data between the PC application and the card is always done using APDUs. In this way, it is not relevant to a PC/SC application whether it's communicating with a contact-type or contactless microcontroller smartcard.

As far as the PC application knows, it's always communicating with a contact card. The PC/SC driver of a contactless reader supports both contact protocols, i.e. T=0 and T=1. After a reset, contact cards transmit an ATR (Answer To Reset). For this reason, the ATS (Answer To Select) is mapped to an ATR. In this way, it isn't necessary to extend the WinSCard API or the PC/SC specification.

11.1.2 Contactless Memory Cards

The PC/SC specification currently includes contactless memory cards compliant with ISO/IEC 14443-3 Type A, ISO/IEC 14443-3 Type B and ISO/IEC 15693. A PC/SC reader need not necessarily support all card types. Which types of contactless cards a PC/SC reader actually supports depends almost entirely on the reader IC used. It is thus advisable to check the reader's documentation.

Since we have dealt extensively with the MIFARE Ultralight and MIFARE 1K/4K cards in previous chapters, these two types will be used for the following examples. In any case, the operating principle is the same for other card types. Although a contactless memory card does not have an ATR, the PC/SC driver always generates a pseudo-ATR for contactless memory cards. The structure of this ATR is specified in detail and allows the distinction between different types of cards (see Section 11.4.4.5).

11.1.2.1 PC/SC-Compliant APDUs

Regardless of card type, communication with a contactless memory card takes place through APDUs. The PC/SC standard specifies the following pseudo-APDU commands: Get Data, Load Keys, General Authenticate, Verify, Read Binary and Update Binary. Additionally to Part 3 of the PC/SC Specification, MIFARE Value Format commands have been specified for a MIFARE 1/4K card.

The PC/SC-compliant APDU commands and their functions will be looked at briefly below. For space reasons, detailed descriptions of all command APDU options, as well as the response APDUs, have been omitted. These can be found either in Part 3 of the PC/SC standard or in manufacturer documentation.

Get Data
Depending on the value in P1, either the card serial number or the ATS historical bytes of an ISO/IEC 14443 Type A card are returned.

Table 11.1. Command APDU: Get Data.

CLA	INS	P1	P2	Lc	Data	Le
0xFF	0xCA	1.)	0x00	–	–	0x00

[1] P1 = 0x00 → UID is returned. P1 = 0x01 → All historical bytes from the ATS, without CRC, are returned.

Load Key
Load Key loads either a reader key or a card key (e.g. MIFARE Key A or Key B) into the reader. A reader key is optional and serves to encrypt sensitive data (for example, a card key) between the PC application and the reader.

In two Parameter bytes, P1 and P2, this command offers numerous options, which not all PC/SC reader manufacturers implement in the same manner (see Section 11.4.4.6).

Table 11.2. PC/SC APDU: Load Key.

CLA	INS	P1	P2	Lc	Data	Le
0xFF	0x82	1.)	2.)	3.)	Key	–

[1] Key Structure; [2] Key number; [3] Key Length

General Authenticate
Depending on the card type, the relevant authentication command is sent to the card: in the case of a MIFARE 1K/4K card, this would be the card commands Auth. Key A or Auth. Key B.

Table 11.3. *PC/SC APDU: General Authenticate.*

CLA	INS	P1	P2	Lc	Data	Le
0xFF	0x86	0x00	0x00	0x05	0x01, Addr.MSB, Addr.LSB, Key type, Key Nr.	–

Verify
Used for authentication of cards that control access using a PIN (e.g. ISO/IEC 15936-compliant tags).

Table 11.4. *PC/SC APDU: Verify.*

CLA	INS	P1	P2	Lc	Data	Le
0xFF	0x20	1.)	2.)	3.)	Pin	–

[1] PIN Structure; [2] Reference Data; [3] PIN Length

Read Binary
The card data is read using Read Binary. For the MIFARE 1K/4K card, the two parameter bytes, P1 and P2, are used to select a block. With MIFARE Ultralight cards, a page is selected.

Table 11.5. *PC/SC APDU: Verify.*

CLA	INS	P1	P2	Le
0xFF	0xB0	1.)	2.)	xx

[1] Address MSB; [2] Address LSB

Update Binary
Card data are written using Update Binary. Addressing a MIFARE block or a page is also done using the two Parameter bytes, P1 and P2.

Table 11.6. *PC/SC APDU: Update Binary.*

CLA	INS	P1	P2	Lc	Data	Le
0xFF	0xD6	1.)	2.)	3.)	data	–

[1] Address MSB; [2] Address LSB; [3] Data Length

11.1.3 Answer To Reset (ATR)

11.1.3.1 Contact-type Card Activation Sequence

In order for readers of this book with no knowledge of contact-type smartcards to understand the term ATR, this section briefly discusses the activation of a contact microcontroller smartcard.

A contact-type reader uses a contact unit to make an electrical connection to the card. In addition, the contact unit uses a small mechanical switch to determine whether a card is present or not. When a card is inserted into the reader, the electrical connections are activated in a fixed, predefined sequence.

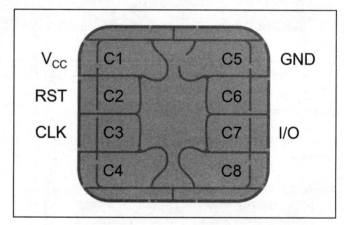

Figure 11.1. Contact field pin assignment on a contact-type smartcard module.

Vcc and GND are used for power input. Modern smart cards support up to three different voltage classes (Class A: 5 V, Class B: 3 V and Class C: 1.8 V). RST is the reset input (active low) of the smart card's microcontroller. Half-duplex communication using the master-slave principle takes place via the bidirectional I/O pin. The CLK input was previously used to supply the CPU clock and also as a reference clock for the UART. Modern smartcard controllers use an internally-generated CPU clock whose frequency is a multiple of the reader-generated clock (3.5712 MHz to 8 MHz).

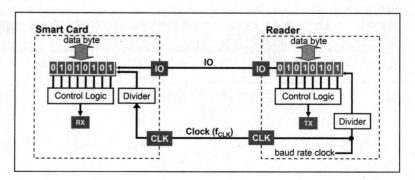

Figure 11.2. Although the data transfer is asynchronous, the shift clock for the card UART is generated by the reader.

These days, the reader-generated clock is used merely as a shift register clock for the asynchronous ISO/IEC 7816 UART. For the standard data transmission rates (9,600 kbit/s to 115,200 kbit/s), a reference clock of 3.5712 MHz is defined. Since most contact-type readers have a special ISO/IEC 7816 UART, most readers have a higher clock rate and thus non-standard data rates.

Figure 11.3 shows the activation sequence of a contact-type microprocessor card. The very first command to the card is a Chip Reset. This could be either a power-on reset (cold reset) or, if the card was already in operation, a reset signal (warm reset). In either case, the card responds with an Answer To Reset (ATR). The ATR holds the supported transmission protocols (usually T=0 and/or T=1) and transmission parameters. An ATR has the same purpose as an ATS and is similarly structured.

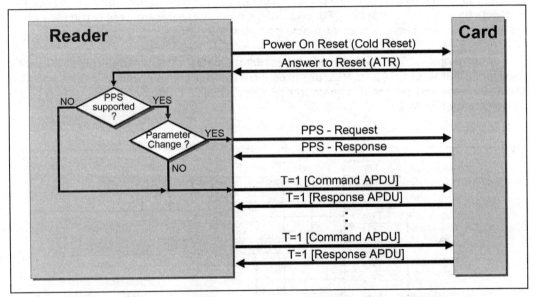

Figure 11.3. Activation sequence of a contact-type smartcard microcontroller.

The PPS sequence (Protocol and Parameter Selection) has the same purpose as with an ISO/IEC 14443 Type A card. With a PPS request, the reader can ask the card to use a higher data transfer rate. A PPS request is permitted only after an ATR. If a PC/SC reader is being used to communicate with a contact-type card, these details need not be taken into account. The WinSCard API function, SCardGetAttrib(), can simply be used to get the transmission parameters used by the reader. Some PC/SC readers' protocol parameters (T=0, T=1 and T=CL) can be configured using the SCardControl() WinSCard API function, or with special Escape APDUs.

Following the optional PPS sequence, the APDUs are exchanged. These are transmitted using either the T=0 or the T=1 protocol.

11.1.3.2 ATR Structure of a Contact-Type Smartcard

The ATR is defined in Part 3 of the ISO/IEC 7816 standard. The structure of the individual data items is quite extensive and requires a detailed knowledge of the T=0 and T=1 transmission protocols. Therefore, this section focuses only on the most important parameters. Figure 11.4 shows the basic structure of the data elements. At first glance, it may appear that this is a description of the ATS.

The first character (initial character) is called TS, and identifies what bit encoding the card uses. value of 0x3B means Direct Convention (high level → logic '1'; low level → logic '0'; LSb-first). Inverse convention (high level → logic '0'; low level → logic '1'; MSb-first) is represented by 0x3F.

The upper nibble of the Format Character, T0, determines whether the Global Interface Characters, TA1, TB1, TC1 and TD1 are present in the data string (bx = 0 → Tx1 not present; bx = 1 → Tx1 present). For example, one of these parameters is the maximum data transmission rate. The lower nibble encodes the number of historical bytes.

The upper nibble of TD2 (Structural Character) indicates whether the Protocol-Specific Interface Characters, TA2 to TC2, are present. Whether these parameters are valid for the T=0 or the T=1 protocol is given in the lower nibble. A value of 0 means T=0 and a value of 1 means T=1.

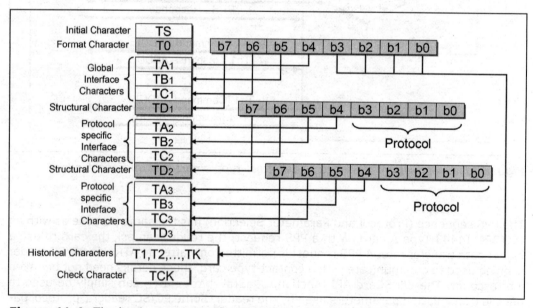

Figure 11.4. The basic structure of an ATR's data elements.

Because contact-type cards partially support both transmission protocols, the TD2 character follows. This has the same purpose as TD1, but for a different protocol. The

ATR data structure is not limited to the two transmission protocols, and, with the TD3 byte, another protocol can be specified. For a long time, the historical bytes' content was not defined in any standard. However, the bytes are still used for information storage about chip hardware, memory size, ROM version or operating system version today. ISO/IEC 7816 Part 4 regulates the structure of the historical bytes. The last byte, TCK (Check Character) is used only as a backup. The checksum is an XOR of all the characters from T0 to TK. If the ATS indicates that the T=0 protocol is used, there is no TCK byte.

11.1.3.3 Contactless Smartcard Pseudo-ATR Structure

The ATR data string (see Figure 11.5) of a contactless card is constructed identically. The indication as to whether it's a microcontroller smartcard or a contactless memory card lies in the historical bytes.

The value of the first character (TS) is always 0x3B (Direct Convention). The ATR contains no Global Interface Characters, nor Protocol-specific Interface Characters. The absence of any Interface Character is indicated in bytes T0, TD1 and TD2. All contactless cards emulate a contact-type card that supports transmission protocols T=0 and T=1 (see lower nibble of TD1 and TD2). Since the T=1 protocol is indicated, the TCK character, or XOR checksum, is appended at the end.

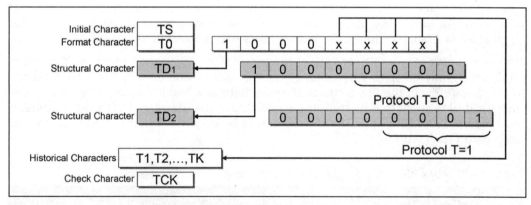

Figure 11.5. Contactless ATR data elements.

Contactless Microcontroller Smartcard Pseudo-ATR Structure

The ATR historical bytes are defined differently between ISO/IEC 14443 Type A and Type B cards. For Type A cards, the historical bytes match the historical bytes of the ATS exactly. With Type B cards, the historical bytes consist of parts of the ATQB (application data and protocol information) as well as the MBLI (Maximum Buffer Length Index) value.

11 PC/SC READERS

Table 11.7. ATR for contactless smartcards generated by a PC/SC reader.

Card Type	Initial	T0	TD1	TD2	Historical bytes	TCK
ISO/IEC 14443 Type A	0x3B	8n	80	01	The historical bytes of the ATS.	
ISO/IEC 14443 Type B					Bytes 1 – 4: ATQB application data Bytes 5 – 7: ATQB protocol info Byte 8, high nibble: MBLI Byte 8, low nibble: 0 (r.f.u.)	x

N – number of historical bytes; x – XOR checksum.

Contactless Memory Card Pseudo-ATR Structure

The historical bytes of a memory card are encoded according to ISO/IEC 7816 Part 4 and contain three data fields: a one-byte Category Indicator, one or more data objects in Compact-TLV format and an optional Status Indicator.

For contactless memory cards, the value of the Category Indicator is 0x80. Basically, this value means that an optional Compact TLV-formatted data object may be present. For contactless memory cards, an internationally-recognized Application Identifier (AID) follows the Character Indicator byte. An Application Identifier actually identifies a smartcard application. For contactless memory cards, the application identifier identifies the type of card. Currently, the PC/SC specification identifies 52 different types of contactless memory card.

The Application Identifier (AID) consists of two parts: the first part is a 5-byte Registered Application Provider ID (RID). This part is unique and is issued by an international or national authority. The upper nibble of an international RID is always 0xA, and for national RIDs it is 0xD. The PC/SC Workgroup's RID is 0xA0 00 00 03 06.

The second part of the AID consists of the Proprietary Application Identifier Extension (PIX), which may be up to 7 bytes long. The PIX is freely defined by the manufacturer. In our specific case, the PIX consists of an SS byte and two NN bytes. The contactless smartcard type supported by the card is encoded in the Standard byte (SS). A value of 0x03 means ISO/IEC Part 3. The two-byte long Byte for Card Name (NN) encodes the exact card type (see Table 11.8).

Table 11.8. MIFARE card pseudo-ATR generated by a PC/SC reader.

Card Type	1.)	T0	TD1	TD2	Historical bytes						r.f.u.	TCK
					2.)	3.)	4.)	AID	SS	NN		
MIFARE 1K	0x3B	8F	80	01	80	4F	C0	A000000306	03	0001	00000000	6A
MIFARE 4K	0x3B	8F	80	01	80	4F	C0	A000000306	03	0002	00000000	69
MIFARE Ultralight	0x3B	8F	80	01	80	4F	C0	A000000306	03	0003	00000000	68

[1] Initial; [2]: Category Indicator; [3]: Application Identifier Presence Indicator (Tag); [4]: Length

The AID is encoded in the Compact TLV format. The first character, T (Tag), is used to identify the data object. The second character, L (Length) encodes the number of data bytes to follow (V for value).

With a PC/SC reader that supports the PC/SC 2.0 standard, one can deploy most memory card types. Unfortunately, not all of the new card types, such as the MIFARE Ultralight C, are listed. For these, the majority of PC/SC readers return the ATR of a MIFARE Ultralight card.

11.2 The Microsoft WinSCard API

The WinSCard API makes the direct exchange of data between readers and cards possible by using APDUs. The WinSCard API's C functions are declared in the header file, `winscard.h`, and the return codes are declared in `winerror.h`.

Table 11.9 lists the main WinSCard API functions. A detailed description of all function parameters is available in the MSDN Library Documentation at *http://msdn.microsoft.com/en-us/library/aa374731(v=VS.85).aspx*.

Table 11.9. The main WinSCard API functions.

Function	Description
`SCardEstablishContext()`	Sets up a connection to and returns the handle to the Resource Manager context.
`SCardListReaders()`	Returns a list of smartcard readers available on the system, in the form of a C multi-string. Individual strings are separated by the null character (0x00). The end of the multi-string is represented by an additional null character.
`SCardGetStatusChange()`	Blocks program execution until a change in card status.
`SCardConnect()`	After successful card activation, returns a card handle, thus creating a logical connection to the card.
`SCardGetAttrib()`	Returns a range of different reader properties. The desired property is selected using a function parameter. The ATR, the reader's name, the manufacturer name and various communication parameters are among the available properties.
`SCardControl()`	Used to control and configure the reader hardware. For example, some PC/SC readers support the ability to turn the RF field on and off and to limit the maximum Baud rate.

Function (continued)	Description (continued)
SCardTransmit()	Sends a command APDU to the card and returns the response APDU.
SCardDisconnect()	Closes the logical connection to the card.
SCardReleaseContext()	Closes the connection to the resource manager.

11.2.1 WinSCard API Programming

The Windows API functions are written entirely in the C programming language. In this way, one may simply call the WinSCard API functions from C/C++. All other programming languages, such as Java, C#, and Visual Basic, require a wrapper. The wrapper in this case acts as an interface between the programming language and the WinSCard API.

The main WinSCard API functions are demonstrated using a console application. This is written in C. Even if you develop PC/SC applications in another language, it is advantageous to know the WinSCard API functions in detail. Later, we'll discuss programming the PC/SC reader in Java. Finally, the C# wrapper developed in the scope of this book is illustrated with some examples.

Figure 11.6 shows the basic program flow of a PC/SC application. Using the `SCardEstablishContext()` function, a connection to the resource manager is established. The function returns a handle to the resource manager context. A handle is nothing more than a number that the operating system creates to reference some other data (see Figure 11.7). In our case, this data is the resource manager. The handle to the resource manager (hContext) is used to differentiate between several PC/SC applications running simultaneously. The handle is passed as a parameter to the `SCardListReaders()`, `SCardGetStatusChange()` and `SCardConnect()` functions.

The `SCardListReaders()` function returns a list of available PC/SC readers on the system, in the form of a C multi-string. The individual strings and the end of the multi-string are terminated with the null character (0x00). Figure 11.8 shows the content of the C multi-string. In this case, three dual-interface readers are connected to the PC.

One of the readers is a dual-interface reader, which supports both contact-type and contactless cards. In order for the interfaces to be individually selectable in the PC application, a separate string is returned for every available interface.

In the following example, the multi-string is printed to the console and the user is prompted to select a reader. The string of the user-selected reader is passed to the two functions, `SCardGetStatusChange()` and `SCardConnect()`. Use of the `SCardGetStatusChange()` function is optional. This feature allows you to react to different card states. In our example, application execution is blocked until a contact-type card is placed in the slot or a contactless card appears in the reader's field.

11.2 THE MICROSOFT WINSCARD API

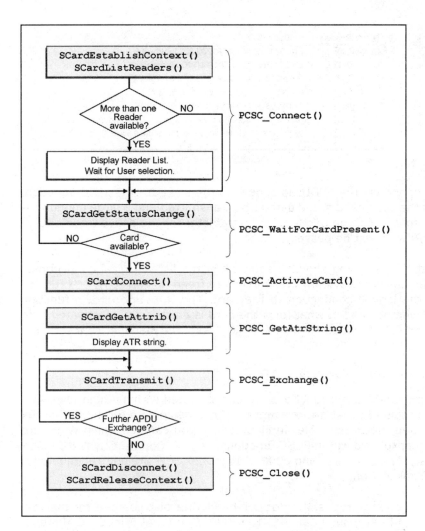

Figure 11.6.
PC/SC application basic program flow.

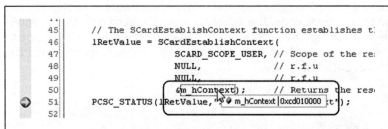

Figure 11.7.
A handle contains a code number that the operating system generates.

The logical connection to a card is created by calling the SCardConnect() function. After successful activation of the card, this function returns the card handle (hCard). The ATR can be retrieved using either SCardGetStatusChange() or SCardGetAttrib(). With some readers, the ATR can also be retrieved using the SCardControl() function.

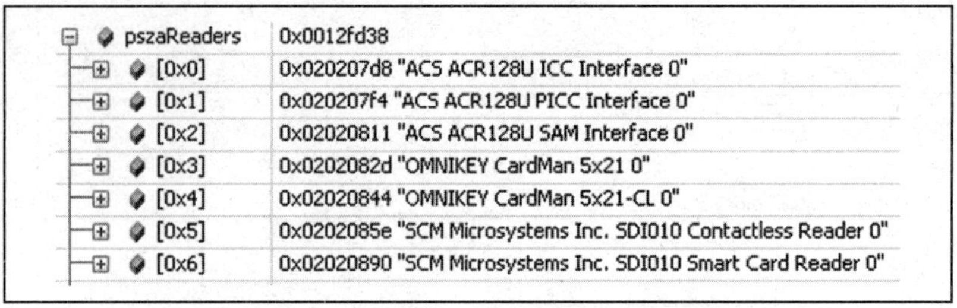

Figure 11.8.
The SCardList-
Readers() func-
tion returns all
readers that are
connected to the
PC, in the form of
a C multi-string.

	pszaReaders	0x0012fd38
	[0x0]	0x020207d8 "ACS ACR128U ICC Interface 0"
	[0x1]	0x020207f4 "ACS ACR128U PICC Interface 0"
	[0x2]	0x02020811 "ACS ACR128U SAM Interface 0"
	[0x3]	0x0202082d "OMNIKEY CardMan 5x21 0"
	[0x4]	0x02020844 "OMNIKEY CardMan 5x21-CL 0"
	[0x5]	0x0202085e "SCM Microsystems Inc. SDI010 Contactless Reader 0"
	[0x6]	0x02020890 "SCM Microsystems Inc. SDI010 Smart Card Reader 0"

The actual data exchange via the APDUs is done with the `SCardTransmit()` function, which simply needs the card handle, the data to be transmitted and a large enough receive buffer passed to it. Also, the `pioSendPci` parameter for the active transfer protocol (T=0, T=1, or Raw), must be passed.

Before closing an application, you should disable all of the listed logical connections. This ensures that the operating system's resources are freed up correctly. `SCardDisconnect()` closes the logical connection to the card. The `dwDisposition` function parameter controls whether and in what form the card is electrically deactivated. The `SCardReleaseContext()` function closes the logical resource manager connection.

11.2.1.1 Programming the WinSCard API in C

In this example, the WinSCard API functions are encapsulated within a high-level API. All of the functions in this API that are relevant to PC/SC begin with the `PCSC_` prefix and are implemented in the `pcsc.c` file. In Figure 11.6, you can see which WinSCard API functions are encapsulated within PCSC functions. `PCSC_Connect()`, `PCSC_ActivateCard()`, `PCSC_Exchange()` and `PCSC_Disconnect()` are quite adequate for developing a PC/SC application.

In addition to the function prototypes, the `pcsc.h` header file also contains the macros, `PCSC_STATUS`, `PCSC_ERROR` and `PCSC_EXIT_ON_ERROR`. All WinSCard API functions return an error code. The `PCSC_STATUS` macro generates output to the console in any event. In addition to the error code, the `PCSC_STATUS` macro passes the WinSCard API function name as a string. The `PCSC_ERROR` macro only prints a message to the console in the event of an error. The `PCSC_EXIT_ON_ERROR` macro ends the console application in the event of an error.

Listing 11.1. *pcsc.h.*

```
#include <winscard.h>

#ifndef PCSC_H_INCLUDE

    #define PCSC_H_INCLUDE

    #define PCSC_STATUS(lRetValue, msg)                         \
```

```c
      if(lRetValue == SCARD_S_SUCCESS)            \
      {                                            \
         printf("\n " msg ": %s",                  \
         SCardGetErrorString(lRetValue));          \
      }                                            \
      else                                         \
      {                                            \
         printf("\n " msg ": Error 0x%04X %s",     \
         lRetValue,SCardGetErrorString(lRetValue));\
         return lRetValue;                         \
      }

   #define PCSC_ERROR(lRetValue, msg)              \
      if(lRetValue != SCARD_S_SUCCESS)             \
      {                                            \
         printf("\n " msg ": Error 0x%04X %s",     \
         lRetValue,SCardGetErrorString(lRetValue));\
         return lRetValue;                         \
      }

   #define PCSC_EXIT_ON_ERROR(lRetValue)  \
      if(lRetValue != SCARD_S_SUCCESS)    \
      {                                   \
         while(!_kbhit());                \
         return 0;                        \
      }

   LONG PCSC_Connect(LPTSTR sReader );
   LONG PCSC_ActivateCard(void);
   LONG PCSC_Exchange(LPCBYTE pbSendBuffer ,DWORD cbSendLength ,
   LPBYTE pbRecvBuffer ,LPDWORD pcbRecvLength );
   LONG PCSC_Disconnect(void);
   LONG PCSC_WaitForCardPresent(void);
   LONG PCSC_WaitForCardRemoval(void);
   LONG PCSC_GetAtrString(LPBYTE atr, LPINT atrLen);
   LONG PCSC_GetVentorName();
   CHAR* SCardGetErrorString(LONG lRetValue);

#endif
```

As can be seen from the header file, this example uses typical Windows C style. For example, instead of an `unsigned char*`, the `LPBYTE` type is used, and instead of `long`, we use the `LONG` type. The following type definitions are found in the Win32 library:

```c
typedef char CHAR;
typedef short SHORT;
typedef long LONG;
typedef int INT;
```

11 PC/SC READERS

For variables and function parameters, so-called 'Hungarian notation' is used, in which the variable name consists of a lowercase type prefix and a variable name beginning with a capital letter. The prefix is always lowercase and specifies the variable's type. For example, `pmszReaders`: the letter `p` denotes a pointer in C, while a C string is denoted with `sz` (string, zero-terminated). If the letter `m` is not the first character, it stands for 'multi'. Thus, the prefix `pmsz` indicates a pointer to a C multi-string.

Listing 11.2. Excerpt from pcsc.c

```
#include <winscard.h>
#include <stdio.h>
#include <conio.h>
#include <string.h>
#include "pcsc.h"
#include "util.h"

#define RcvLenMax 300l                  // Max. APDU Buffer length.
```

The PCSC function set a few parameters using global variables. These include the resource manager handle, `m_hContext`, and the card handle, `m_hCard`. The user-selected reader name is stored in `m_szSelectedReader`. The active transmission protocol returned by `SCardConnect()` is stored in the `m_dwActiveProtocol` variable and is used as a parameter for the `SCardTransmit()` function.

```
SCARDCONTEXT m_hContext;           // Resource manager handle
SCARDHANDLE m_hCard;               // Card Handle
CHAR m_szSelectedReader[256];      // Selected card reader name.
DWORD m_dwActiveProtocol;          // Active protocol (T=0, T=1 or undefined).
```

The `PCSC_Connect()` function alls the WinSCard API functions, `SCardEstablishContext()` and `SCardListReaders()`. Should multiple cards be available, the user is prompted to select one of them in the console window.

Alternatively, an already-known reader name can be passed to the `PCSC_Connect()` function in the `szReader` parameter. In this case, the reader is used without prior prompting. Reader selection is then suppressed. In this instance, the `SCardListReaders()` function will not be called.

The `PCSC_WaitForCardPresent()` function blocks the application until a card is present in or at the selected reader.

```
LONG PCSC_WaitForCardPresent(void)
{
    SCARD_READERSTATE sReaderState;
    LONG lRetValue;
    sReaderState.szReader = m_szSelectedReader;
    sReaderState.dwCurrentState = SCARD_STATE_UNAWARE;
    sReaderState.dwEventState = SCARD_STATE_UNAWARE;
```

```
    //The SCardGetStatusChange function blocks execution until
                                                        the current
    //availability of the cards in a specific set of readers changes.
    lRetValue = SCardGetStatusChange(
            m_hContext,    // Resource manager handle.
            30,            //Max. amount of time
                        (in milliseconds) to wait for an action.
            &sReaderState, // Reader state
            1);            // Number of readers
    PCSC_STATUS(lRetValue,"SCardGetStatusChange");

    // Check if card is already present
    if((sReaderState.dwEventState & SCARD_STATE_PRESENT) ==
                                            SCARD_STATE_PRESENT)
    {
        printf(": Card present...\n");
    }
    else
    {
        printf(": Wait for card...\n");
        do
        {
            lRetValue = SCardGetStatusChange(m_hContext,30,&sReaderState,1);
            PCSC_ERROR(lRetValue, "SCardGetStatusChange");
            Sleep(100);
        }
        while((sReaderState.dwEventState & SCARD_STATE_PRESENT) == 0);
    }

    return lRetValue;
}
```

A call to the `PCSC_ActivateCard()` function activates the card. This is done exclusively by calling the `SCardConnect()` function. The `dwShareMode` parameter determines whether access to the card is exclusive or shared (see Section 10.4.1.4). The preferred transmission protocol is defined in the `dwPreferredProtocols` parameter (T=0, T=1, or Raw). Should the T=0 protocol be selected, for example, and the card supports only T=1, the `SCardConnect()` function returns the error code `SCARD_E_PROTO_MISMATCH` and the card is not activated. Generally, we use either `SCARD_PROTOCOL_T0` or `SCARD_PROTOCOL_T1` as a parameter.

A contactless reader emulates both the T=0 and T=1 protocols. So, you won't find a constant definition, `SCARD_PROTOCOL_TCL` in the `winsmcrd.h` header file. The protocol actually used by the reader is returned in the `dwActiveProtocol` parameter.

```
LONG PCSC_ActivateCard(void)
{
    LONG RetValue;

    //Establishes a connection to a Smartcard contained by a specific
```

11 PC/SC READERS

```
                                                    reader.
    lRetValue = SCardConnect(
                m_hContext,             // Resource manager handle.
                m_szSelectedReader,     // Reader name.
             SCARD_SHARE_EXCLUSIVE, // Share Mode.
         SCARD_PROTOCOL_Tx,       // Preferred protocols (T=0 or T=1).
         &m_hCard,                // Returns the card handle.
         &m_dwActiveProtocol);    // Active protocol.
    PCSC_STATUS(lRetValue,"SCardConnect");

    switch(m_dwActiveProtocol)
    {
       case SCARD_PROTOCOL_T0:
          printf(": Card Activated via T=0 protocol");
          break;

       case SCARD_PROTOCOL_T1:
          printf(": Card Activated via T=1 protocol");
          break;

       case SCARD_PROTOCOL_UNDEFINED:
          printf(": ERROR: Active protocol unnegotiated or unknown");
          lRetValue = -1;
          break;
    }

    return lRetValue;
```

`PCSC_Exchange()` is used to exchange APDUs. As already mentioned, it's necessary to include the active protocol in the `pioSendPci` parameter passed to the `SCardTransmit()` function.

```
    LONG PCSC_Exchange(LPCBYTE pbSendBuffer ,
                       DWORD cbSendLength ,
                       LPBYTE pbRecvBuffer,
                       LPDWORD pcbRecvLength )
    {
       LPCSCARD_IO_REQUEST ioRequest;
       LONG lRetValue;

       switch (m_dwActiveProtocol)
       {
          case SCARD_PROTOCOL_T0:
             ioRequest = SCARD_PCI_T0;
             break;

          case SCARD_PROTOCOL_T1:
             ioRequest = SCARD_PCI_T1;
             break;
```

```
            default:
                ioRequest = SCARD_PCI_RAW;
                break;
        }

        *pcbRecvLength = RcvLenMax;

        // APDU exchange.
        lRetValue = SCardTransmit(
                    m_hCard,            // Card handle.
                    ioRequest,          // Pointer to the send protocol header.
                    pbSendBuffer,       // Send buffer.
                    cbSendLength,       // Send buffer length.
                    NULL,               // Pointer to the rec. protocol header.
                    pbRecvBuffer,       // Receive buffer.
                    pcbRecvLength);     // Receive buffer length.

        PCSC_STATUS(lRetValue,"SCardTransmit");

        printHexString("\n --> C-Apdu: 0x",(LPBYTE)pbSendBuffer, cbSendLength);
        printHexString(" <-- R-Apdu: 0x",pbRecvBuffer, *pcbRecvLength);
        printf(" SW1SW2: 0x%02X%02X\n\n",pbRecvBuffer[*pcbRecvLength - 2],
        pbRecvBuffer[*pcbRecvLength - 1]);

        return lRetValue;
    }
```

The PCSC_Disconnect() function deactivates the card and closes all logical connections.

```
    LONG PCSC_Disconnect(void)
    {
        long lRetValue;

        // Terminates the Smartcard connection.
        lRetValue = SCardDisconnect(
                    m_hCard,                    // Card handle.
                    SCARD_UNPOWER_CARD);        // Action to take on the card
        // in the connected reader on close.
        PCSC_STATUS(lRetValue,"SCardDisconnect");

        // Release the Resource Manager Context.
        lRetValue = SCardReleaseContext(m_hContext);
        m_hContext = 0;

        return lRetValue;
    }
```

The `PCSC_ GetAtsString()` returns the ATR.

```c
LONG PCSC_GetAtrString(LPBYTE atr, LPINT atrLen)
{
    LPBYTE pbAttr = NULL;
    DWORD cByte = SCARD_AUTOALLOCATE;
    LONG lRetValue;

    // Gets the current reader attributes for the given handle.
    lRetValue = SCardGetAttrib(m_hCard,     // Card handle.
                SCARD_ATTR_ATR_STRING,  // Attribute identifier.
                (LPBYTE)&pbAttr,        // Attribute buffer.
                &cByte);                // Returned attribute length.
    PCSC_ERROR(lRetValue,"SCardGetAttrib (ATR_STRING)");

    printHexString("\n Atr: 0x",(LPBYTE)pbAttr, cByte);

    if(atr != NULL)
    {
        copyByte(atr, pbAttr, cByte);
        *atrLen = cByte;
    }

    // Releases memory that has been returned from the resource manager
    // using the SCARD_AUTOALLOCATE length designator.
    lRetValue = SCardFreeMemory( m_hContext, pbAttr );
    PCSC_ERROR(lRetValue, "SCardFreeMemory");

    return lRetValue;
}
```

The main program is shown in the `main.c` file. The example outputs a contactless card's serial number to the console. The card's serial number is retrieved using the `GetData` pseudo-APDU command (see Section 11.1.2.1). The same example can safely be tried with a contact-type reader. In this case, the `GetData` response APDU is `0x9000`.

The `getClessCardType()` auxiliary function gets the card type of a contactless memory card by doing a simple byte-wise comparison of the reader-generated pseudo-ATR with the known pseudo-ATRs of MIFARE 1K, MIFARE 4K and MIFARE Ultralight cards.

Listing 11.2. *Excerpt from pcsc.c*

```c
#include <conio.h>
#include "stdio.h"
#include "pcsc.h"
#include "util.h"
#include "clessCardType.h"

int main(int argc, char* argv[])
{
    // Get Data: CLA = 0xFF, INS = 0xCA, P1 = 0x00, P2 = 0x00, Le = 0x00
    BYTE baCmdApduGetData[] = { 0xFF, 0xCA, 0x00, 0x00, 0x00};
```

```c
    BYTE baResponseApdu[300];
    DWORD lResponseApduLen = 0;
    BYTE atr[40];
    INT atrLength;
    LONG lRetValue;

    system("cls");
    printf("PCSC API Example - Read Card Serial Number (UID)...\n\n");

    lRetValue = PCSC_Connect(NULL );
    PCSC_EXIT_ON_ERROR(lRetValue);

    lRetValue = PCSC_ActivateCard();
    PCSC_EXIT_ON_ERROR(lRetValue);

    lRetValue = PCSC_GetAtrString(atr, &atrLength);
    PCSC_EXIT_ON_ERROR(lRetValue);

    // Send pseudo APDU to retrieve the card serical number (UID)
    lRetValue = PCSC_Exchange(baCmdApduGetData, (DWORD)
                                        sizeof(baCmdApduGetData),
                              baResponseApdu, &lResponseApduLen);

   if( baResponseApdu[lResponseApduLen - 2] == 0x90 &&  // Verify if
                                                        status word
        baResponseApdu[lResponseApduLen - 1] == 0x00)   // SW1SW2 is
                                                        equal 0x9000.
    {
        // Contactless card detected.
        // Retrieve the card serical number (UID) form the response APDU.
        printHexString("Card Serial Number (UID): 0x", baResponseApdu,
                                                   lResponseApduLen - 2);

        if( getClessCardType(atr) == Mifare1K)
        {
            printf("Card Type: MIFARE Classic 1K");
        }
        else if( getClessCardType(atr) == Mifare4K)
        {
            printf("Card Type: MIFARE Classic 4K");
        }
        else if( getClessCardType(atr) == MifareUL)
        {
            printf("Card Type: MIFARE Ultralight");
        }
    }
    PCSC_Disconnect();

    printf("\n");
    return 0;
}
```

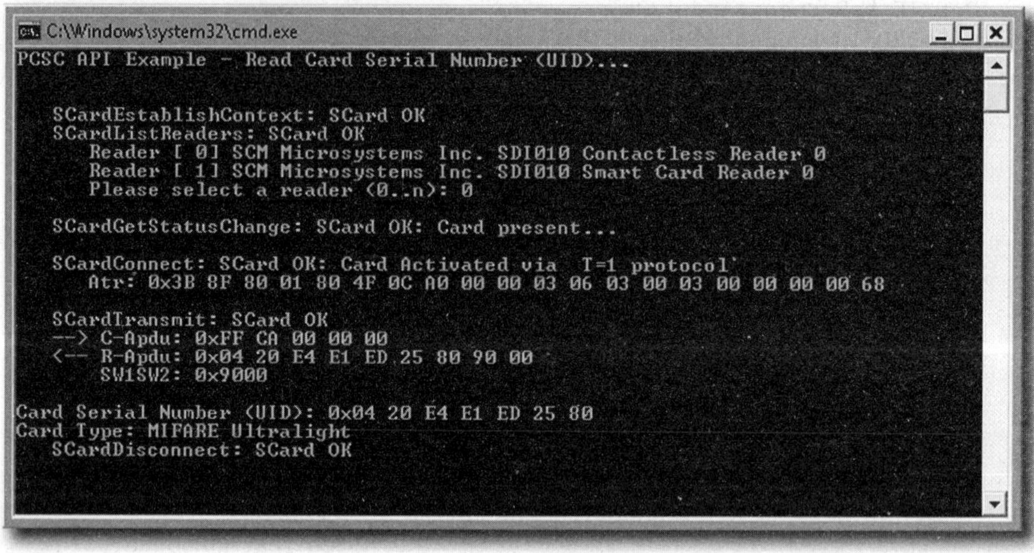

11 PC/SC READERS

```
C:\Windows\system32\cmd.exe
PCSC API Example - Read Card Serial Number (UID)...

    SCardEstablishContext: SCard OK
    SCardListReaders: SCard OK
        Reader [ 0] SCM Microsystems Inc. SDI010 Contactless Reader 0
        Reader [ 1] SCM Microsystems Inc. SDI010 Smart Card Reader 0
        Please select a reader (0..n): 0

    SCardGetStatusChange: SCard OK: Card present...

    SCardConnect: SCard OK: Card Activated via T=1 protocol
        Atr: 0x3B 8F 80 01 80 4F 0C A0 00 00 03 06 03 00 03 00 00 00 00 68

    SCardTransmit: SCard OK
    --> C-Apdu: 0xFF CA 00 00 00
    <-- R-Apdu: 0x04 20 E4 E1 ED 25 80 90 00
        SW1SW2: 0x9000

Card Serial Number (UID): 0x04 20 E4 E1 ED 25 80
Card Type: MIFARE Ultralight
    SCardDisconnect: SCard OK
```

Figure 11.9. Console output of the first PC/SC example.

11.3 Java and PC/SC

The Java programming language supports many ways of integrating PC/SC readers. The Open Card Framework (OCF) has been briefly introduced in Section 10.3 already and will not be discussed further. In the following section, the JPC/SC Java API and the Java™ SmartCard I/O will be discussed briefly.

11.3.1 JPC/SC Java API

JPC/SC (Java PC/SC) was implemented by IBM as part of the MUSCLE project (Movement for the Use of SmartCards in a Linux Environment). JPC/SC can be used in both Linux and Windows. This library consists largely of a Java WinSCard API wrapper, which the Java Native Interface (JNI) uses. The Java Native Interface provides the ability to integrate platform-specific functions (e.g. WinSCard API) into a Java library. However, such a Java program would not be platform-independent, as this would require the native library to be available on all platforms.

Listing 11.4. JPC/SC: Extract from Test.java.

```
/* Ask user to choose a reader, wait for card in reader, print its ATR,
 * connect to card, ask for apdu and send it. */
private static void interactivetest(){
    System.out.println("EstablishContext(): ...");
    Context ctx = new Context();
    ctx.EstablishContext(PCSC.SCOPE_SYSTEM, null, null);

    String[] sa = ctx.ListReaders();
```

```java
       if (sa.length == 0){
          System.out.println("No reader detected. " +
          " Make sure reader is avalaiable and start again.");

          System.exit(1);
       }
       System.out.println("Wait for card in a certain reader ...");
       System.out.println("Pick reader ...");
       String rn = Util.stdinPickReader(ctx);
       System.out.println("Please, insert card into reader " + rn + " ...");
       State state = null;
       do{
          state = ctx.WaitForCard(rn, PCSC.INFINITE); // block until card
                                                                is inserted
       }while(state == null);

       System.out.println("Card inserted into " + state.szReader);
       System.out.println("Card ATR is " + Apdu.ba2s(state.rgbAtr));
       System.out.println("Connect to card " + state.szReader);
       Card card = ctx.Connect(state.szReader, PCSC.SHARE_EXCLUSIVE,
                               PCSC.PROTOCOL_T0|PCSC.PROTOCOL_T1);
       while(true){
          String l = Util.readLine("Type APDU to send or q to leave: ");
          if (l.toUpperCase().equals("Q"))
             break;
          byte[] ba = null;
          try{
             ba = Apdu.s2ba(l);
          }catch(Exception e){
             System.out.println("invalid APDU: " + e.getMessage());
             continue;
          }

          try{
             byte[] response = card.Transmit(ba, 0, ba.length);
             System.out.println("Response: " + Apdu.ba2s(response));
          }catch(Exception e){
             System.out.println("sending APDU failed: " + e.getMessage());
             continue;
          }
       }
       card.Disconnect();
       ctx.ReleaseContext();
    }
```

The JPC/SC Package is open source and can be obtained for free from the MUSCLE website (http://www.musclecard.com/middle.html). Listing 11.4 demonstrates the use of the library. Upon closer inspection, there is a strong resemblance to the previous C example.

11.3.2 Java Smartcard I/O API

Since Java Version 1.6, the Smartcard I/O API has been available. This API supports communication with PC/SC readers and smartcards.

The following classes are included in the `javax.smartcardio` package: ATR, CommandAPDU, ResponseAPDU, TerminalFactory, CardTerminals, CardTerminal, Card, CardChannel, CardPermission, CardException, CardNotPresentException and TerminalFactorySpi. The documentation is available here: *http://download.oracle.com/javase/6/docs/jre/api/security/smartcardio/spec/javax/smartcardio/package-summary.html*.

Because the Smartcard I/O API has similar functions to the JPCSC library, only the API should be used in new projects. Listing 11.5 should show the general idea. In this example, the contactless card's serial number is retrieved with the help of the pseudo-APDU command, `GetData`, and is output to the console.

Listing 11.5. *SmartCardReader.java.*

```java
import java.util.List;
import javax.smartcardio.*;

public class SmartCardReader {
  public static void main(String[] args) {
    try{
      // show the list of available terminals
      TerminalFactory factory = TerminalFactory.getDefault();
      List<CardTerminal> terminals = factory.terminals().list();
      System.out.println("Terminals: " + terminals);
      CardTerminal terminal = terminals.get(0); // get the first terminal

      Card card = terminal.connect("T=1"); // establish a connection
      System.out.println("card: " + card); // with the card

      ATR atr = card.getATR();              // get the ATR
      byte[] baAtr = atr.getBytes();

      System.out.print("ATR = 0x");
      for(int i = 0; i < baAtr.length; i++ ){
        System.out.printf("%02X ",baAtr[i]);
      }

      // Get Data: CLA = 0xFF, INS = 0xCA, P1 = 0x00, P2 = 0x00, Le = 0x00
      byte[] cmdApduGetUid = new byte[]{ (byte)0xFF, (byte)0xCA,
                          (byte)0x00, (byte)0x00, (byte)0x00};
      CardChannel channel = card.getBasicChannel();
      ResponseAPDU respApdu = channel.transmit(
                          new CommandAPDU(cmdApduGetUid));
      if(respApdu.getSW1() == 0x90 && respApdu.getSW2() == 0x00){
        byte[] baCardUid = respApdu.getData();
        System.out.print("Card UID = 0x");
```

```
        for(int i = 0; i < baCardUid.length; i++ ){
            System.out.printf("%02X ", baCardUid [i]);
        }
    }
    card.disconnect(false);
} catch (CardException e) {
    e.printStackTrace();
}
      }
   }
}
```

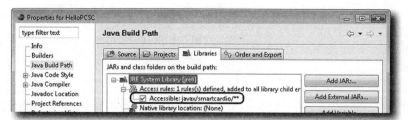

Figure 11.10.
In the Eclipse environment, access to the javax.* package must be shared.

11.4 The CSharpPCSC Wrapper for .NET

The .NET Framework class library also supports smartcards in Version 4.0. In this book, a CSharp PC/SC wrapper will be developed, and the main WinSCard API functions will be embedded in the class. In addition to the `WinSCard` class, the `PCSCReader` class offers a high-level API. This includes the `Connect()`, `ActivateCard()`, `Exchange()` and `Disconnect()` methods, as well as a range of useful properties.

The following C# program example can be compiled and tested with Microsoft Visual Studio C# Express Edition, SharpDevelop or even with Smart Card Magic.NET. Developing the program in Smart Card Magic.NET is done in the same manner as presented in Section 7.6.2. The utility classes presented in Chapter 8, `CmdApdu`, `RespApdu`, `ByteArray`, `HexEncoding`, `HexFormatting`, `HexUtil`, `MifareClassicKeys`, `MifareClassicUtil`, `TripleDESCrypto` and `Rng`, are used again in the following examples.

11.4.1 How Does One Create an API Wrapper?

The .NET Framework's Platform Invocation Service allows the option to call unmanaged functions from any .NET language. For this purpose, the functions must be encapsulated in a native DLL (dynamic link library). Using Platform Invoke (P/Invoke), all WinSCard API functions are available in the C# programming language.

An external function is declared with the `DLLImport` attribute, in which the constructor contains the DLL's file name, excluding extension. The `DLLImport` attribute has optional parameters for controlling the function call. In the source code, the API method is defined below the attribute, using the `static` and `extern` modifiers. An external method is always declared within a class. There are often data types in API function declarations that have no direct equivalent in C#. If such a function is called from C#, the C# data type will be converted into an API data type – a process is called marshalling. Correctly declaring a C# data type for a Windows API function is surely the most complex task. *http://www.pinvoke.net*, a Wiki website is very helpful for finding the correct API declaration. The external `SCardTransmit()` function is included in C# as follows:

```
[DllImport( "WinScard.dll" )]
public static extern int SCardTransmit( IntPtr hCard,
                                        ref ScardIoRequest pioSendPci,
                                        byte[] pbSendBuffer,
                                        int cbSendLength,
                                        IntPtr pioRecvPci,
                                        byte[] pbRecvBuffer,
                                        ref int pcbRecvLength );
```

11.4.2 The `WinSCard` Class

The `WinSCard` class, in the `GS.SCard` namespace, encapsulates the most commonly used WinSCard API functions in the form of a class. Functions not yet implemented in the WinSCard API can be added to the open source library if needed. All constants for the `WinSCARD` API from the `winscard.h` header file, as well as the error codes from `SCardErr.h` are declared in several enumerations (namespaces `GS.SCard.Const` and `GS.SCard.ReturnCodes`). For example, the WinSCard API function, `SCardEstablishContext()`, expects one of three values as its first parameter. The three valid constants, SCARD_SCOPE_USER, SCARD_SCOPE_TERMINAL and SCARD_SCOPE_SYSTEM are declared in the `winscard.h` header file. The corresponding enumeration looks as follows:

```
public enum SCARD_SCOPE : uint
{
    User = 0,
    Terminal = 1,
    System = 2
}
```

The C# enumerations always use the same names as the corresponding C constants. The use of enumerations instead of simple C# constants has the advantage of only allowing valid constant values to be passed to a method.

```
public void ExtablishContext(SCARD_SCOPE dwScope)
```

The following tables show the main properties and methods of the `SCard` class. More information can be found in the `CSharpPCSC.chm` help file. This is opened either by double-clicking on the file, or in the Smart Card Magic.NET menu option, *Help → PC/SC*.

Table 11.10. The properties of the WinSCard class.

Property	R/W	Description
`byte[] Atr`	R	Returns the ATR of a contact-type smartcard or the pseudo-ATR of a contactless card.
`string AtrString`	R	The name of the active reader.
`string ConnectedReaderName`	R	The name of the active reader.
`string[] ReaderNames`	R	Returns all of the available smartcard readers on the system, in the form of an array of strings.
`string ReaderName`	R	The reader name of the previously selected reader.
`bool IsRMContextEstablished`	R	Returns true if an active connection to the resource manager has been established.
`bool IsCardContextEstablished`	R	Returns true if an active connection to a smartcard has been established.
`bool IsCardPresent`	R	Returns true if a card is currently connected to the reader.
`bool TraceSCard`	R	Enables or disables the SCard trace output.

Table 11.12. Main WinSCard API functions.

Function	Description
`void EstablishContext( SCARD_SCOPE dwScope)`	Creates a connection to the resource manager context. Valid values of the optional `dwScope` parameter are: `SCARD_SCOPE.User` (default), `SCARD_SCOPE.Terminal` and `SCARD_SCOPE.System`.
`void EstablishContext()`	
`string[] ListReaders()`	Returns the available smartcard readers in the system in the form of an array of strings. This method has the same purpose as the `ReaderNames` property.
`void AddReaders( ComboBox comboBox)`	Adds all available smartcard readers in the system to a ComboBox.
`void Connect(string szReader, SCARD_SHARE_MODE dwShareMode, SCARD_PROTOCOL dwPrefProtocol)`	Creates a logical connection to the card after successful card activation. The ATR is then displayed in the Trace Window. `szReader` parameter: reader name. Valid values for optional parameter `dwShareMode`: `SCARD_SHARE_MODE.Shared` (default), `SCARD_SHARE_MODE.Exclusive` and

Function (continued)	Description (continued)
`void Connect( string szReader)`	`SCARD_SHARE_MODE.Direct`. Valid values for optional parameter `dwPrefProtocol`: `SCARD_PROTOCOL.Tx` (default), `SCARD_PROTOCOL.T0`, `SCARD_PROTOCOL.T1`, `SCARD_PROTOCOL.Raw`, `SCARD_PROTOCOL.Default` and `SCARD_PROTOCOL.Undefined`.
`void WaitForCardPresent()`	Program execution is blocked until a card is available at the reader. This method repeatedly calls `SCardGetStatusChange()`.
`void WaitForCardPresent( string szReader)`	
`void Reconnect()`	This function allows the forcing of a card power-on reset. Valid values for the optional parameter, `disconnectAction`, are: `SCARD_DISCONNECT.Unpower` (default), `SCARD_DISCONNECT.Leave`, `SCARD_DISCONNECT.Reset`, `SCARD_DISCONNECT.Unpower` and `SCARD_DISCONNECT.Eject`.
`void Reconnect( SCARD_DISCONNECT disconnectAction)`	
`void Transmit(byte[] sendBuffer, int sendLength, byte[] responseBuffer, ref int responseLength)`	Sends a command APDU to the card and returns the response APDU. The size of the read buffer is returned in the `responseLength` parameter. In the same parameter, the number of bytes received from the card is returned.
`void GetAttrib( SCARD_ATTR AttrId, byte[] responseBuffer, ref int responseLength)`	With the help of the `SCardGetAttrib()` function, a range of different reader properties can be read. All WinSCard API `AttrId` values are defined in the `SCARD_ATTR` enumeration.
`void GetAttrib( SCARD_ATTR AttrId)`	This variant of `GetAttrib()` merely outputs the `responseBuffer` array to the Trace Window.
`bool GetCardPresentState( string szReader)`	Returns true if a contact-type card is in the reader or a contactless card is in the field of specified reader `szReader`.
`void WaitForCardRemoval()`	As long as the reader is available, program execution is blocked. This method repeatedly calls `sCardGetStatusChange()`.
`void WaitForCardRemoval( string szReader)`	

Function (continued)	Description (continued)
`void Disconnect()`	Closes the logical connection to a card. The `disposition` parameter determines whether the card should first be electrically deactivated. Valid values for this optional parameter are:
`void Disconnect(SCARD_DISCONNECT disposition)`	`SCARD_DISCONNECT.Unpower` (default), `SCARD_DISCONNECT.Leave`, `SCARD_DISCONNECT.Reset` and `SCARD_DISCONNECT.Eject`.
`void ReleaseContext()`	Closes the connection to the resource manager.

11.4.3 The `PCSCReader` Class

The `PCSCReader` class, from the `GS.PCSC` namespace, offers the `Connect()`, `ActivateCard()`, `Exchange()` and `Disconnect()` methods. All of these methods are partially overloaded, making it easier to program a PC/SC-compliant reader.

Like the Elektor RFID Reader, the embedded class *Mifare* implements the two interfaces called IMifareUltraLight and IMifareClassic. The class *Mifare* is extensively covered in Chapter 8.

Table 11.14. Properties of the PCSCReader class.

Property	R/W	Description
`SCard`	R/W	With the `SCard` property, one has full access to all properties and methods from the `WinSCard` class.

Table 11.15. Methods in the PCSCReader class.

Method	Description
`void Connect()`	Calls the `WinSCard()`, `EstablishContext()` and `WinSCard.ListReaders()` methods. Should several readers be available on the PC, the user is prompted to select one of them via the console.
`void Connect(SCARD_SCOPE dwScope)`	
`void Connect(string szReader)`	By using the `szReader` parameter, a specific reader can be selected and the user-selection prompt suppressed.
`void Connect(string szReader, SCARD_SCOPE dwScope)`	

11 PC/SC READERS

Method (continued)	Description (continued)
void ActivateCard()	This ActivateCard() method first calls the WinSCard.WaitForCardPresent() method. Program execution is blocked until a card is available in or at the selected card reader. When a card is detected, it is activated using the WinSCard.Connect() method and, finally, the ATR is output to the Trace Window.
void ActivateCard(SCARD_PROTOCOL dwPrefProtocol)	
void ActivateCard(SCARD_SHARE_MODE dwShareMode, SCARD_PROTOCOL dwPrefProtocol)	
void Disconnect()	Calls the WinSCard.Disconnect() and WinSCard.ReleaseContext() methods, thus freeing system resources.
void Disconnect(SCARD_DISCONNECT disposition)	

The Exchange() method is used to send and receive APDUs and is overloaded about 11 times. In Table 11.6, all variants used in the following examples are listed. The expected response from the card can be passed in the optional parameter, expectedSW1SW2. Should the card return a different status word, an ApduException is triggered. This feature has already been described in Section 8.8.4.

Table 11.16. Methods in the PCSCReader class.

Method	Description
void Exchange(byte[] sendBuffer, int sendLength, byte[] responseBuffer, ref int responseLength, ushort? expectedSW1SW2);	In the sendBuffer parameter, the C-APDU is passed as a byte array and a sufficiently large receive buffer is passed in the responseBuffer parameter. The receive buffer size is returned in the responseLength parameter. In the same parameter, the method also returns the number of bytes received from the card.
RespApdu Exchange(string cmdApdu);	The C-APDU can be simply passed as a hexadecimal string. These two variants of the Exchange() method return an object of type RespApdu.
RespApdu Exchange(string cmdApdu, ushort? expectedSW1SW2);	
RespApdu Exchange(CmdApdu cmdApdu);	In these two variants, the C-APDU is passed as CmdApdu object. In this case, the Exchange() method also returns an object of type RespApdu.
RespApdu Exchange(CmdApdu cmdApdu, ushort? expectedSW1SW2);	

446

11.4.4 Program Examples

The following examples cover the essential aspects of programming contactless PC/SC readers, and also show the correct use of the CSharpPCSC library. All examples can be ported with relatively little effort into the C and Java programming languages. The framework presented in Sections 11.2 and 11.3 are used for this purpose. Additionally, GS.CSharpPCSC.dll can be used in any .NET language without any additional porting.

11.4.4.1 "Hello Contactless Card"

In these two examples, which also retrieve the contactless card's serial number using the GetData pseudo-APDU command, the most important methods of the WinSCard and PCSCReader classes are presented.

Listing 11.6. WinScardExample.cs.

```csharp
using System;
using GS.PCSC;
using GS.SCard;
using GS.SCard.Const;

namespace GS.PCSCExample
{
    public class WinScardExample
    {
        public static void Main(MainForm script, string[] args)
        {
            WinSCard scard = new WinSCard();

            try
            {
                scard.EstablishContext();
                scard.ListReaders();
                Console.WriteLine( "Available PC/SC Readers:" );

                for (int i = 0; i < scard.ReaderNames.Length; i++)
                {
                    Console.WriteLine( string.Format( " Reader {0}: {1}", i,
                                        scard.ReaderNames[i] ) );
                }
                Console.Write( "Please select a reader (0...n): " );
                scard.Connect( scard.ReaderNames[int.Parse( Console.ReadLine() )] );
                Console.WriteLine( "ATR: 0x" + scard.AtrString );
                script.TraceWriteLine("Get Card UID ...");
                byte[] cmdApdu = { 0xFF, 0xCA, 0x00, 0x00, 00};
                byte[] respApdu = new byte[256];
                int respLength = respApdu.Length;
                scard.Transmit(cmdApdu, cmdApdu.Length, respApdu, ref respLength);
            }
            catch (WinSCardException ex)
```

11 PC/SC READERS

```
         {
            Console.WriteLine( ex.WinSCardFunctionName + " 0x" +
            ex.Status.ToString( "X08" ) + " " + ex.Message );
         }
         finally
         {
            scard.Disconnect();
            scard.ReleaseContext();
         }
      }
   }
}
```

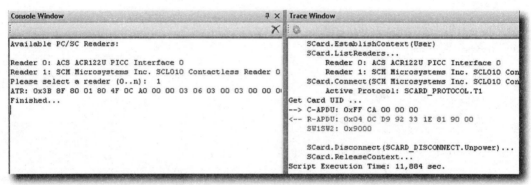

Figure 11.11. Example: WinScardExample.cs – Console and Trace outputs.

Using the PCSCReader class achieves the same thing with much less source code.

Listing 11.7. PCSCReaderExample.cs.

```
using System;
using GS.Apdu;
using GS.PCSC;
using GS.SCard;
using GS.SCard.Const;

namespace GS.PCSCExample
{
   public class PCSCReaderExample
   {
      public static void Main(MainForm script, string[] args)
      {
         PCSCReader reader = new PCSCReader();

         try
         {
            reader.Connect();
            reader.ActivateCard();
```

```
          RespApdu cardResponse = reader.Exchange("FF CA 00 00 00");
        }
        catch (WinSCardException ex)
        {
          Console.WriteLine( ex.WinSCardFunctionName + " Error 0x" +
                ex.Status.ToString( "X08" ) + ": " + ex.Message );
        }
        catch (Exception ex)
        {
          Console.WriteLine( ex.Message );
        }
        finally
        {
          reader.Disconnect();
        }
      }
    }
  }
```

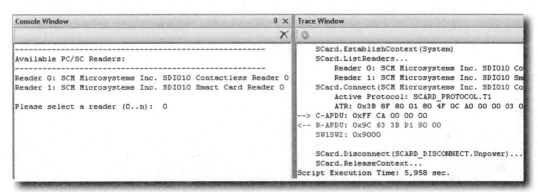

Figure 11.12. Example: PCSCReaderExample.cs – Console and Trace outputs.

In Figure 11.12, a dual-interface reader was used, so the `SCardListReaders()` function returned two strings. Therefore, the console prompts the user to select one of them. This manual selection is especially time consuming and annoying during program development. This problem is easily remedied. One need only pass the reader name as a string to the `Connect()` method. This is most easily copied from the console output.

```
    reader.Connect("SCM Microsystems Inc. SDI010 Contactless Reader 0");
```

11.4.4.2 Determine All Installed PC/SC Drivers

In the next example, a list of all installed PC/SC readers is generated. Up to now, the `SCardListReaders()` WinSCARD API function has been used to get a list of available (i.e. connected to the PC) smartcard readers. In order to retrieve a list of all installed PC/SC readers (drivers), one merely has to pass a null value to the `hContext` function

11 PC/SC READERS

parameter, instead of a reference to the resource manager (handle). For good reasons, all handles are declared as private in the CSharp wrapper and are thus protected from external access. Therefore, we have no way to set this value directly. However, the resource manager handle is set to null when the object is created or when the `ReleaseContext()` method is called. In the simplest implementation, it is sufficient to call the `ListReaders()` method. If, instead, the `EstablishContext()` method is called before `ListReaders()`, the list of readers connected to the PC can be retrieved.

Listing 11.8. *WinSCardListInstalledReads.cs.*

```
using System;
using GS.PCSC;
using GS.SCard;
using GS.SCard.Const;

namespace GS.PCSCExample
{
    public class ListInstalledReaders
    {
        public static void Main(MainForm script, string[] args)
        {
            try
            {
                WinSCard scard = new WinSCard();
                scard.ListReaders();
            }
```

Figure 11.13. *The list of all installed PC/SC readers is output to the Trace Window.*

```
Trace Window
SCard.ListReaders...
    Reader 0: ACS ACR122 0
    Reader 1: ACS ACR122U PICC Interface 0
    Reader 2: ACS ACR128U ICC Interface 0
    Reader 3: ACS ACR128U PICC Interface 0
    Reader 4: ACS ACR128U SAM Interface 0
    Reader 5: ARYGON CL Reader  0
    Reader 6: ARYGON CL Reader 00000000000000C3 0
    Reader 7: ARYGON CL Reader 0000000001000234 0
    Reader 8: ARYGON CNT Reader  0 0
    Reader 9: ARYGON CNT Reader  1 0
    Reader 10: Broadcom Corp Contacted SmartCard 0
    Reader 11: Broadcom Corp Contactless SmartCard 0
    Reader 12: Duali DE-620 Contact Reader 0
    Reader 13: Duali DE-620 Contactless Reader 0
    Reader 14: FEIG ELECTRONIC GmbH ID CPR40.xx-U Slot:CL 388757901
    Reader 15: FEIG ELECTRONIC GmbH ID CPR40.xx-U Slot:SC1 388757901
    Reader 16: FEIG ELECTRONIC GmbH ID CPR40.xx-U Slot:SC2 388757901
    Reader 17: Gemalto Prox-DU Contact_10500977 0
    Reader 18: Gemalto Prox-DU Contactless_10500977 0
    Reader 19: NXP Pegoda 2 X 0
    Reader 20: NXP Pegoda N CL 0 0
    Reader 21: NXP Pegoda X 0 0
    Reader 22: OMNIKEY AG Smart Card Reader USB 0
    Reader 23: OMNIKEY CardMan 5x21 0
    Reader 24: OMNIKEY CardMan 5x21 1
    Reader 25: OMNIKEY CardMan 5x21-CL 0
    Reader 26: OMNIKEY CardMan 5x21-CL 1
    Reader 27: OMNIKEY Smart Card Reader USB 0
    Reader 28: Raisonance ContactLAB 0
    Reader 29: REINER SCT cyberJack RFID basis 0
    Reader 30: SCM Microsystems Inc. SCL010 Contactless Reader 0
    Reader 31: SCM Microsystems Inc. SCL3711 reader & NFC device 0
    Reader 32: SCM Microsystems Inc. SDI010 Contactless Reader 0
    Reader 33: SCM Microsystems Inc. SDI010 Smart Card Reader 0
    Reader 34: SCM Microsystems Inc. SDI011G Contactless Reader 0
    Reader 35: SCM Microsystems Inc. SDI011G Contactless Reader 1
    Reader 36: SCM Microsystems Inc. SDI011G Smart Card Reader 0
    Reader 37: SCM Microsystems Inc. SDI011G Smart Card Reader 1
Script Execution Time: 0,048 sec.
```

```
                catch (WinSCardException ex)
                {
                }
                catch (Exception ex)
                {
                }
            }
        }
    }
```

11.4.4.3 Getting the Reader and Card Properties

Using the WinSCard API function, `SCardGetAttrib()`, one can determine a number of different reader and card properties. Selecting the properties is done using the numerical value in the `dwAttr` parameter. The required values are defined as constants in the WinSCard API header file, `WinSmCrd.h`.

The C# `SCard` class has a method of the same name, `GetAttrib()`. The most important constants are declared in the `SCARD_ATTR` enumeration. Not all PC/SC readers support the reading of all properties defined in the WinSCard API. The properties used in Listing 11.9 should be supported by all PC/SC readers. The `GetAttrib()` method in the `SCard` class, is overloaded twice. In this case, only one value is received from the `SCARD_ATTR` enumeration. In this way, the reader response (see Figure 11.14) cannot be used in the program, but is output to the Trace Window. The second variant of the `GetAttrib()` method allows the value to be used further in the program.

Listing 11.9. *PCSCGetAttribute.cs.*

```
    using System;
    using GS.Apdu;
    using GS.Util.ByteArray;
    using GS.PCSC;
    using GS.SCard;
    using GS.SCard.Const;

    namespace GS.PCSCGetAttribute
    {
        public class PCSCExample
        {
            public static void Main(MainForm script, string[] args)
            {
                PCSCReader reader = new PCSCReader();

                try
                {
                    reader.Connect();
                    reader.ActivateCard();
                    reader.SCard.GetAttrib(SCARD_ATTR.ATR_STRING);
```

```
                reader.SCard.GetAttrib(SCARD_ATTR.DEVICE_SYSTEM_NAME_A);
                reader.SCard.GetAttrib(SCARD_ATTR.VENDOR_IFD_VERSION);
                byte[] baRecBuffer = new byte[256];
                int recLength = baRecBuffer.Length;
                reader.SCard.GetAttrib(SCARD_ATTR.MAX_DATA_RATE,
                    baRecBuffer, ref recLength );
            }
            catch (WinSCardException ex)
            {
                Console.WriteLine( ex.WinSCardFunctionName + " Error 0x" +
                       ex.Status.ToString( "X08" ) + ": " + ex.Message );
            }
            catch (Exception ex)
            {
                Console.WriteLine( ex.Message );
            }
            finally
            {
                reader.Disconnect();
            }
        }
    }
}
```

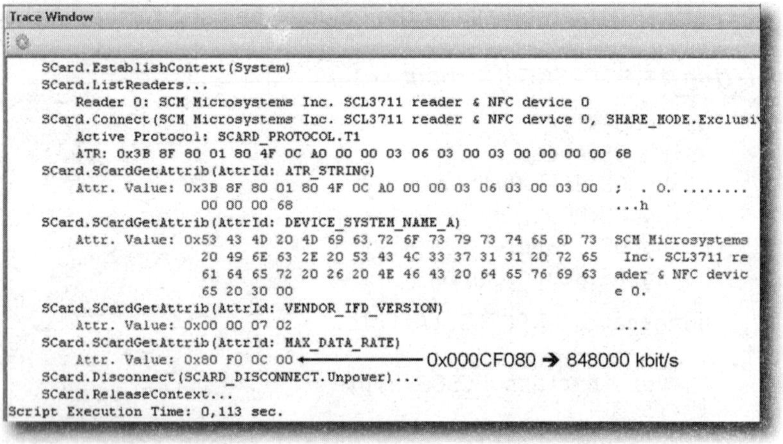

Figure 11.14.
When interpreting numerical values, the byte order must be considered.

11.4.4.4 Testing the Reading Range

In Terminal mode, the Elektor RFID Reader displays a card UID on the LCD. Thus, one can easily see at what distance the reader is able to read the card. The next example is similar in function to the Elektor RFID Reader program in Listing 8.5 (`GetReading-Distance.cs`), making it possible to determine the reading range of any contactless reader.

After the card is activated for the first time, the following sequence is repeated in an endless loop: the `SCard.WaitForCardPresent()` method blocks the application until a card is present; `SCard.Reconnect()` ensures that the card is activated. The new activation is necessary so that the card can be moved an arbitrary distance from the reader and returned into the reader field. It's obviously also possible to switch to another card or even card type.

As soon as the card is activated using `SCard.Reconnect()`, the `GetData` C-APDU retrieves the card UID and outputs it to the console. In addition, a beep sounds. `SCard.WaitForCardRemoval()` blocks the application until the same card is found in the reader's field. Should the card be removed, the text, "`No card present…`", is written to the console, and the entire sequence is repeated.

Listing 11.10. *PCSCGetAttribute.cs.*

```
using System;
using GS.Util.Hex;
using GS.Apdu;
using GS.PCSC;
using GS.SCard;
using GS.SCard.Const;

namespace GS.PCSCExample
{
    public class PCSCGetReadingDistance
    {
        public static void Main(MainForm script, string[] args)
        {
            PCSCReader reader = new PCSCReader();

            try
            {
                reader.Connect();
                Console.Clear();
                Console.WriteLine("No card present...");
                reader.ActivateCard();

                while(true)
                {
    reader.SCard.WaitForCardPresent();
    reader.SCard.Reconnect(SCARD_DISCONNECT.Unpower);
    Console.Clear();
    RespApdu respApdu = reader.Exchange("FF CA 00 00 00");// Get UID

                    if( respApdu.SW1SW2 == 0x9000)
                    {
                        Console.WriteLine("UID = 0x" +
                        HexFormatting.ToHexString(respApdu.Data, true));
```

```
                    Console.Beep();
                }
                else
                {
                    Console.WriteLine("Please use a PC/SC2.01 compliant
contactless Reader!");
                }

            reader.SCard.WaitForCardRemoval();
            Console.Clear();
            Console.WriteLine("No Card present...");
        }
        catch (WinSCardException ex)
        {
            Console.WriteLine( ex.WinSCardFunctionName + " Error 0x" +
                    ex.Status.ToString( "X08" ) + ": " + ex.Message );
        }
        catch (Exception ex)
        {
            Console.WriteLine( ex.Message );
        }
        finally
        {
            reader.Disconnect();
        }
    }
  }
}
```

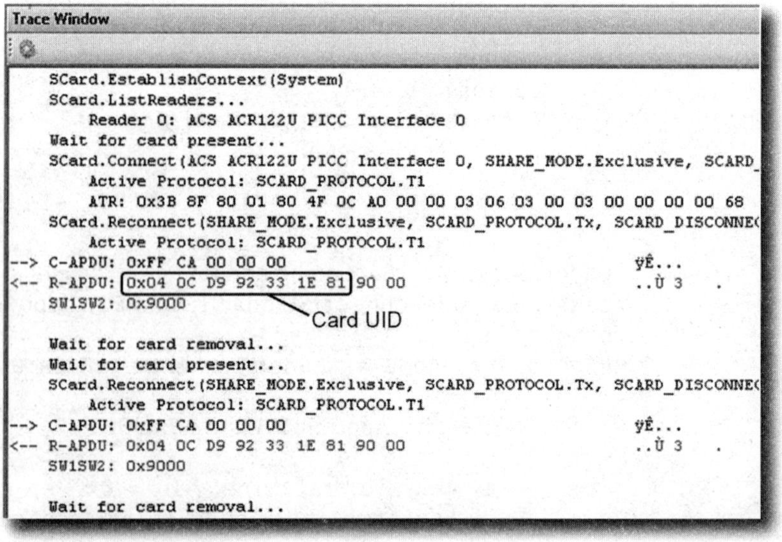

Figure 11.15. With double size UIDs, the GetData pseudo-APDU command does not return a Cascade Tag Byte (0x88).

```
Trace Window

    SCard.EstablishContext(System)
    SCard.ListReaders...
        Reader 0: ACS ACR122U PICC Interface 0
    Wait for card present...
    SCard.Connect(ACS ACR122U PICC Interface 0, SHARE_MODE.Exclusive, SCARD_
        Active Protocol: SCARD_PROTOCOL.T1
        ATR: 0x3B 8F 80 01 80 4F 0C A0 00 00 03 06 03 00 03 00 00 00 00 68
    SCard.Reconnect(SHARE_MODE.Exclusive, SCARD_PROTOCOL.Tx, SCARD_DISCONNEC
        Active Protocol: SCARD_PROTOCOL.T1
--> C-APDU: 0xFF CA 00 00 00                                        ÿÊ...
<-- R-APDU: [0x04 0C D9 92 33 1E 81] 90 00                          ..Ù 3   .
    SW1SW2: 0x9000
                                        Card UID
    Wait for card removal...
    Wait for card present...
    SCard.Reconnect(SHARE_MODE.Exclusive, SCARD_PROTOCOL.Tx, SCARD_DISCONNEC
        Active Protocol: SCARD_PROTOCOL.T1
--> C-APDU: 0xFF CA 00 00 00                                        ÿÊ...
<-- R-APDU: 0x04 0C D9 92 33 1E 81 90 00                            ..Ù 3   .
    SW1SW2: 0x9000

    Wait for card removal...
```

> **Note** Currently, all available PC/SC readers support communication with only one contactless card at a time.

11.4.4.5 Determining the Type of Contactless Memory Card

The preceding example (Listing 11.10) is very easy to extend to the detection of contactless memory cards. For this purpose, all that is necessary is to byte-wise compare the pseudo-ATR produced by the reader with the ATR values of all memory cards to be tested. The definition of a MIFARE 1K card is as follows:

```
static byte[] baAtrMifare1K = new byte[]{
                    0x3B, // Initial
                    0x8F, // T0
                    0x80, // TD1
                    0x01, // TD2
                    0x80, // Category Indicator
                    0x4F, // Appl. Id. Precence Indicator
                    0x0C, // Tag Length
                    0xA0,0x00,0x00,0x03,0x06, // AID
                    0x03, // SS
                    0x00,0x01, // NN - Mifare 1K
                    0x00,0x00,0x00,0x00, // r.f.u.
                    0x6A }; // TCK
```

The ATRs for the MIFARE 4K and the MIFARE Ultralight are also defined in the `CLess-CardTypeIdentification.cs` source file. Byte-by-byte comparison of the result is most easily achieved using the `ByteArray.DataAreEqual()` static method in the `GS.Util.ByteArray` namespace.

Listing 11.11. Extract from PCSCClessCardTypeIdentification.cs.

```
    RespApdu respApdu = reader.Exchange("FF CA 00 00 00"); // Get Card UID

    if( respApdu.SW1SW2 == 0x9000)
    {
       if(ByteArray.DataAreEqual( reader.SCard.Atr , baAtrMifare1K ))
       {
          Console.WriteLine("Card Type: MIFARE 1K");
       }
       else if(ByteArray.DataAreEqual( reader.SCard.Atr , baAtrMifare4K ))
       {
          Console.WriteLine("Card Type: MIFARE 4K");
       }
       ...
```

11.4.4.6 MIFARE Classic 1K/4K and MIFARE Ultralight

The PC/SC pseudo-APDUs, `Get Data`, `Load Keys`, `General Authenticate`, `Verify`, `Read Binary` and `Update Binary` were presented in Section 11.1.2.1. These APDUs enable the reading and writing of MIFARE Ultralight and MIFARE Classic 1K/4K cards.

The following example implements the `Authentication`, `Read` and `Write` MIFARE commands.

The declarations of the `ClassicAuth()`, `Read()` and `Write()` methods in Listing 11.12 are identical to the corresponding methods in the MifareClassic interface, as well as to the embedded class *Mifare*.

Listing 11.12. *PCSCMIFAREExample.cs.*

```
using System;
using GS.Util.Hex;
using GS.ISO14443_Reader;
using GS.Apdu;
using GS.PCSC;
using GS.SCard;
using GS.SCard.Const;

namespace GS.PCSCExample
{
    public class PCSCMifareClassic
    {
        static MainForm script;
        PCSCReader reader;
        public PCSCMifareClassic()
        {
            reader = new PCSCReader();
        }
```

Before an Auth. Key A or Auth. Key B command can be sent to the card, it is necessary to load the MIFARE key into the reader's memory using the `Load Key` pseudo-APDU. Depending on the reader used and its IC, the MIFARE key may be stored in either non-volatile or volatile memory. Basically all readers support writing to a volatile memory, i.e. RAM.

The `Load Key` command, however, is not implemented in the same way by all manufacturers; the encoding of the APDU `Load Key` commands differs slightly. In the source code, there are two variants, which have been tested with PC/SC readers from manufacturers Advanced Card Systems, Ltd., Gemalto, Omnikey (HID Global) and SCM Microsystems.

The actual authentication is performed with the pseudo-APDU command, `General Authenticate`. Depending on the content of the C-APDU, the reader firmware will send either Auth. Key A or Auth. Key B to the card. Upon successful authentication, the reader returns the status word `0x9000`. In every other case, the `Exchange()` method throws an exception for this example. Whenever it is called, the `ClassicAuth()` method sends the key in the `key` parameter to the reader. If the key is stored in RAM, as in this example, the implementation should present no problems. However, if non-volatile memory (EEPROM / Flash) is used to store one or more keys, one must bear in mind that this memory can only withstand a limited number or write cycles.

```
int ClassicAuth(byte block, string key, MFKeyType mfKeyType)
{
    script.TraceWriteLine(string.Format(" Authenticate MIFARE
                                        Classic Block {0}", block));
    byte mfKeyValue;

    if(mfKeyType == MFKeyType.KeyA)
    {
        mfKeyValue = 0x60;
    }
    else
    {
        mfKeyValue = 0x61;
    }

    // Load the MIFARE key into the reader.
    // The key is stored in volatile memory.
    // Note: The command parameter are reader type specific.
    // Works with ACS, Gemalto and Omnikey readers.
    string cmdApduLoadKeys = string.Format("FF 82 20 {0:X2} 06"
                                            + key,mfKeyValue);

    // Works with SCM readers
    //string cmdApduLoadKeys = string.Format("FF 82 00 {0:X2} 06"
                                            + key, mfKeyValue);

    reader.Exchange(cmdApduLoadKeys, 0x9000);
    string cmdApduGeneralAuth = string.Format("FF 86 00 00 05 01
                                            00 {0:X2}{1:X2} 00",
                                            block, mfKeyValue);
    reader.Exchange(cmdApduGeneralAuth, 0x9000);

    return 0;
}
```

A MIFARE block is read using the `Read Binary` APDU.

```
int Read(byte block, out byte[] data)
{
    script.TraceWriteLine(string.Format(" Read MIFARE Block
                                        {0}",block));
    string cmdApduReadBinary = string.Format("FF B0 00 {0:X2}
                                            10", block);
    RespApdu respApdu = reader.Exchange(cmdApduReadBinary, 0x9000);
    data = respApdu.Data;
    return 0;
}
```

`Write Binary` is used to write MIFARE blocks.

```
public int Write(byte block, string data)
{
```

```
            return Write(block, HexEncoding.GetBytes(data));
        }

        public int Write(byte block, byte[] data)
        {
            script.TraceWriteLine(string.Format(" Write MIFARE Block
                                                          {0}",block));
            int len = 0;

            if (data != null)
            {
                if (data.Length <= 16)
                {
                    len = data.Length;
                }
                else
                {
                    len = 16;
                }
            }
            else
            {
                data = new byte[16];
            }

        byte[] baData = new byte[16];
        Array.Copy(data, 0, baData, 0, len);
        string sData = HexFormatting.ToHexString(baData);
        string cmdApduUpdateBinary = string.Format("FF D6 00 {0:X2} 10" +
                                                          sData, block);
            reader.Exchange(cmdApduUpdateBinary, 0x9000);

            return 0;
        }
        public static void Main(MainForm script, string[] args)
        {
            PCSCMifareClassic readerScript = new PCSCMifareClassic();
            PCSCMifareClassic.script = script;
            readerScript.Run();
        }
```

The main program tests the `ClassicAuth()`, `Read()` and `Write()` methods.

```
        public void Run()
        {
            try
            {
                reader.Connect();
                reader.ActivateCard();
                ClassicAuth(4, "FF FF FF FF FF FF", MFKeyType.KeyA);

                // Write the hex data 0x11 22 33 44 to MIFARE block 4
                Write(4, "11 22 33 44 ");
```

```csharp
            // Write the string "Hello MIFARE" to MIFARE block 5
            Write(5, "\"Hello MIFARE\"");
            byte[] data;
            MifareRead(4, out data); // Read MIFARE block 4
            MifareRead(5, out data); // Read MIFARE block 5
        }
        catch (WinSCardException ex)
        {
            Console.WriteLine( ex.WinSCardFunctionName + " Error 0x" +
                ex.Status.ToString( "X08" ) + ": " + ex.Message );
        }
        catch (Exception ex)
        {
            Console.WriteLine( ex.Message );
        }
        finally
        {
            reader.Disconnect();
        }
    }
  }
}
```

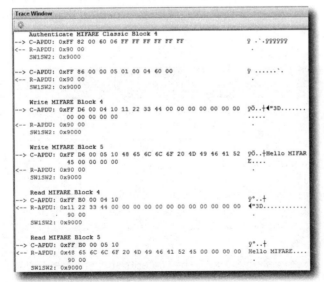

Figure 11.16.
Reading and writing a MIFARE 1K or 4K card is done using pseudo-APDUs.

11.4.4.7 MIFARE DESFire EV1

The MIFARE DESFire EV1 was presented in Section 8.7. The card's entire instruction set is encoded in the proprietary DESFire native command structure. Because communication with a PC/SC reader always takes place in the ISO/IEC 7816-3 APDU format, the card supports the ability to wrap the DESFire native commands in an APDU structure.

11 PC/SC READERS

Some PC/SC readers also feature a proprietary APDU command, which transfers data via the air interface in the native DESFire format. This has the advantage that the commands and card responses are generally shorter. In both cases, the T=CL protocol is used.

Conversion of DESFire native commands into the ISO 7816 APDU structure.
Figure 11.17 shows the principle of converting a DESFire native command into an ISO/IEC 7816-compliant APDU structure. The value of the CLA byte is always 0x90, indicating a vendor-specific command. The value of the INS byte is the same as the DESFire's command byte. P1, P2 and Le are always 0x00.The optional native DESFire parameter bytes and the data field are represented in the data field of the command APDU.

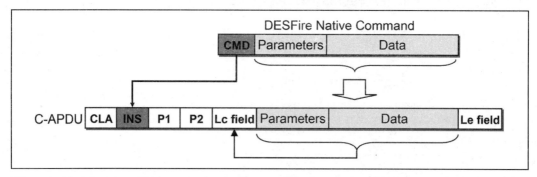

Figure 11.17. A handle contains a code number that the operating system generates.

Table 11.17. DESFire ISO/IEC 7816 wrapped command APDU.

CLA	INS	P1	P2	Lc	Data	Le
0x90	1.)	00	00	2.)	DESFire native command parameter + data	0x00

[1] INS: DESFire native command code,
[2] Le: Number of DESFire command parameter and data bytes.

Conversion of Native DESFire Responses into the ISO 7816 APDU Structure.
When a MIFARE DESFire EV1 card receives the first command in APDU format, it responds in APDU format. The data field in both formats contains the card data excluding status information.

The value of the first status byte, SW1, is always 0x91.The second status byte, SW2, holds the DESFire native status byte.

Table 11.18. DESFire ISO/IEC 7816 wrapped response APDU.

Data	SW1	SW2
The card's response data.	0x91	1.)

[1] SW2: DESFire native status byte.

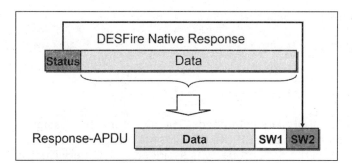

Figure 11.18.
DESFire ISO/IEC 7816
wrapped response APDU.

Creating a DESFire ISO/IEC 7816 Wrapper

In Listing 8.32, reading and writing a standard data file was presented. The following program example is functionally identical, and shows how DESFire native commands are converted into an APDU structure.

The easiest is to follow the example using a MIFARE DESFire EV1 card, which was already used for the program example in Section 8.7.6. Alternatively, the required DESFire application can be created with the `PCSCMifareDesFireCreateApplication.cs` script. This script is functionally equivalent to Listing 8.31 and is thus not printed to save space.

Listing 11.13. *PCSCMifareDesStdDataFileRdWr.cs.*

```
using System;
using GS.Util.Hex;
using GS.Apdu;
using GS.PCSC;
using GS.SCard;
using GS.SCard.Const;
namespace GS.PCSCExample
{
    public class PCSCMifareDesStdDataFileRdWr
    {
        public static void Main(MainForm script, string[] args)
        {
            PCSCReader reader = new PCSCReader();
```

In this example, the DESFire ISO/IEC 7816 wrapper is realized with the `CmdApdu` class. The `wrappedApud` object is used to send all commands to the card. Therefore, once the `wrappedApud` object has been created, all command APDU static values can be initialized as per Table 11.17.

```
            try
            {
                CmdApdu wrappedApdu = new CmdApdu();

                wrappedApdu.CLA = 0x90;
```

11 PC/SC READERS

```
                wrappedApdu.P1   = 0x00;
                wrappedApdu.P2   = 0x00;
                wrappedApdu.Le   = 0x00;

                reader.Connect();
                reader.ActivateCard();
```

The next two steps are identical for all DESFire commands. The INS property is assigned the value of the DESFire native command byte. If the DESFire native command has no data field (parameter and data), then the data property is assigned null. Otherwise, the Data property is assigned the value of the DESFire native command data field. The `Hex-Encoding.GetBytes()` method from the `GS.Util.Hex` namespace converts a string into a byte array. The string data is interpreted as hexadecimal numbers. The optional space characters are ignored during conversion.

```
                //*** Select Application ***
                wrappedApdu.INS = 0x5A; // DESFire Command Code:
                                                            Select Application
                wrappedApdu.Data = HexEncoding.GetBytes("33 22 11");
                                                    // AID = 0x112233
                script.TraceWriteLine("\n Select Application:
                                                    AID = 0x112233");
                reader.Exchange(wrappedApdu, 0x9100);

                //*** Write Data ***
                wrappedApdu.INS = 0x3D; // DESFire Command Code: Write Data
                wrappedApdu.Data = HexEncoding.GetBytes(
                            "01" + // DESFire Native File No. = 0x01
                            "00 00 00" + // File Offset
                            "05 00 00" + // Length = 5 Byte
                            "11 22 33 44 55");
                                        // Data = 0x1122334455
                script.TraceWriteLine(" Write Data: File No. = 0x01; " +
                                        "Offset = 0x00 00 00");
                reader.Exchange(wrappedApdu, 0x9100);

                //*** Read Data ***
                wrappedApdu.INS = 0xBD;
                                // DESFire Command Code: Write Data
                wrappedApdu.Data = HexEncoding.GetBytes("01" +
                                        // DESFire Native File No. = 0x01
                            "00 00 00" + // File Offset
                            "0A 00 00");
                                        // Length = 10 Bytes (0x00000A)
                script.TraceWriteLine(" Read Data: File No. = 0x01; " +
                                        "Offset = 0x00 00 00;
                                            Length = 10 Bytes");
                reader.Exchange(wrappedApdu, 0x9100);
            }
            catch (WinSCardException ex)
```

```
            {
                Console.WriteLine( ex.WinSCardFunctionName + " Error 0x" +
                    ex.Status.ToString( "X08" ) + ": " + ex.Message );
            }
            catch (Exception ex)
            {
                Console.WriteLine( ex.Message );
            }
            finally
            {
                reader.Disconnect();
            }
        }
    }
}
```

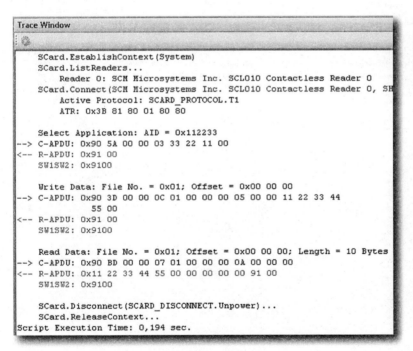

Figure 11.19.
Writing and reading data to and from a DESFire EV1 standard data file.

12
List of Abbreviations

2K3DES . . . . Two-Key-Triple-DES
3DES . . . . . . Triple-DES
3K3DES . . . . Three-Key-Triple-DES

AC CL n . . . . Anti-collision Cascade Level n, ISO/IEC 14443 Type A
AC . . . . Access Condition
AC . . . . Anti-collision
ACK . . . . Acknowledge
AES . . . . Advanced Encryption Standard
AFI. . . . . Application Family Identifier
AID . . . . Application Identifier
APDU . . . . . . Application Protocol Data Unit
API. . . . . Application Programming Interface
ASCII . . . . . . American Standard Code for Information Interchange
ASK . . . . Amplitude Shift Keying
ATQA . . . . . . Answer To Request, Type A
ATQB . . . . . . Answer To Request, Type B
ATR . . . . Answer To Reset
ATS . . . . Answer To Select
ATTRIB . . . . PICC selection command, Type B

BCC . . . . Block Check Character
BCD . . . . Binary Coded Digit
BPSK . . . . . . Binary Phase Shift Keying
BSI . . . . Bundesamt für Sicherheit in der Informationstechnik

C# . . . . CSharp
C-APDU . . . . Command APDU
CBC . . . . Cipher Block Chaining
CC . . . . Common Criteria
CCID . . . . . . Chip/Smartcard Interface Device
CID . . . . Card IDentifier
CIU . . . . Contactless Interface Unit
CLA . . . . Class Byte
CLK . . . . Clock
CL n . . . . Cascade Level n, ISO/IEC 14443 Type A
CMAC . . . . . . Cryptic Message Authentication Code
COS . . . . Card Operating System
CRC . . . . Cyclic Redundancy Check
CT . . . . Cascade Tag, Type A
CT-API . . . . . Card Terminal API

D	Divisor
DES	Data Encryption Standard
DF	Dedicated File
DIF	Dual Interface
DLL	Dynamic Link Libraries
DPA	Differential Power Analysis
DR	Divisor Receiver
DS	Divisor Send
DUT	Device Under Test
ECB	Electronic Code Book
ECC	Elliptic Curve Cryptosystems
EEPROM	Electrically-Erasable Programmable Read Only Memory
EF	Elementary Files
EGT	Extra Guard Time, Type B
EMC	Electromagnetic Compatibility
EMD	Electro Magnetic Disturbance
EMV	Europay, MasterCard, Visa
EOF	End Of Frame
etu	Elementary Time Unit
fc	Carrier Frequency
FCC	Federal Communications Commission
FCI	File Control Information
FDT	Frame Delay Time, Type A
FID	File Identifier
FIFO	First-In-First-Out
FO	Frame Option
fs	Subcarrier Frequency
FSC	Frame Size for proximity Card
FSCI	Frame Size for proximity Card Integer
FSD	Frame Size for proximity coupling Device
FSDI	Frame Size for proximity coupling Device Integer
FWT	Frame Waiting Time
FWTI	Frame Waiting Time Integer
GND	Ground
GSM	Global System for Mobile communications
HF	High Frequency
HLTA	Halt command, Type A
HLTB	Halt command, Type B
I/O	Input/Output
I2C	Inter-IC bus
I-Block	Information Block
ICAO	International Civil Aviation Organization
ID	IDentification number, Type A
INF	Information field
INS	Instruction byte
ISO	International Organization for Standardization
IV	Init Vector

JNI	Java Native Interface
Key A, B	MIFARE Classic Key A, B
Lc	Length command
Le	Length expected
LF	Low Frequency
LSB	Least-Significant Bit
MAC	Message Authentication Code
MAD	MIFARE Application Directory
MBL	Maximum Buffer Length, Type B
MBLI	Maximum Buffer Length Index, Type B
MF	Master File
MSB	Most Significant Bit
N	Number of anti-collision slots, Type B
NAD	Node ADdress
NAK	Not AcKnowledge
NDA	Non-Disclosure Agreement
NFC	Near Field Communication
NIST	National Institute of Standards and Technology
NRZ	Non-Return to Zero encoding
NVB	Number of Valid Bits, Type A
OCF	Open Card Framework
OSI	Open System Interconnection
OTP	One-Time Programmable
Pn	Parameter Byte n
PC	Personal Computer
PC/SC	Personal Computer/Smartcard
PCB	Protocol Control Byte
PCD	Proximity Coupling Device
PICC	Proximity Integrated Circuit Card
PIN	Personal Identification Number
PIX	Proprietary application Identifier eXtension
PKI	Public Key Infrastructure
POR	Power-On Reset
PPS	Protocol and Parameter Selection
PUPI	Pseudo-Unique PICC Identifier, Type B
R	slot number chosen, Type B
RAM	Random Access Memory
R-APDU	Response APDU
RATS	Request for Answer To Select
R-Block	Receive Ready Block
REQA	Request command, Type A
REQB	Request command, Type B
RFID	Radio Frequency IDentification
RFU	Reserved for Future Use
RID	Random ID
RID	Registered application provider IDentifier

LIST OF ABBREVIATIONS

RND . . . . Random number
ROM . . . . Read-Only Memory
RSA . . . . Rivest, Shamir und Adleman
RST . . . . Reset

SAK . . . . Select AcKnowledge, Type A
SAM . . . . Secure Application Module
S-Block . . . . . Supervisory Block
SCL . . . . I2C Serial CLock line
SDA . . . . I2C Serial Data line
SELECT . . . . . Select command, Type A
Select CLn . . Select Cascade Level n
SFGI . . . . Start-up Frame Guard Time Integer
SFID . . . . Short File IDentifier
SFR . . . . Special Function Register
SIM . . . . Subscriber Identity Module
SL n . . . . Security Level n
SM . . . . Secure Messaging
SOF . . . . Start Of Frame
SPA . . . . Simple Power Analysis
SPI . . . . Serial Peripheral Interface
SW1SW2 . . . . Status Word

T=0 . . . . Contact byte-oriented Transmission Protocol (ISO/IEC 7816-3)
T=1 . . . . Contact block-oriented Transmission Protocol (ISO/IEC 7816-3)
T=CL . . . . . . Contactless Transmission Protocol ISO/IEC 14443-4
TL . . . . Length byte
TLV . . . . Tag Length Value, Data Format
TPDU . . . . . . Transport Protocol Data Unit
TR0 . . . . Guard Time, Type B
TR1 . . . . Synchronization Time, Type B
TR2 . . . . Frame delay Time PICC to PCD, Type B
Type A . . . . . ISO/IEC 14443 Part 3 Type A
Type B . . . . . ISO/IEC 14443 Part 3 Type B

UART . . . . . . Universal Asynchronous Receiver and Transmitter
UHF . . . . Ultra High Frequency
UID . . . . Unique Identifier, Type A
USB . . . . Universal Serial Bus
UTF . . . . Universal Transformation Formats

WTX . . . . Waiting Time eXtension
WTXM . . . . . Waiting Time eXtension Multiplier
WUPA . . . . . . Wake Up, Type A
WUPB . . . . . . Wake Up, Type B

XML . . . . eXtensible Markup Language
XOR . . . . eXclusive-OR

13 Bibliography

Axelson Jan: *Serial Port Complete: COM Ports, USB Virtual COM Ports and Ports for Embedded Systems*; 2nd Edition, Lakeview Research LLC, 2007

Baldischweiler Michael: *Keil C51 / Philips LPC900: Hardware – Software – Toolchain*; Electronic Media Giesler & Danne GmbH & Co. KG, 2004

Bayer Jürgen: *Visual C# 2008 Kompendium: Windows-Programmierung mit dem .NET Framework 3.5. Inkl. WPF und LINQ*; Markt und Technik Verlag, 2008

Beutelspacher Albrecht, Neumann Heike B., Schwarzpaul Thomas: *Kryptografie in Theorie und Praxis: Mathematische Grundlagen für Internetsicherheit, Mobilfunk und elektronisches Geld*; 2nd Edition, Vieweg+Teubner, 2009

Bienert Renke: *Mit Funk-Chips in die Zukunft: Ein kleiner RFID-Überblick*; Elektor Issue 9/2006 (Germany)

Doberenz Walter, Gewinnus Thomas: *Visual C# 2008: Grundlagen und Profiwissen*; Carl Hanser Verlag, 2008

Doberenz Walter, Gewinnus Thomas: *Visual C# 2008: Kochbuch*; Carl Hanser Verlag, 2008

Ertel Wolfgang: *Angewandte Kryptographie*; Carl Hanser Verlag, 3. Auflage, 2007

Freeman Eric, Freeman Elisabeth, Sierra Kathy, Bates Bert: *Entwurfsmuster von Kopf bis Fuß*; O'Reilly Verlag, 2005

Freitag Horst: *Einführung in die Zweitortheorie*; 4th Edition, B.G. Teubner Stuttgart, 1990

Frischalowski Dirk: *Visual C# 2008: Master Class. Einstieg für Anspruchsvolle*; Addison-Wesley, 2008

Gilbert H. Owyang: *Foundations for Microwave Circuits*; Springer Verlag, 1989
Hilyard Jay, Teilhet Stephen: *C# Kochbuch*; O'Reilly Verlag, 2006

Hoffman Michael H.W.: *Hochfrequenztechnik – Ein systemtheoretischer Zugang*; SpringerVerlag, 1997

Klaus Finkenzeller: *RFID-Handbuch: Grundlagen und praktische Anwendungen von Transpondern, kontaktlosen Chipkarten und NFC*; 5th Edition, Carl Hanser Verlag, 2008

13 BIBLIOGRAPHY

Mayes Keith E, Konstantinos Markantonakis: *Smart Cards, Tokens, Security and Applications*; Springer Verlag, 2009

Meinke H.H., Gundlach F.W., Lange Klaus, Löcherer Karl-Heinz: *Taschenbuch der Hochfrequenztechnik, Band 1 – 3*; Springer Verlag, 1999

Ossmann Martin: *Experimenteller RFID-Reader*; Elektor Issue 9/2006 (Germany)

Petzold Charles: *Windows Forms-Programmierung mit Visual C sharp 2005*; Microsoft Press, 2006

Rankl Wolfgang, Wolfgang Effing: *Handbuch der Chipkarten: Aufbau – Funktionsweise– Einsatz von Smart Cards*; 5th Edition, Carl Hanser Verlag, 2008

Rankl Wolfgang: *Chipkarten-Anwendungen: Entwurfsmuster für Einsatz und Programmierung von Chipkarten*; Carl Hanser Verlag, 2006

Rothammel Karl, *Antennenbuch*; Franckh'sche Verlagshandlung, 1984

Schalk H. Gerhard: *Die Elektor-RFID-Karte: 13,56-MHz-Karte mit MIFARE Ultralight-IC von Philips*; Elektor Issue 9/2006 (Germany)

Schalk H. Gerhard: Elektor Issue 9/2006 (Germany); Elektor Issue 9/2006 (Germany)

Schalk H. Gerhard: *Der Chip im Pass*; Elektor Issue 9/2006 (Germany)

Scheider Bruce: *Angewandte Kryptographie: Protokolle, Algorithmen und Sourcecode in C.*; Pearson Studium, 2005

Volpe Francesco P., Safinaz Volpe: *Chipkarten Grundlagen, Technik, Anwendungen*; Heinz Heise Verlag, 1996

Voß Herbert: *Kryptografie mit Java: Grundlagen und Einführung zur kryptografischen Programmierung mit Java*; Franzis Verlag, 2006

14 Index

.NET Framework 189
3DES *see triple DES*
3-pass mutual authentication 129, 322, 324

A

AC CL1 40, 142, 230, 233, 393
AC CL2 41, 142, 230, 234, 394
AC CL3 41, 230
access condition 151, 281, 290
Access Control Manager 398
access control systems 396
    offline systems 396
    online systems 396
ACK 58, 149, 345
Acknowledge *see ACK*
active mode 72
ACTIVE state 41, 42
adjustment 74
Advanced Encryption Standard *see AES*
AES 122, 124, 155, 356
AFI 48
AID 156, 361
algorithm 109
amplitude modulation 29
Answer to Request, Type A *see ATQA*
Answer to Request, Type B *see ATQB*
Answer to Reset *see ATR*
Answer to Select *see ATS*
antenna
    coil 76, 93
    current 78
    design 73, 159, 173
    matching 162
    Q factor 82
anti-collision 33
    algorithm 38, 41, 44, 46, 233, 387
    Cascade Level 1 *see AC CL1*
    Cascade Level 2 *see AC CL2*
    Cascade Level 3 *see AC CL3*

APDU 337, 368, 456
    C-APDU 340, 368
        CLA Byte 369
        INS Byte 369
        length field Lc 369
        length field Le 369
        Parameter Byte P1 369
        Parameter Byte P2 369
    Command APDU *see C-APDU*
    commands
        External Authenticate 378
        General Authenticate 420, 456
        Get Challenge 378
        Get Data 420, 456
        Internal Authenticate 378
        Load Key 420, 456
        Read Binary 373, 421, 456
        Select 372
        Update Binary 375, 421, 456
        Verify 421, 456
    extended APDU 370
    R-APDU 340, 371
        data field 371
        Status Word SW1SW2 371
    Response APDU *see R-APDU*
    short APDU 369
API 411
application
    interface *see API*
    identifier *see AID*
    identifier extension *see PIX*
    layer 337, 368
    programming interface
        see API
    protocol data unit
        see APDU
asymmetric cryptography 120
ATQA 40, 42, 228, 231
ATQB 46
ATR 419, 422, 423, 455
    pseudo 425, 426, 455

ATS 51, 340, 419
attack 112
    brute force 273
    cloning memory contents 271
    combined 120
    denial-of-service 116
    differential power analysis (DPA) 117
    eavesdropping 113
    laser 119
    logical 112
    man-in-the-middle 116
    physical 116
    power analysis 117
    relay 114
    replay 114
    reverse engineering 118
    side channel attack 117
    simple power analysis (SPA) 117
    temperature and frequency 120
    unauthorized manipulation of data 113
    unauthorized reading of data 112
ATTRIB command 48
authentication 113, 129, 155, 282, 322, 324, 378, 421
authenticity 277, 324

B

Backup Data File 157, 360
backup management 286, 289, 310, 316
balun 97
bandwidth 77, 83
BCC 39, 394
Biot-Savart law 79
bit-frame anti-collision
    see anti-collision
block 151, 277
    chaining see chaining
    check character see BCC
    cipher 123
    number 56
BPSK 34
brute force 273
BSI 125
Bundesamt für Sicherheit in der Informationstechnik 125

C

C# 189, 212, 215, 223, 236, 428
    CSharpPCSC 441
        PCSCReader 443
        WinSCard 441
    Platform Invoke 441
C/C++ 189, 212, 428
calibration 68, 102
    coil 65
card 137, 227
    activation see ISO/IEC 14443 Part 3
    antennas 97
    emulator 398
    identifier see CID
    operating system see cos
    serial number see UID
    terminal API see CT-API
    type identification 253
    specific keys see key diversification
Cascade Bit 43
Cascade Tag see CT-API
CBC 127, 131, 155, 275, 326
chaining 57, 58, 343
challenge-response 324
check-in 24
check-out 24
chip card 137, 227
CID 50, 56, 58, 338
Cipher Block Chaining see CBC
circuit board 94
CLA Byte 369
Class 1 36
coil design 174
cold reset 423
collision 41
    detection 33
combined attacks 120
Commit Perso 154
Common Criteria (CC) 135
communication protocol 49, 419
compensated antenna coil 96
compensation 102
conductor twisting 94
contact card 49, 337
    activation 422
    contact unit 422

memory card 413
microcontroller smartcard 424
contactless
    interface *see ISO/IEC 14443*
    memory card 419, 455
        pseudo ATR 426
        type identification 455
    microcontroller smartcard 419
        pseudo ATR 425
    smartcard *see smartcard*
COS 338, 355
coupling 28
    factor 80
CRC 56, 344, 355, 390, 393
cryptography 109, 120
CSharp *see C#*
CSharpPCSC 441
    PCSCReader 443
    WinSCard 441
CT 39, 229
CT-API 411, 412
Cyclic Record File 158, 360
cyclic redundancy check *see CRC*

D

Data Encryption Standard *see DES*
Data File 157, 360
data link layer 337
data
    corruption 310, 313
    field 371
    integrity 111
    security 109
    transmission 29
Dedicated File *see DF*
denial-of-service attack 116
derived key *see key diversification*
DES 122, 124, 356
    single DES 273, 277
    triple DES 125, 273, 277, 324
Deselect 60, 340
DESFire EV1 *see DESFire EV1*
device under test 66
DF 359, 378
DIF card 139, 38
Differential Power Analysis *see DPA*
DIN 36

dipole 75
directory names 359
Divisor Receiver *see DR*
Divisor Send *see DS*
DLL 212, 441
door opener 396
double size UID 39, 44, 228, 258
downlink 29, 52
DPA 117
DR 52, 55, 338
DS 52, 55, 338
dual interface
    card *see DIF card*
    smartcard reader 411
dynamic keys 132
dynamic link libraries *see DLL*

E

eavesdropping 113
ECB 127
eddy current losses 76
EEPROM 310
    erase cycle 310
    program cycle 310
EF 359, 378
electrical shielding 94
electromagnetic disturbance 61
electronic code book *see ECB*
electronic identity card *see identity card*
electronic passport *see passport*
electronic wallet 289, 290, 310, 316
Elektor RFID Reader 68, 187, 202, 227, 396
    Access Control Manager 398
    firmware 208
        access control systems 399
        software architecture 208
        update 203
        version control 205
    library 189, 203, 212
        APDU 374, 441
        card activation 236
        MIFARE Classic 294
        MIFARE ClassicUtil 287, 294, 441
        MIFARE DESFire EV1 362
        MIFARE Ultralight 263

MIFARE Ultralight C 327
reader selection 243
T=CL 345
modes 205
Access Control mode 205, 398
PC Reader mode 207
Terminal mode 205
programming 212
RC522 386
transmission protocol 210
USB driver installation 202
elementary files *see EF*
elliptic curves 121
EMC filter 174
EMD 61, 167
employee identification cards 25
encryption 109
methods 109, 124
energy transfer 77
energy transmission 27
error detection and handling 59, 344
extended APDU 370
extended sectors 281
External Authenticate 378

F

far field 75
FDT 44, 61
Felica 72
ferrite 90
FID 361
field strength 79, 89
FIFO 390
file
identifier *see FID*
system 156, 356, 378
Flash Magic 203
Format Byte T0 52
frame delay time *see FDT*
frame size for proximity
card *see FSC*
card integer *see FSCI*
coupling device *see FSD*
coupling device integer *see FSDI*
Frame Waiting Time *see FWT*
Frame Waiting Time Integer *see FWTI*
FSC 52, 56, 338, 343

FSCI 52
FSD 51, 56, 338, 343
FSDI 51
FT232R 190, 197, 385
configuration 199
driver 198, 201, 202
EEPROM 199
FTDI 190
Future Technology Devices *see FTDI*
FWT 53, 338, 344, 53

G

General Authenticate 420, 456
Get Challenge 378
Get Data 420, 456

H

half duplex 337, 422
HALT state 42, 60, 340
Halt, Type A *see HLTA*
Helmholtz arrangement 65
higher data rates 31, 156
Historical Bytes 54, 338
HLTA 42, 60, 142, 234, 340
HLTB 48

I

I Block 56, 340, 343
I²C 160, 195, 307, 385
ICC 413
ID-1 98
identification 18
identity card 26, 406
personalization 406
reading the card data 405
IDLE state 40, 46, 230, 232
IFD 414
handler 414
impedance
analyzer 175
measurement 101
measuring device 101
INF 56, 59
Information Block *see I Block*
Information Field *see INF*

init vector (IV) 114, 127, 326
initiator 72
inlay 22
input impedance 73
INS Byte 369
Interface Byte TA 52
Interface Byte TB 53
Interface Byte TC 53
interference 183
Inter-Integrated Circuit see I²C
internal authenticate 378
ISO/IEC 10373-6 64, 96
ISO/IEC 14443 22, 27, 29, 35, 64, 84, 156, 227, 412, 413, 419
ISO/IEC 14443 Part 3: Card Activation 43, 154, 228
  Answer to Request, Type A see ATQA
  anti-collision algorithm 233
  CT 229
  HALT state 60, 340
  HLTA 60, 340
  Type A 29, 31, 33, 34, 35, 38, 84
    AC CL1 40, 142, 230, 233
    AC CL2 41, 142, 230, 234
    AC CL3 42, 230
    ACTIVE state 41, 42
    anti-collision algorithm 38, 41, 44
    ATQA 40, 42, 228, 231
    BCC 39
    Block Check Character see BCC
    card-type detection 253
    Cascade Level 1 see AC CL1
    Cascade Level 2 see AC CL2
    Cascade Level 3 see AC CL3
    FDT 44, 61
    Frame Delay Time see FDT
    HALT state 42
    Halt, Type A see HLTA
    HLTA 42, 142, 234
    IDLE state 40, 230, 232
    Number of Valid Bits see NVB
    Random ID see RID
    READY state 40, 41, 232
    REQA 142, 230
    REQB 230
    RID 38, 112, 133, 206
    SAK 41, 42, 228, 233

    Select Acknowledge see SAK
    Cascade Level 1 (CL1) 41, 142, 230, 233
    Cascade Level 2 (CL2) 41, 142, 230, 234
    Cascade Level 3 (CL3) 42
    UID see UID
    Unique Identifier see UID
    Wake Up, Type A see WUPA
    WUPA 42, 230
ISO/IEC 14443 Part 3: Card Activation Type B 30, 32, 34, 35, 38, 46, 85
  AFI 46
  Answer to Request, Type B see ATQB
  anti-collision algorithm 46
  Application Family Identifier 48
  ATQB 46
  ATTRIB command 48
  HLTB 48
  IDLE state 46
  number of anti-collision slots 48
  PARAM byte 48
  pseudo-unique PICC identifier see PUPI
  PUPI 49
  random number 46, 47, 48
  READY_DECLARED state 48
  REQB 46
  Request Command, Type B see REQB
  Slot MARKER 48
  Slot Number Chosen 48
  time slot 46, 47, 48
  Wake Up, Type B see WUPB
  WUPB 47
ISO/IEC 14443 Part 4: T=CL 49, 56, 156, 234, 337, 368
  ACK 58, 149, 345
  Acknowledge see ACK
  Answer to Select see ATS
  ATS 51, 340, 419
  block number 56
  block type 56
  Card Identifier see CID
  chaining 57, 58, 343
  CID 50, 56, 58, 338

CRC 56
Cyclic Redundancy Check *see CRC*
Deselect 60, 340
Divisor Receiver *see DR*
Divisor Send *see DS*
DR 52, 55, 338
DS 52, 55, 338
error detection and handling 59, 344
Format Byte T0 52
Frame Size for Proximity
    Card *see FSC*
    Card Integer *see FSCI*
    Coupling Device *see FSD*
    Coupling Device Integer
      see FSDI
Frame Waiting Time *see FWT*
Frame Waiting Time Integer
    see FWTI
FSC 52, 56, 38, 43, 52
FSD 51, 56, 338, 343
FSDI 51
FWT 53, 338, 343
FWTI 53
Historical Bytes 54, 338
I Block 56, 340, 343
INF 56, 59
Information Block *see I Block*
Information Field *see INF*
Interface Byte TA 52
Interface Byte TB 53
Interface Byte TC 53
Length Byte TL 52
multi-card activation 341, 352
NAD 53, 56, 338
NAK 58, 149, 345
Node Address *see NAD*
Not Acknowledge *see NAK*
PCB 56, 58, 343
PPS 54, 55, 339
presence check 59
protocol activation 49, 338, 350
Protocol and Parameter Selection
    see PPS
Protocol Control Byte *see PCB*
protocol deactivation 60, 338, 350
R Block 58, 341, 344
RATS 50, 338

Receive Ready Block
    see R Block
Request for Answer to Select
    see RATS
S Block 59, 341
SFGI 53
SFGT 53
Startup Frame Guard Time
    see SFGT
Startup Frame Guard Time Integer
    see SFGI
Supervisory Block *see S Block*
WTX 60, 61, 344
WTXM 61
ISO/IEC 15693 72, 412, 413, 419
ISO/IEC 7816 49, 324, 337, 368
    activation 422
    Answer to Reset *see ATR*
    ATR 419, 422, 423, 455
        pseudo 425, 426, 455
    cold reset 423
    PPS 423
    T=0 49, 338, 368
    T=1 49, 338, 368
    UART 422
    -warm reset 423
ISO/IEC Part 3: Card Activation
    Type A
        AC CL1 393
        AC CL2 394
        anti-collision algorithm 387
        BCC 394
        NVB 393
        REQA 390
        SAK 394
        select
            Cascade Level 1 (CL1) 393
            Cascade Level 2 (CL2) 394
        WUPA 390

J

Java 189, 412, 428
    card applet 137
    Native Interface (JNI) 438
    Smartcard I/O API 440
JPC/SC 438

INDEX 14

K

Kerckhoff's principle 122
Key A *see MIFARE Classic Key A*
Key B *see MIFARE Classic Key B*
keys
    diversification 132, 273, 294
    management 132, 294
    pair 120
    rights *see MIFARE Classic Access Condition*

L

laser attacks 119
Length Byte TL 52
Length Field Lc 369
Length Field Le 369
Linear Record File 158, 360
little-endian 287
Load Keys 420, 456
load modulation 33, 37
load resistance 74
load simulation 67
load switching 33
Lock Bits 144, 147, 262
logical attacks 112
low-pass filter 177

M

MAC 111, 114, 116, 131, 155, 355
magnetic antennas 75
magnetic coupling 77
magnetic field 27
main coil 66
Manchester encoding 33
man-in-the-middle attack 116
Master File *see MF*
Master Key 133, 273, 276
Master Key *see Master Key*
master-slave principle 337, 386, 422
matching 86
    network 178
maximum field strength 37
measurement coil 67
memory card 137, 188, 413, 419, 455

Message Authentication Code *see MAC*
metal enclosures 89
metal surfaces 87, 205
MF 359, 381
MFRC522 63, 85, 148, 159, 190, 194, 307, 385
    block diagram 195
    FIFO 385
    I²C 160, 195, 385
    oscillator 161
    quartz oscillator 161
    Serial Peripheral Interface 160
    SFR
        AnalogTestReg 172
        CollLevel 168
        CWGsPReg 165, 184
        DemodReg 169, 387
        FIFODataReg 391
        FIFOLevelReg 390, 391
        GsNReg 164, 184
        MinLevel 168
        ModGsPReg 165
        ModWidthReg 164
        RFCfgReg 169, 184, 387
        RxGain 169
        RxModeReg 64, 167
        RxMultiple 64, 167
        RxNoErr 63, 167
        RxSelReg 168
        RxThresholdReg 168, 184, 387
        RxWait 64
        TauRcv 169
        TauSync 169
        TxASKReg 164
        TxControlReg 163
        TXModeReg 163, 170
    SPI 160
    test pins
        AUX1 171
        AUX2 171
        MFOUT 170
    test signals 169
        ADC_I 173
        ADC_Q 173
    Tx output 161
microcontroller smartcard 137, 227, 419, 422

477

14 INDEX

Microsoft
   Visual C# 2012 Express Edition 189
   WinSCard API see PC/SC WinSCard API
MIFARE 140, 227, 456
   1K see MIFARE Classic
   4K see MIFARE Classic
   application directory see MAD
   card-type detection 253
   Classic 119, 124, 140, 148, 150, 227, 277, 456
      Access Condition 151, 277, 290
      Auth. Key A 282, 456
      Auth. Key B 282, 456
      blocks 151, 277
      Crypto 1 281
      Decrement 150, 283, 289, 292
      extended sectors 281
      Increment 150, 283, 289, 292
      instruction set 282
      Key A 152, 281, 282, 290, 456
      Key B 152, 281, 282, 291, 456
      key rights see Access Condition
      memory organization 151, 281, 282
      Read 150, 283, 456
      Restore 150, 283, 289, 292
      Sector 151, 281
      Sector Trailer 151, 281, 291
      Standard Sector 281
      Transfer 150, 283, 289, 290
      Value Format 150, 289, 316
      Write 150, 283, 286, 456
   clone 140
   DESFire EV1 141, 156, 227, 355, 459
      Application Master Key 157
      command structure
         chaining 358
         DESFire Native APDU 357, 460
         ISO/IEC 7816 APDU 356
      commands 356
      Record File 158
      directory names 361
      file system 156, 356, 378
         Application Identifier AID 361
         Backup Data File 157, 360
         Cyclic Record File 360
         Dedicated File DF 359, 378
         Elementary File EF 359, 378
         File Identifier FID 361
         Linear Record File 360
         Master File MF 359
         Standard Data File 157, 360
         Value File 360
            Limited Credit 158
            Linear Record File 158
            Value File 158
   hack 141
   MAD 291
   Magic 188
   memory organization 156
   MFRC522 see MFRC522
   Mini see MIFARE Classic
   Plus 115, 141, 152
      authentication 155
      Perso 154
      memory organization 152
      Multi-Block Read 155
      Multi-Block Write 155
      Multisector Authentication 155
      Plus S 153
      Plus X 153
      Proximity Check 152
      Security Level 0 153
      Security Level 1 154
      Security Level 2 154
      Security Level 3 154
      security levels 153
      Write Perso 153
   protocol 154
   Ultralight 141, 227, 258, 451
      memory organization 144, 258
      Read 142, 259
      security functions 273
      Write 143, 259
   Ultralight C 141, 145, 227, 322, 456
      Authent Bytes 148
      Authenticate 146, 326
      authentication 148, 324
      Counter 148
      instruction set 324
      Lock Bits 147
      memory organization 147, 322
      triple DES keys 333
      Compatibility Write 143, 259

instruction set 259
    Lock Bits 144, 262
MifareWnd 291
Mikron 140
Miller encoding 30
minimum field strength 37
miniVNA 101, 181
mismatch 74
modulation index 38, 85, 86
Motorola 160
Multi-Block Read 155
Multi-Block Write 155
multi-card activation 341, 352
MultiSector Authentication 155
mutual authentication 24, 129, 322
mutual authentication *see 3-pass mutual authentication*
mutual inductance 78

N

NAD 53, 56, 338
NAK 58, 149, 345
natural resonance frequency 107
NDA 227, 233, 356
near field 74
Near-Field Communication *see NFC*
NFC 70, 71, 72, 412
    active mode 72
    initiator 72
    modes of operation 71
    peer-to-peer 159
    PN512 159
    target 72
NIST 124
Node Address *see NAD*
non-disclosure agreement *see NDA*
Not Acknowledge *see NAK*
NRZ encoding 30, 34
Number of Valid Bits *see NVB*
NVB 393
NXP Semiconductors 70, 140, 160, 176

O

OCF 411, 412
one-time programmable *see OTP*
Open Card Framework *see OCF*

Open System Interconnection *see OSI*
oscillator 161
OSI 337, 368
OTP 144, 258

P

P89LPC936 190, 193, 203, 306, 385, 399, 401
    EEPROM 399, 401
padding 125
paper cards 261
Parameter Byte P1 369
Parameter Byte P2 369
passport 22, 26, 115
PC/SC 411
    APDU
        General Authenticate 420, 456
        Get Data 420, 456
        Load KeyS 420, 456
        Read Binary 421, 456
        Update Binary 421, 456
        Verify 421, 456
    application 417
    architecture 413
    contact card
        activation 422
        memory card 413
        microcontroller smartcard 422
    contactless
        memory card 419, 455
            pseudo ATR 426
            type identification 455
    CSharpPCSC 428, 441
        classes
            PCSCReader 443
            WinSCard 441
    driver 414, 419, 449
    ICC 414
    IFD 414
    Java 428
    Java Smartcard I/O API 440
    JPC/SC 438
    microcontroller smartcard 419
        pseudo ATR 425
    MIFARE 456
        Classic 456
        DESFire EV1 459

    Ultralight 456
    Ultralight C 456
  reader 305
  Resource Manager 414
  service provider 416
  specification 411, 413, 419
  transaction 416
  WinSCard API 428, 438, 441
PCB 56, 58, 343
PCD 23, 227
peer-to-peer 159
Pegoda reader 291
personal identification number see PIN
personalization 333
phase shift keying 34
Philips Semiconductors 70, 140
physical attacks 116
PICC 22, 227
pickup coil 66
PIN 115, 421
PIX 361, 426
Platform Invocation 441
PN512 159
POR 38, 144, 199, 208, 230, 390, 419, 423
power analysis 117
power supply 173, 192, 422
power-on reset see POR
PPS 54, 55, 339, 423
prepaid see prepaid system
prepaid system 310
presence check 59
privacy protection 112
private key 120
protection mechanism 109
protection objectives 109
protocol activation 49, 338, 350
protocol analyzer 304
Protocol and Parameter Selection see PPS
Protocol Control Byte see PCB
protocol deactivation 60, 338, 350
proximity card see PICC
Proximity Check 152
proximity coupling device see PCD
Proxispy Protocol Analyzer 305
pseudo-unique PICC identifier see PUPI
public key 120

public transport 24, 261
PUPI 49, 112

Q

Q factor 84, 87, 176
quartz oscillator 161

R

R Block 58, 341, 344
radiation losses 76
radio frequency 17
Raisonance 306
Random ID see RID
random number 47, 48, 129, 324
range 185, 205, 249, 452
range measurement 69, 205, 249
RATS 50, 338
RC522 see MFRC522
Read Binary 373, 421, 456
reader 23, 159
  antenna 87
    device see reader IC
    Elektor RFID Reader see Elektor RFID Reader
    IC 159, 190, 194, 385
    MFRC522 IC see MFRC522
reading distance 80
reading range see range
READY state 40, 41, 232
READY_DECLARED state 48
Receive Ready Block see R Block
Record File 158
records 360
ReferencePICC 66, 86
Registered Application Provider Identifier 361, 426
relay attack 114, 130
REQA 142, 230, 390
REQB 46, 230
Request Command, Type B see REQB
Request for Answer to Select see RATS
resonant circuit 73, 84
resonant frequency 73, 83, 86, 98
Resource Manager 414

rest level 63
rest period 62
retroactive effect 86
reverse engineering 118
RF reset *see POR*
RFID 17
    access control system 396
        offline systems 396
        online systems 396
    system 19, 22
RFSIM99 179
RID 38, 112, 133, 206, 228
Rivest, Shamir, Adleman *see RSA*
rolling back 310, 316
RS-232 communication protocol 210
RSA 121

S

S Block 59, 341
SAK 41, 42, 228, 233, 394
SAM 23, 110, 135, 273
sector 151, 281
Sector Trailer 151, 281, 291
Secure Application Module *see SAM*
security 109
    unctions 144, 147
    level *see MIFARE Plus: Security Level*
    sensors 120
Select 372
    Acknowledge *see SAK*
    Cascade Level 1 (CL1) 41, 142, 230, 233, 393
    Cascade Level 2 (CL2) 142, 230, 233, 394
    Cascade Level 3 (CL3) 42, 230
sense coils 66
sensitivity 165, 168
Serial Peripheral Interface 160
service provider 416
session keys 130
SFGI 53
SFGT 53
SFR *see MFRC522 SFR*
SharpDevelop 189
short APDU 369
side channel attacks 116
sidebands 84

signature 111, 120
SIM card 135, 139, 337
Simple Power Analysis *see SPA*
single DES 124
single size UID 38, 42, 228
skin effect 76
Slot MARKER 48
Smart Card Magic.NET 190, 214
    breakpoint 222
    card activation 231
    ISO/IEC 14443-4 Transmission Protocol Window 340
    MIFARE Classic Keys Window 284
    MIFARE Classic Window 283, 289
    MIFARE Ultralight C Window 260, 326
    script 215
        breakpoint 222
        compiler 215
        compiling and running 219
        Console Window 219, 220
        creation 216
        Editor 220
        File Explorer 216
        Trace Window 219
        user input 220
    user input 220
smartcard 19, 22, 36, 137, 140, 227, 350, 355, 368
    module 422
Smith chart 102, 104, 105, 106
Sony 70
Sony Felica 72
SPA 117, 120
Special Function Register *see MFRC522 SFR*
SPI 160
Standard Data File 158, 360
standard sector 281
Startup Frame Guart Time *see SFGT*
Startup Frame Guart Time Integer *see SFGI*
storage card 227
stream cipher 123
subcarrier modulation 33
Supervisory Block *see S Block*
symmetric cryptography 122
    Cipher Block Chaining (CBC) *see CBC*

Electronic Code Book (ECB) *see ECB operation mode* 127
symmetrical antenna 92
symmetrical feed 75

T

T=0 49, 337, 368, 419
T=1 49, 337, 368, 419
T=CL 368, 419
T=CL *see ISO/IEC 14443 Part 4: T=CL*
tag 19, 100, 137
target 72
temperature and frequency attacks 120
test equipment 65
test methods 64
test PCD assembly 65
threat 109
ticket 24, 261, 268, 293
time slot 46, 47, 48
TLV format 54, 362, 426
token 100
tolerances 173
Topaz 233
TPDU 338
transaction 316, 416
    buffer 310
    time 303
    protocol 338
transformer 28, 77
    model 77
    principle 32
transmission layer *see Data Link Layer*
transmission protocol 49, 337
    T=0 *see T=0*
    T=1 *see T=1*
    T=CL *see T=CL*
Transport Protocol Data Unit *see TPDU*
triple DES 124, 125
triple size UID 39, 228
tuning 68

U

UART 160, 385, 422
UID 38, 41, 112, 133, 397
10-byte UID 39, 228
4-byte UID 228, 394
7-byte UID 39, 228, 258
double size UID 39, 44, 228, 258
Random ID *see RID*
RID 38, 112, 133, 206, 228
single size UID 38, 42, 228
triple size UID 39, 228
Ultralight *see MIFARE Ultralight*
Ultralight C *see MIFARE Ultralight C*
unauthorized manipulation of data 113
unauthorized reading of data 112
Unique Identifier *see UID*
Update Binary 375, 421, 455
uplink 29, 52
UPM Raflatac 100
USB 192, 197, 385
    CCID 415
    driver 415
    enumeration 192, 199
    Product ID 199
    USB-to-RS232 converter *see FT232R*
    Vendor ID 199

V

Value Block *see Value Format*
Value File 158, 360
Value format 150, 289, 316
Value Operations *see Value Format*
vector network analyzer 101
Verify 421, 455
Visual Basic.NET 212
Visual C# 2012 Express Edition 189, 223
    Elektor RFID Reader Library 225
vna/J 102

W

Wake Up, Type A *see WUPA*
Wake Up, Type B *see WUPB*
warm reset 423
WG8 36
wideband transformer 97
winding count 80
WinSCard API *see PC/SC WinSCard API*

wrapper 428, 438
Write Perso 153
WTX 60, 61, 344
WTXM 61
WUPA 42, 231, 390

WUPB 47

Z

Zplots 102, 181